Traffic Signal Systems 2018

Transportation Research Record: Journal of the Transportation Research Board

Volume 2672, Issue 18

ISBN: 978-1-5108-7823-5

Printed from e-media with permission by:

Curran Associates, Inc.
57 Morehouse Lane
Red Hook, NY 12571

Some format issues inherent in the e-media version may also appear in this print version.

Published by Sage Publications USA

ISBN (Print) 978-1-5108-7823-5
ISBN 2018 Printed Set (All Issues) 978-1-5108-7735-1

Printed by Curran Associates, Inc. (2019)

Additional copies of this publication are available from:

Curran Associates, Inc.
57 Morehouse Lane
Red Hook, NY 12571 USA
Phone: 845-758-0400
Fax: 845-758-2633
Email: curran@proceedings.com
Web: www.proceedings.com

Volume 2672 Number 18 December 2018

TRR
JOURNAL OF THE TRANSPORTATION RESEARCH BOARD

Contents

Traffic Signal Systems 2018

Articles

TRB

TRANSPORTATION RESEARCH BOARD

CORE SPONSORS

Transportation Departments of the 50 States and the District of Columbia

Federal Government

U.S. Department of Transportation
- Federal Aviation Administration (FAA)
- Federal Highway Administration
- Federal Motor Carrier Safety Administration (FMCSA)
- Federal Railroad Administration (FRA)
- Federal Transit Administration
- National Highway Traffic Safety Administration (NHTSA)
- Office of the Assistant Secretary of Transportation for Research and Technology (OST-R)

U.S. Air Force Civil Engineer Center (AFCEC)
U.S. Coast Guard (USCG)
U.S. Department of Energy
U.S. Department of the Interior (DOI)
U.S. Environmental Protection Agency (EPA)

Nongovernmental Organizations

American Public Transportation Association (APTA)
Association of American Railroads (AAR)
California Air Resources Board

The National Academies of
SCIENCES · ENGINEERING · MEDICINE

The **National Academy of Sciences** was established in 1863 by an Act of Congress, signed by President Lincoln, as a private, nongovernmental institution to advise the nation on issues related to science and technology. Members are elected by their peers for outstanding contributions to research. Dr. Marcia McNutt is president.

The **National Academy of Engineering** was established in 1964 under the charter of the National Academy of Sciences to bring the practices of engineering to advising the nation. Members are elected by their peers for extraordinary contributions to engineering. Dr. C. D. Mote, Jr., is president.

The **National Academy of Medicine** (formerly the Institute of Medicine) was established in 1970 under the charter of the National Academy of Sciences to advise the nation on medical and health issues. Members are elected by their peers for distinguished contributions to medicine and health. Dr. Victor J. Dzau is president.

The three Academies work together as the National Academies of Sciences, Engineering, and Medicine to provide independent, objective analysis and advice to the nation and conduct other activities to solve complex problems and inform public policy decisions. The Academies also encourage education and research, recognize outstanding contributions to knowledge, and increase public understanding in matters of science, engineering, and medicine.

Learn more about the National Academies of Sciences, Engineering, and Medicine at **www.national-academies.org.**

The **Transportation Research Board** is one of seven major programs of the National Academies of Sciences, Engineering, and Medicine. The mission of the Transportation Research Board is to increase the benefits that transportation contributes to society by providing leadership in transportation innovation and progress through research and information exchange, conducted within a setting that is objective, interdisciplinary, and multimodal. The Board's varied activities annually engage about 7,000 engineers, scientists, and other transportation researchers and practitioners from the public and private sectors and academia, all of whom contribute their expertise in the public interest. The program is supported by state transportation departments, federal agencies including the component administrations of the U.S. Department of Transportation, and other organizations and individuals interested in the development of transportation.

Learn more about the Transportation Research Board at **www.TRB.org**.

Article

TRR
JOURNAL OF THE TRANSPORTATION RESEARCH BOARD

Transportation Research Record
2018, Vol. 2672(18) 1–14

DOI: 10.1177/0361198118782753
journals.sagepub.com/home/trr
SAGE

An Arterial-Based Transit Signal Priority Control System

Hyeonmi Kim[1], Yao Cheng[1], and Gang-Len Chang[1]

Abstract

To contend with the negative impacts of intersection-based transit system priority (TSP) control on side-street traffic, and also to enhance its effectiveness from the perspective of the whole traffic flow on the arterial, this study presents an integrated arterial-based TSP system to promote bus operations. The proposed system employs an off-line transit-based progression system to produce the base signal plan, in which is embedded a feature to minimize the negative effects of local TSP implementation on non-priority movements at critical intersections. With its specially designed evaluation function, the proposed system is capable of minimizing the ineffective activation of local TSP which often results in undesirable patterns of traffic arriving at the next intersection downstream and excessive delays for all vehicles. Extensive evaluation with simulation experiments confirms that the proposed system, integrating a local TSP with the transit-based signal progression system, can indeed circumvent the deficiencies of conventional TSP controls and produce the expected benefits to both passenger cars and transit vehicles over the entire arterial road.

Transit signal priority (TSP) has long been viewed as an effective strategy to reduce delays for transit vehicles at urban intersections, and a large body of studies on the subject has been conducted by transportation researchers over the past few decades. Despite the significant contributions of various studies, however, implementation of either passive or active TSP control in urban networks remains quite limited in practice. Among the various reasons for limited usage of TSP control, the negative impacts on the side-street traffic and the downstream arrival patterns of the subject intersections deserve the most attention.

With respect to the first issue of inevitable negative impacts on the side streets, it can be challenging to justify the effectiveness of TSP control in many urban networks, especially its implementation at those intersections experiencing heavy traffic volumes or high transit frequencies from all approaches, despite the design of various constraints on the TSP activation frequency and/or with some remedies for vehicles delayed by such operations. The second issue that may degrade a TSP's benefits from the perspective of all traffic on an arterial is due mostly to the signal control system at the arterial level, which facilitates passenger car flows but not transit vehicles even with proper signal offsets. Hence, reduced queue delays for transit vehicles at one intersection due to TSP activation are likely to be transferred into excessive waiting time at the downstream intersection, because of their early arrival with other concurrently discharged passenger cars, which leads to longer queues. Certainly, embedding the TSP within a full-scale adaptive signal control could minimize the negative impacts observed in practice. However, the costs associated with the deployment, operation, and maintenance of an effective adaptive network signal control system are beyond the acceptable level of expenditure for most responsible traffic agencies. This is evidenced by the fact that so far adaptive network signal control systems have mostly been deployed for demonstration and not day-to-day operation.

In response to such concerns, this study presents an integrated arterial-based signal priority control system which first employs an off-line module to yield the optimal offsets for bus progression, and then activates real-time TSP control only at critical intersections. It is expected that the implementation of TSP under such a design can yield benefits for the entire arterial without costly investments for all its intersections, and the negative impacts due to real-time TSP activations on side-street traffic can be minimized.

[1]Department of Civil and Environmental Engineering, University of Maryland, College Park, MD

Corresponding Author:
Address correspondence to Yao Cheng: chengyao09@gmail.com

Literature Review

Depending on the mechanism for the priority decision, one can classify existing TSP controls into rule-based and model-based strategies. The former category would provide transit priority with either predetermined or varying priority duration, based on a set of decision rules. An example of studies in this category is that of Ludwick who simulated a bus route with unconditional TSP under multiple scenarios defined with different bus headways and bus stop locations (*1*). That study showed a reduction of 15% to 20% in bus travel time, yielding the expected benefits to other vehicles using the same phase but incurring negative impacts on the side street. Garrow and Machemehl further evaluated unconditional TSP with a simulation to test its performance under different intervals for green extension and different volumes from the cross street (*2*). They found that such a system needs to limit the activation of TSP frequency in order to minimize the negative impact on the side street of high traffic volume. To mitigate such impacts, some researchers have promoted the use of conditional priority control based on the scheduled delay of detected buses, preset maximal TSP activation frequency, and the state of priority control in the previous cycle (*3*). In this regard, Evans and Skiles, and the U.K. Department of Transport explored the strategy of allowing the red truncation only when the previous green has not been extended (*4*, *5*). Moghimidarzi et al. developed a rule-based TSP for application in a self-organizing traffic signal control, where the priority decision based on stochastic dwell time can be reversed if the updated prediction of the expected remaining dwell time indicates an ineffective green extension (*6*). Some studies recommend that TSP be activated only to those buses behind schedule or deviating from the required headway (*7–9*).

Using mathematical formulations and quantified performance index, some researchers have proposed model-based TSP strategies to reduce bus delays or total person delays based on bus locations, operation conditions, and passenger car volumes. An example is the studies by Lin et al., who developed a model to determine the optimal priority duration with the objective of minimizing the bus passenger waiting time at the downstream bus stop, but not to increase the total passenger delays (*10*). Head et al. presented a model to deal efficiently with multiple priority requests with mixed integer programming formulations (*11*). Using the adaptive control logic, Chang et al. suggested that the activation of priority decisions be based on a performance index which includes vehicle delay, bus schedule delay, and passenger delay under the absence or presence of automatic vehicle location systems (*12*). Hu et al. adopted connected vehicle technologies in TSP control, taking into account the coordination between the TSP intersection and its adjacent ones (*13*). Christofa et al. presented a real-time signal control system on an arterial to minimize total person delay experienced by drivers and passengers on vehicles. With the objective of maintaining passenger car progression, their method recognizes the adherence of buses to their schedules and can provide priority to transit vehicles even when they travel on conflicting routes (*14*). Xu et al., Hu et al., and Ye and Xu further conducted studies to accommodate conflicting TSP requests from different routes using optimization formulations (*15–17*). Yu et al. combined adaptive TSP strategies and person-based signal timings with consideration of uncertain bus arrival times when determining green extensions (*18*). Zeng et al. proposed a person-based adaptive TSP using connected vehicle information which could better predict the arrival time of vehicles (*19*). Zamanipour et al. presented a model for multimodal signal priority control, considering coordination between intersections and multiple priority requests (*20*). He et al. developed a request-based mixed-integer linear program to address the conflicting issues between actuated coordination and multimodal priority control using information from connected vehicles (*21*). Accounting for more factors, including average passenger delay, maximum queue length, and emissions, Ding et al. established a multi-objective TSP strategy that utilized the predicted bus dwell time (*22*). Delgado et al. integrated at-station bus holding control with TSP, considering high-frequency and capacity-constrained services (*23*). Zeng et al. explicitly accounted for the randomness of bus arrival times by considering the dwell time and vehicle queues to produce TSP timing (*24*).

To address the same issue from different perspectives, some researchers have studied passive TSP strategies which would neither interrupt side-street traffic nor need to detect bus arrivals. Examples of such strategies include: shortening cycle length, metering vehicles, and splitting phases at the isolated intersection level (*25*). Since most TSP controls may yield benefits only at the target intersection, the impacts of such operations on the patterns of downstream arriving traffic and the progression effectiveness of the entire arterial may not be desirable outcomes. Hence, facilitating bus progression over the whole arterial could be a viable strategy to serve as the basis for any further on-line TSP control. As an example of studies in this regard, Lin et al. presented an arterial progression strategy with the objective of maximizing the weighted summation of bus and passenger car bandwidths (*26*). Lin et al. further proposed a model to facilitate the platoon of buses and passenger cars travelling concurrently along arterials by balancing the travel delays of both modes (*27*). Dai et al. developed a model to minimize bus dwell time based on the predetermined maximum and minimum of bus bands and the general vehicle bands (*28*). Dai et al. further categorized

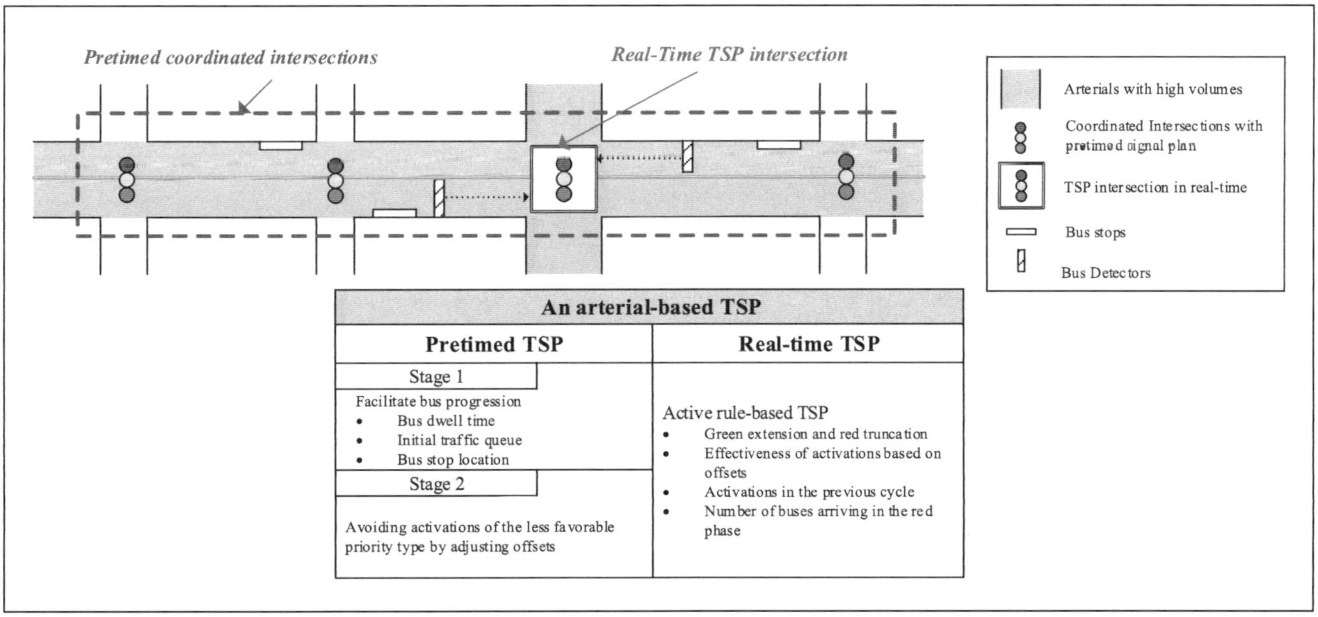

Figure 1. Structure of the proposed integrated TSP system.

intersections into different groups, based on the location of bus stops, to construct a progression model for buses considering the average bus dwell time (*29*). To address the stochastic nature of bus dwell time, Cheng et al. proposed a model to optimize the offsets for a bus-based progression for arterials with heavy transit flows, following the core logic of MAXBAND (*30*). Note that, despite the extensive literature on passive TSP design, some vital issues remain to be addressed, such as the impacts of general traffic queues on bus progression, and integration of passive TSP with potential active TSP activations.

As a viable option to circumvent some deficiencies of TSP control, there have been some proposals to integrate TSP with an adaptive network traffic signal control system. For example, a rule-based TSP has been integrated with a split-cycle offset optimization technique (SCOOT) (*31*). A real-time, hierarchical, optimized, distributed, and effective system (RHODES) has also been used as a base plan for TSP (*32*).

System Development

This study proposes an integrated TSP system for an arterial with heavy transit volumes, where the pre-timed bus-based progression model serves to produce a base plan to facilitate bus flows, and the real-time TSP model functions to supplement the bus progression signal plan at the critical intersections. Figure 1 shows the structure of the proposed system, and critical issues associated with the development of each component will be illustrated in this section.

Pre-timed TSP Module

The pre-timed TSP module aims to facilitate bus movements and, more importantly, to minimize the need to deploy a costly real-time TSP system at all intersections. It executes the function via two stages, where Stage 1 optimizes the offsets to maximize the bus bandwidths, considering bus dwell time, bus stop location, and initial traffic queues at intersections. Stage 2 then proceeds to enhance the progression design so as to reduce the frequency of activating the less favorable TSP strategy at critical intersections, and also to minimize the impacts on the non-priority movements under the same progression bandwidth from Stage 1. Key notations used in the proposed model are defined below, and represented as a diagram in Figure 2.

Stage 1: Design a Set of Bus-Based Progression Bands along an Arterial. With the identified information such as phase plans, traffic flow patterns, and bus dwell time, one can directly extend MAXBAND (*33*) with the following objective function for offset optimization:

$$Max b + k\bar{b} \quad (1)$$

$$(1-k)\bar{b} \geq (1-k)kb \quad (2)$$

where k is a weight factor for bus progression bands in two directions that represents the target ratio of inbound to outbound bandwidths, approximated with the ratio of total inbound to total outbound bus frequency; and $b(\bar{b})$ is the outbound (inbound) bus progression bandwidth (cycles). Equation 2 ensures that the bandwidth for the less favored direction would remain within a reasonable

Notation	Description
	Parameters
N	Number of intersection
r_i	Outbound (inbound) red time at intersection *i* (cycles)
e_i, u_i $(\bar{e}_i, \bar{u}_i)$	Lower and upper limits on outbound(inbound) speed (feet/second)
$g_i, h_i(\bar{g}_i, \bar{h}_i)$	Lower and upper limits on change in outbound(inbound) speed (feet/second)
$v_i(\bar{v}_i)$	Average bus running speed between intersection *i* (*i* + 1) and intersection *i* + 1 (*i*) (feet/cycles)
$f_i(\bar{f}_i)$	Average bus dwell time between intersections *i* (*i* + 1) and *i* + 1 (*i*) (cycles)
d_i	Distance between intersections *i* (*i* + 1) and *i* + 1 (*i*) (feet)
$l_i(\bar{l}_i)$	Distance from intersection *i* (i + 1) to a bus stop between intersections *i* (*i* + 1) and *i* + 1 (*i*) (feet)
$\tau_i(\bar{\tau}_i)$	Queue clearance time at intersection *i* (cycles)
$\varsigma_i(\bar{\varsigma}_i)$	Maximum queue length at intersection *i* (feet)
$S_i(\bar{S}_i)$	The distance from intersections *i*-1(*i* + 1) to the end of maximum queue at intersection *i* (feet)
z	Inverse of cycle length (1/second)
k	Weight factor
M	Large number
	Variables
$b(\bar{b})$	Outbound (inbound) bus bandwidth (cycles)
$b_m(\bar{b}_m)$	Maximized outbound (inbound) bus bandwidth from Stage-1 (cycles)
θ_i	Offset of intersection *i* (cycles)
$w_i(\bar{w}_i)$	Interference variables, equals to the time period from right(left) side of red at intersection *i* to left (right) edge of outbound(inbound) bus band(cycles)
m_i	Loop integer variable for intersection *i* for the outbound (inbound) of bus band
$t_i(\bar{t}_i)$	Average bus running time from intersection *i* (*i* + 1) to intersection *i* + 1 (*i*) (cycles)
$g(\cdot)(\bar{g}(\cdot))$	A function of left boundary of an outbound(inbound) bus band
$a_i(\bar{a}_i)$	The time when the queue vanishes at intersection *i* (cycles)

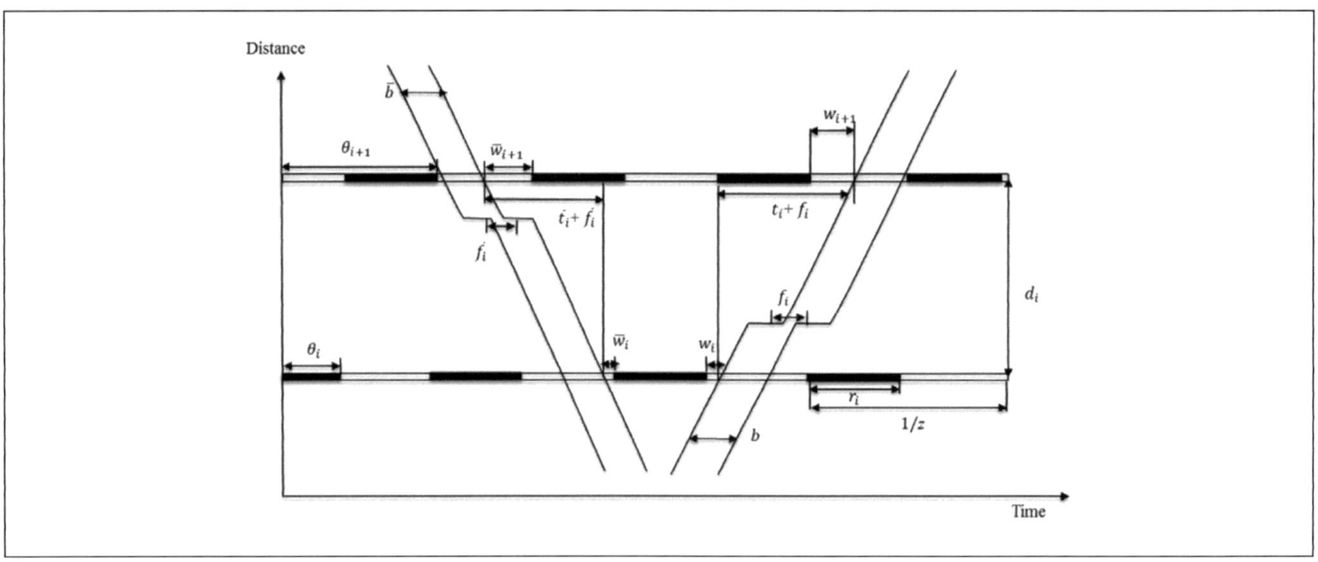

Figure 2. Notations used in the proposed model.

range. The following constraints are specified to ensure the progression for buses:

Interference constraints: The first group, named the directional interference constraints, is to ensure that bus progression bands use the available green time at intersection *i*. Such constraints can be specified as follows:

$$w_i + b \leq (1 - r_i) \ i = 1, \ldots, \mathrm{N} \quad (3)$$

$$\bar{w}_i + \bar{b} \leq (1 - \bar{r}_i) \ i = 1, \ldots, \mathrm{N} \quad (4)$$

where $w_i(\bar{w}_i)$ denotes the time period from the right (left) side of the red phase at intersection *i* to the left (right) boundary of the outbound (inbound) bus band (cycles); and $r_i(\bar{r}_i)$ denotes outbound (inbound) red time at intersection *i* (cycles).

Loop integer constraints: The second group of constraints is specified to ensure that the boundary lines of the progression bands pass through the green phases under the specified offsets. Each constraint functions to limit the difference between the left (right) boundaries of the outbound (inbound) bands for each pair of neighboring intersections to equal the travel time on that link (33). Hence, for a pair of neighboring intersections, i and $i+1$, one can express the loop integer constraints as follows:

$$(w_i + \bar{w}_i) - (w_{i+1} + \bar{w}_{i+1}) + (t_i + f_i + \bar{t}_i + \bar{f}_i) - m_i = (r_{i+1} - r_i) i = 1, \ldots, \mathrm{N} - 1 \tag{5}$$

The bus travel times between intersections i $(i+1)$ and $i+1$ (i) specified in the loop integer constraints are the sum of bus running time, $t_i(\bar{t}_i)$, and dwell time at a bus stop, $f_i(\bar{f}_i)$. For those links without a bus stop, the average dwell time is set to zero.

Constraints for the progression speed and changes of speed: The third group of constraints is to obtain progression speed by taking into account the lower and upper speed limits along the target link and its neighboring links. For a pair of neighboring intersections, i and $i+1$, one can express the progression speed constraints as follows:

$$\left(\frac{d_i}{u_i}\right) \times \mathrm{z} \leq \mathrm{t_i} \leq \left(\frac{d_i}{e_i}\right) \times \mathrm{z}\ i = 1, \ldots, \mathrm{N} - 1 \tag{6}$$

$$\left(\frac{\bar{d}_i}{\bar{u}_i}\right) \times \mathrm{z} \leq \bar{\mathrm{t}}_\mathrm{i} \leq \left(\frac{\bar{d}_i}{\bar{e}_i}\right) \times \mathrm{z}\ i = 1, \ldots, \mathrm{N} - 1 \tag{7}$$

$$\left(\frac{d_i}{h_i}\right) \times \mathrm{z} \leq \left(\frac{d_i}{d_{i+1}}\right) \mathrm{t_{i+1}} - \mathrm{t_i} \leq \left(\frac{d_i}{g_i}\right) \times \mathrm{z}\ i = 1, \ldots, \mathrm{N} - 2 \tag{8}$$

$$\left(\frac{\bar{d}_i}{\bar{h}_i}\right) \times \mathrm{z} \leq \left(\frac{\bar{d}_i}{\bar{d}_{i+1}}\right) \bar{\mathrm{t}}_{\mathrm{i+1}} - \bar{\mathrm{t}}_\mathrm{i} \leq \left(\frac{\bar{d}_i}{\bar{g}_i}\right) \times \mathrm{z}\ i = 1, \ldots, \mathrm{N} - 2 \tag{9}$$

where $d_i(\bar{d}_i)$ is the distance between intersections $i(i+1)$ and $i+1(i)$ (feet); z is the inverse of cycle length (1/second); $\mathrm{e_i}, u_\mathrm{i}(\bar{e}_i, \bar{u}_i)$ denotes the lower and upper limits of the outbound (inbound) speed (feet/second); and $\mathrm{g_i}, h_\mathrm{i}(\bar{g}_i, \bar{h}_i)$ denotes lower and upper limits on such changes (feet/second).

Bus stop location and intersection traffic queue constraints: Another important set of constraints is to account for the locations of bus stops and their relationship with intersection traffic queues. To ensure bus progression, the anticipated bus band should not be interrupted by the intersection traffic queues. Taking the outbound direction in Figure 3*a* as an example, the queue vanish point (a_{i+1}, S_{i+1}) needs to be located above the left boundary of the bus progression band to avoid interrupting bus progression.

The left boundary of the bus progression band is determined by the bus stop location and bus travel time (running and dwell times) between the intersections. It can be divided into three time intervals: before the bus stop, at the bus stop, and after the bus stop. Referring to Figure 3, one can formulate the left boundary of the bus band in a time–space diagram, where the origin, (0, 0), is

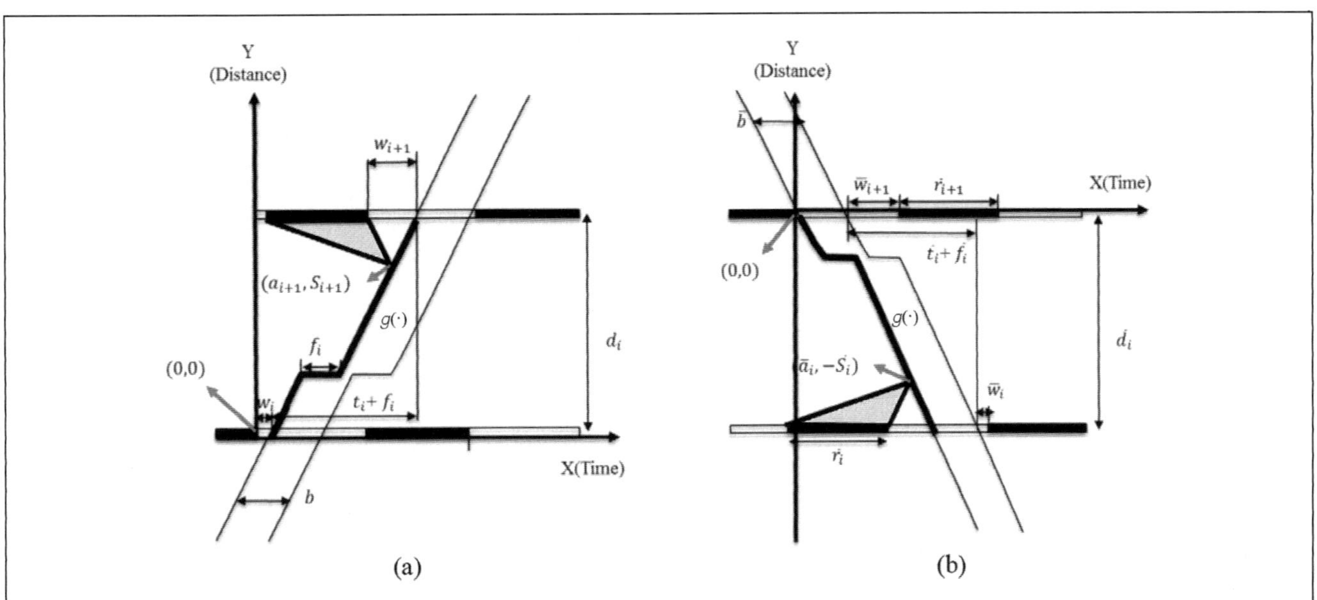

Figure 3. Illustration of a relationship of traffic queue and bus progression band: (*a*) outbound; (*b*) inbound.

the onset of a green phase at the outbound (inbound) upstream intersection i ($i+1$), as follows:

$$g(z_{i+1}) = \begin{cases} v_i \times (z_{i+1} - (w_i)), & |w_i \le z_{i+1} \le w_i + t_i \times l_i/d_i \\ v_i \times t_i \times l_i/d_i, & |w_i + t_i \times l_i/d_i \le z_{i+1} \le w_i + t_i \times l_i/d_i + f_i \quad i = 1, \ldots, \mathrm{N}-1 \\ v_i \times (z_{i+1} - (w_{i1} + f_i)), & |w_i + t_i \times l_i/d_i + f_i \le z_{i+1} \end{cases} \tag{10}$$

$$\bar{g}(z_i) = c\begin{cases} -\bar{v}_i \times (z_i - (1 - \bar{r}_{i+1} - \bar{w}_{i+1})), & |1 - \bar{r}_{i+1} - \bar{w}_{i+1} \le z_i \le 1 - \bar{r}_{i+1} - \bar{w}_{i+1} + \bar{t}_i \times \bar{l}_i/\bar{d}_i \\ -\bar{v}_i \times \bar{t}_i \times \bar{l}_i/\bar{d}_i, & |1 - \bar{r}_{i+1} - \bar{w}_{i+1} + \bar{t}_i \times \bar{l}_i/\bar{d}_i \le z_i \le 1 - \bar{r}_{i+1} - \bar{w}_{i+1} + \bar{t}_i \times \bar{l}_i/\bar{d}_i + \bar{f}_i\, i = 1, \ldots, \mathrm{N}-1 \\ -\bar{v}_i \times (z_i - (1 - \bar{r}_{i+1} - \bar{w}_{i+1} + \bar{f}_i)), & |1 - \bar{r}_{i+1} - \bar{w}_{i+1} + \bar{t}_i \times \bar{l}_i/\bar{d}_i + \bar{f}_i \le z_i \end{cases} \tag{11}$$

where $v_i(\bar{v}_i)$ denotes average bus running speed between intersections i ($i+1$) and $i+1$ (i) (feet/cycle); and $l_i(\bar{l}_i)$ denotes the distance from the upstream intersection to a bus stop between intersections i ($i+1$) and $i+1$ (i) (feet). In Equations 10 and 11, an array of functions represents the location of the left boundary of the bus band for the corresponding subdomains of time interval.

Given the traffic volume at an intersection, one can approximate the average queue length (ς) and average queue clearance time (τ) as follows:

$$\varsigma = \left(V \times r + \frac{V \times r}{s} \times V\right) \times L \tag{12}$$

$$\tau = \left(V \times r + \frac{V \times r}{s} \times V\right)/s \tag{13}$$

where V denotes the traffic volume (vehicles/cycle); r is the red interval (cycles); s is the saturation flow rate (vehicles/cycle); and L is the length of a vehicle (feet). The point where queues vanish in Figure 3, (a_i, S_i) $((\bar{a}_i, \bar{S}_i))$, can then be defined as follows:

$$a_{i+1} = (w_i - w_{i+1}) + t_i + f_i + \tau_{i+1} \quad i = 1, \ldots, \mathrm{N}-1 \tag{14}$$

$$S_{i+1} = d_i - \varsigma_{i+1} \quad i = 1, \ldots, \mathrm{N}-1 \tag{15}$$

$$\bar{a}_i = ((1 - r_{i+1}) - \bar{w}_{i+1}) + \bar{t}_i + \bar{f}_i - ((1 - r_i) - \bar{w}_i) + \bar{\tau}_i \quad i = 1, \ldots, \mathrm{N}-1 \tag{16}$$

$$\bar{S}_i = \bar{d}_i - \bar{\varsigma}_i \quad i = 1, \ldots, \mathrm{N}-1 \tag{17}$$

In a time–space diagram as shown in Figure 3a, Equation 14 indicates that the time when the queues vanish at the downstream intersection, $i+1$ (a_{i+1}), is the sum of the relative offset of intersection $i+1$ from intersection i, $((w_i - w_{i+1}) + t_i + f_i)$, and the queue clearance time at intersection $i+1$ (τ_{i+1}). Equation 15 represents the available storage space of the segment (S_{i+1}) at that time (a_{i+1}), which is the length of the link between intersections i and $i+1$, d_i, with the maximum queue length, ς_{i+1}, subtracted. By the same token, one can set the constraints for the inbound direction as shown in Equations 16 and 17 (see Figure 3b).

Note that the available storage, $S_{i+1}(\bar{S}_i)$, needs to be sufficiently long to prevent the intersection traffic queue from breaking the left boundary of the bus band at the time prior to the complete dissipation of the queue, $g(a_{i+1})(\bar{g}(\bar{a}_i))$, as follows:

$$g(a_{i+1}) \le S_{i+1} \quad i = 1, \ldots, \mathrm{N}-1 \tag{18}$$

$$-\bar{g}(\bar{a}_i) \le \bar{S}_i \quad i = 1, \ldots, \mathrm{N}-1 \tag{19}$$

In brief, the proposed off-line TSP model can be summarized as follows:

$$Max b + k\bar{b}$$

s.t. Equations 2–11
Equations 14–19
$b, \bar{b}, w_i, \bar{w}_i, t_i, \bar{t}_i, a_i, \bar{a}_i \geq 0$ *for* $i = 1,\ldots,\mathrm{N}$
m_i integer *for* $i = 1,\ldots,\mathrm{N}-1$

Stage 2: Enhance the Proposed Off-Line TSP Model by Considering the Effects of Green Extension or Red Truncation. TSP control at a critical intersection may execute a red truncation or green extension to facilitate incoming buses. The negative impacts of executing a red truncation or green extension on non-priority movements differ due to the differences in the resulting green time. Note that Stage 1, with its mathematical nature, may yield multiple sets of offsets for the same total bandwidth, depending on the green intervals of intersections and the travel times between them. Hence, Stage 2 will function to find a solution which can minimize the potential negative impacts caused by TSP activations on the non-priority movements.

To approximate the negative impacts of a TSP activation on a shortened phase, the following index is introduced:

$$I = \frac{VC - s(g - t_s)}{sg - VC} = \frac{st_s}{sg - VC} - 1 \tag{20}$$

where V denotes the highest traffic volume of the movement during the given phase (vehicles/lane/second); g denotes the green interval of the phase (seconds); t_s is the potential extension or truncation duration (seconds); C is the cycle length (seconds); and s denotes the saturation flow rate (vehicles/second). The index, I, indicates the estimated number of cycles for the sacrificed phase to recover after a TSP activation. It represents the magnitude of negative impacts of TSP activation on that phase; the TSP strategy with a higher positive value for I is viewed as less favorable.

More specifically, let I^+ denote the index for the phase after the priority phase and I^- for the phase before. For example, if $I^- > I^+$, the red truncation is expected to cause more negative impacts than under the green extension. Hence, the likelihood of activating a red truncation under this case should be minimized, so the offsets from Stage 1 should be designed such that those buses, if they need to arrive at the TSP intersection during the red phase, are more likely to arrive at the beginning of the red phase rather than the end of the red phase. Hence, one shall maximize $\theta_{k-1} - \theta_k$ for the outbound direction and $\theta_{k+1} - \theta_k$ for the inbound direction ($\theta_i$denotes offset of intersection i, and k denotes the real-time TSP intersection), without compromising the maximum bands obtained from Stage 1. Following the same logic, to reduce the potential activation of a green extension, one should minimize the offset difference from the upstream intersection to the TSP intersection.

In brief, let intersection k be the real-time TSP intersection and $b_m(\bar{b}_m)$ be the maximized bandwidth obtained from Stage 1. The proposed off-line TSP in Stage 2, designed such that each TSP intersection activates the strategy with least negative impact, can be summarized as follows:

$$\text{If red truncation is less favorable}(I^- > I^+), \quad \max \theta_{k-1} - \theta_k + (\theta_{k+1} - \theta_k) \tag{21}$$

$$\text{If green extension is less favorable}(I^- < I^+), \quad \min \theta_{k-1} - \theta_k + (\theta_{k+1} - \theta_k) \tag{22}$$

$$\text{s.t.} b \geq b_m, \bar{b} \geq \bar{b}_m \tag{23}$$

Equations 2–11
Equations 14–19
$b, \bar{b}, w_i, \bar{w}_i, t_i, \bar{t}_i, a_i, \bar{a}_i \geq 0 \; for \; i = 1,\ldots,N$
m_i integer $for \; i = 1,\ldots,N-1$

Real-Time TSP Module

Despite a pre-timed signal plan for bus progression, some buses may still encounter the red phase at some intersections along an arterial, due to the stochastic nature of dwell time and travel time. Considering that a pre-timed bus-oriented base signal plan is expected to reduce the number of buses encountering the red phase along an arterial, a real-time TSP control is implemented only at the critical intersections where major arterials meet to smooth bus progression further with minimum negative impacts on non-priority movements.

The TSP control at a critical intersection may execute either red truncation or green extension if a bus is detected by the detectors located upstream of the TSP intersection. If a bus is detected during the detection interval at the end of green, an extension of a predetermined duration will be granted to the current green phase. If a bus is detected during the red phase, the red phase will end earlier to provide extra green time to the transit vehicle. The real-time TSP in the proposed system is a rule-based strategy and the activation of a green extension and red truncation will be determined with the following rules:

Rule 1: Execute the TSP Only if the Reduced Delay Would Not be Transferred into Extra Waiting Time at the Downstream Intersection. Note that the benefits from a green extension or red truncation at the TSP intersection can be diminished at the downstream intersection under some circumstances. For example, in Case I shown in Figure 4, although an outbound bus is granted a green extension to pass the TSP intersection, it would actually experience longer waiting time at the downstream intersection.

More specifically, if offsets of the TSP intersection (i) and the downstream intersection ($i + 1$) encounter the following condition (see Equation 24), the reduced travel time at the TSP-activated intersection will cause the

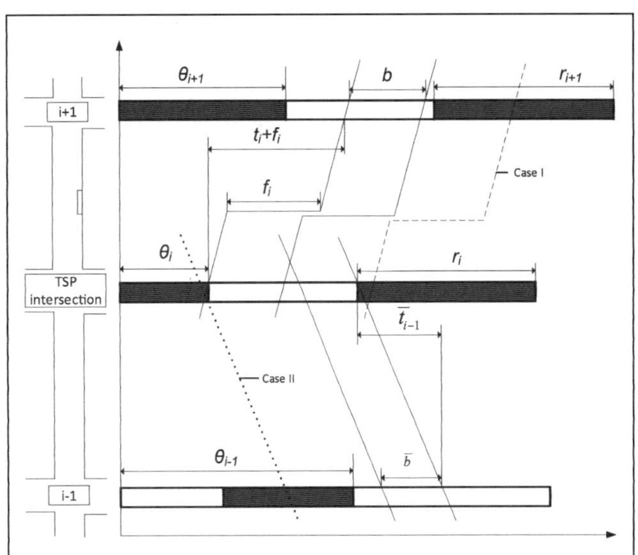

Figure 4. Offsets causing ineffective TSP activations.
Note: Case I is shown with a dashed line and a label. Case II is shown with a dotted line and a label.

target bus to experience extra waiting time at the downstream intersection,

$$\theta_i + (1 - r_i) + t_i + f_i \geq \theta_{i+1} + (1 - r_{i+1}) + \alpha z \quad (24)$$

where α is a positive number denoting a buffer time (e.g., 10 seconds in this study), and z denotes the inverse of cycle length (1/sec). If Equation 24 is satisfied, an outbound bus, receiving a green extension, would not be able to pass the downstream intersection without a stop. By the same token, as shown in Case II in Figure 4, if an inbound bus, receiving red truncation at the TSP intersection (i), should encounter the red phase at the downstream intersection i–1, the priority activation would be viewed as ineffective.

Hence, the logic for assessing the effectiveness of granting a green extension or red truncation in both directions at a TSP operation under the proposed system can be listed as follows:

- If $\theta_i + (1 - r_i) + t_i + f_i \geq \theta_{i+1} + (1 - r_{i+1}) + \alpha z$, green extension for outbound buses is ineffective;
- If $\theta_i + t_i + f_i \leq \theta_{i+1} - \alpha z$, red truncation for outbound buses is ineffective;
- If $\theta_i + (1 - \bar{r}_i) + \bar{t}_{i-1} + \bar{f}_{i-1} \geq \theta_{i+1} + (1 - \bar{r}_{i+1}) + \alpha z$, green extension for inbound buses is ineffective; and
- If $\theta_i + \bar{t}_{i-1} + \bar{f}_{i-1} \leq \theta_{i+1} - \alpha z$, red truncation for inbound buses is ineffective.

In the proposed strategy, the request to activate green extension or red truncation would be ignored if it is assessed as ineffective. Note that, since the activation is triggered by buses from either direction, the priority activations can be implemented independently from one direction to the other. By applying Rule 1, the proposed system can decrease the negative impacts on the non-priority movements by eliminating the inefficient TSP activation.

Rule 2: Do Not Execute the Same Activation Type in Consecutive Cycles. Since the phases for non-priority movements may be interrupted by the TSP activations, they would need time to recover from the interruption. Moreover, consecutive activations of the same priority control may further aggravate the negative impacts on the corresponding non-priority movements (*34*). Therefore, in this study, it is not allowed to activate the type of TSP control (i.e., green extension or red truncation) if it has been activated in the previous cycle.

Rule 3: Activate Red Truncation if at Least Two Buses are Detected in the Detection Interval. To avoid frequent activation of TSP with marginal benefits but significant negative impacts on the side street, in the proposed system the red truncation would only be activated if at least two buses were detected during the detection interval.

Case Study

Test Site

To evaluate the potential of the proposed system for real-world applications, we selected an arterial segment on Dongzhimenwai Road in Beijing, China, for a case study. The simulation network consists of five intersections with two roadside bus stops in each direction. The key features, such as geometric conditions, bus operations, traffic volumes, and signal settings, used in the evaluation analysis are summarized in Figure 5. Note that the bus dwell times at bus stops in each direction are assumed to follow normal distributions with mean and standard deviation (seconds) shown next to the bus stops in Figure 5*a*. For example, dwell times at a bus stop between intersections III and IV in Eastbound follow normal distributions with mean of 30 seconds and standard deviation of 5 seconds. To simulate the stochastic bus entry times to the arterial, a dummy stop is located at the upstream of the first intersection in each direction with a high standard deviation of dwell times of 18 seconds. The maximum and minimum cruising speeds for buses are set as 55 feet/second (38 mph) and 45 feet/second (30 mph). To test the performance under different traffic demands, this study adopts three traffic volume scenarios, as shown in Figure 5*b*. The signal timings are calculated based on the traffic volumes, as shown in Figure 5*c*.

Experimental Design

Since the relative offsets between a TSP intersection and its adjacent intersections may significantly affect the number of priority calls and their effects, TSP strategies may perform differently under different base signal plans along an arterial. For comparison, this study has adopted MULTIBAND (*35*), one of the most rigorous signal progression models, to yield a base signal plan, and to work with unconditional and conditional real-time TSPs. The detector is located 200 feet upstream from the stop line and the detection interval for green extension is within the last 5 seconds of the green phase. The green extension and red truncation durations are set at 10 seconds (*2*), which is longer than the travel time to the stop bar, to make sure that the green extension activation will be effective. The unconditional TSP can execute the green extension and red truncation, as introduced in the real-time TSP section, without rules. The conditional TSP, implemented with MULTIBAND, includes the state of activations in the previous cycle, as stated in Rule 2. The proposed bus-based progression plan without real-time

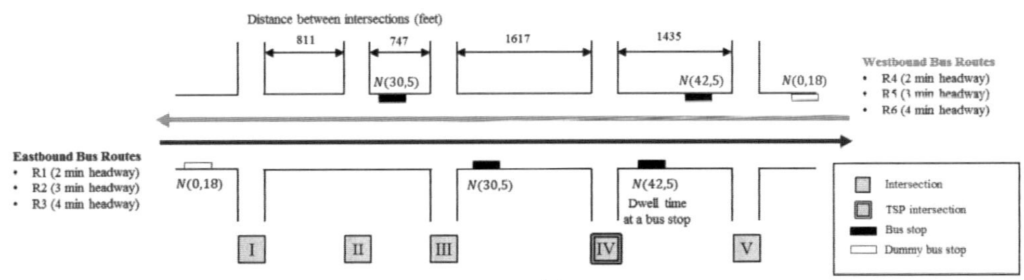

(a)

	Intersections	I			II			III			IV			V		
	Movements	L	T	R	L	T	R	L	T	R	L	T	R	L	T	R
Q1	EB	\	963	65	\	1127	\	51	1076	\	430	722	108	156	737	54
	WB	\	784	\	\	716	111	\	721	67	395	545	97	53	847	19
	NB	46	124	43	\	\	\	89	148	142	243	450	\	114	68	22
	SB	123	166	43	\	\	68	69	143	40	223	494	\	35	76	76
Q2	EB	\	1116	76	\	1306	\	59	1247	\	499	837	126	181	854	63
	WB	\	908	\	\	830	128	\	836	78	457	632	113	61	982	22
	NB	54	144	50	\	\	\	103	172	164	282	522	\	132	79	26
	SB	142	192	50	\	\	79	79	166	46	259	573	\	41	88	88
Q3	EB	\	1208	82	\	1414	\	64	1350	\	540	906	136	196	924	68
	WB	\	983	\	\	898	139	\	905	84	495	684	122	66	1063	24
	NB	58	156	54	\	\	\	112	186	178	305	565	\	143	85	28
	SB	154	208	54	\	\	85	86	180	50	280	620	\	44	95	95

(b)

I			II			III			IV			V		
Q1	Q2	Q3	Q1	Q2	Q3	Q1	Q2	Q3	Q1	Q2	Q3	Q1	Q2	Q3
44	59	73	65	87	108	44	59	73	28	37	46	35	47	58
16	21	27	25	33	42	31	41	51	22	30	37	16	21	26
30	40	50				15	20	25	18	24	30	21	27	34
									22	29	36	19	25	31

- Note that a 3-second yellow interval and a 2-second all-red interval are included in each phase.

(c)

Figure 5. General information associated with the study site: (*a*) Geometry and bus operations; (*b*) Traffic demands (unit: vph); (*c*) Phase plan and signal timings (unit: seconds).

TSP control will also be adopted for comparison to illustrate the effectiveness of the integrated control strategy. In summary, those cases for comparison are set to operate with one of the following models:

- MULTIBAND - TSP 1: Conditional TSP on MULTIBAND signal plan
- MULTIBAND - TSP 2: Unconditional TSP on MULTIBAND signal plan
- BUSBAND: The proposed bus-based progression model without real-time TSP
- BUSBAND with TSP: The proposed real-time TSP on BUSBAND signal plan

With the predetermined cycle length and signal timings, offsets are optimized with each progression model for either cars or buses. To ensure that the negative impacts on the non-priority movement are comparable under two different base plans, the phase sequence at each intersection is also predetermined. Active TSP controls are implemented at intersection IV, which consists of two major arterials and high volumes, for the non-priority movements. This study adopts VISSIM as an unbiased traffic simulation tool, and the active TSP are implemented with COM interface and Visual Basic code. The simulation duration is set to one and a half hours and the averages of five simulation runs are adopted for comparison.

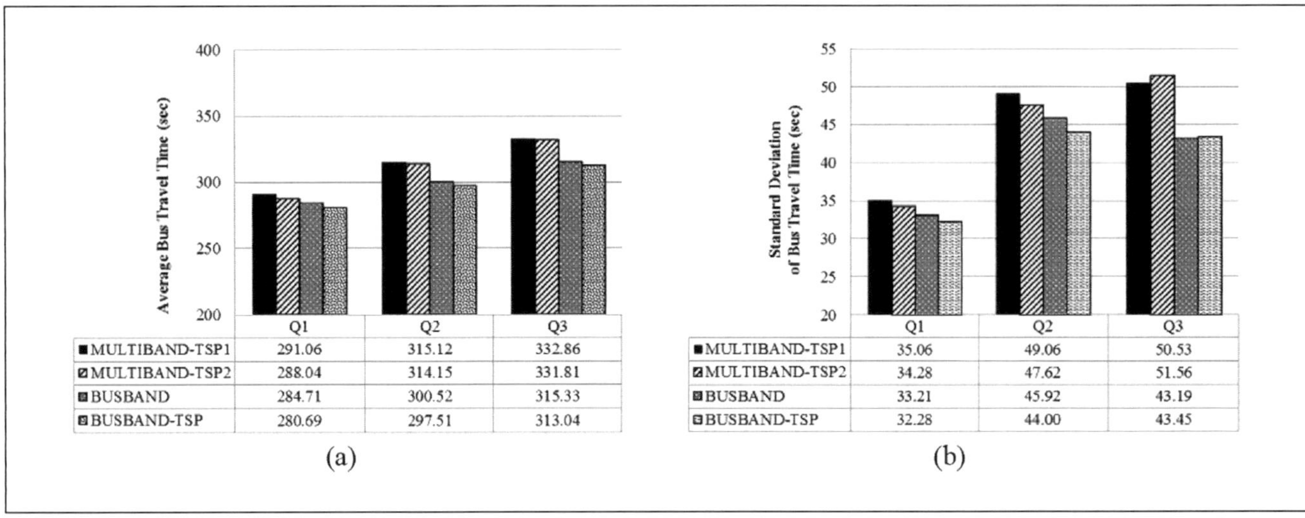

Figure 6. Average bus travel times and standard deviations (seconds): (*a*) Average bus travel time; (*b*) Standard deviations of bus travel time.

Simulation Results

To assess the performance of bus progression and the benefits for the overall network under the proposed arterial-based TSP system, this study has adopted the following measurements of effectiveness (MOE).

Average Bus Travel Times and Standard Deviations along the Arterial. As shown in Figure 6*a*, the proposed BUSBAND integrated with TSP can outperform the TSP strategies implemented on MULTIBAND with respect to reducing travel times under Demand 1 to Demand 3, especially under the higher demand scenarios. Taking the scenario of Demand 2 as an example, while the MULTIBAND with conditional TSP shows an average travel time of 315.12 seconds, BUSBAND supplemented with TSP shows an average travel time of 297.51 seconds, a reduction of 5.59%. BUSBAND with TSP, in comparison with BUSBAND only, has achieved a greater reduction in travel times (from 4.63% to 5.59% under Demand 2).

In addition, Figure 6*b* shows that BUSBAND and BUSBAND-TSP yield significantly lower travel time standard deviations than the MULTIBAND with either conditional or unconditional TSP control, and result in a lower headway variation for buses along the arterial. Taking the scenario of Demand 2 as an example, while MULTIBAND with conditional TSP shows the standard deviation of travel times of 49.06 seconds, BUSBAND with TSP shows the standard deviation of travel times of 44.00 seconds, a reduction of 10.31%.

It is suggested that the proposed arterial-based TSP strategies can benefit not only the bus users but also transit operators since reduced variability of travel time can better ensure the service reliability at bus stops and the terminals.

Delays at the TSP Intersection and along the Arterial and Number of Stops of Buses along the Arterial. Figure 7*a* presents how BUSBAND with TSP outperforms MULTIBAND with unconditional TSP at the target intersection under Demand 2, but shows slightly longer bus delay under Demand 1 and Demand 3. However, over the entire arterial, BUSBAND with TSP produces the lowest average bus delay under three demand scenarios among all signal strategies (Figure 7*c*). Furthermore, BUSBAND-only also outperforms MULTIBAND with conditional and unconditional TSPs with respect to bus delay along the arterial.

Since an active TSP utilizes signal timings of non-priority movements at the TSP intersection, BUSBAND-only shows the lowest average delay of non-priority movements at the TSP intersection, but other real-time TSP strategies show an increase in such delays (Figure 7*b*). Moreover, it is evident that the proposed system is able to reduce the delays to the non-priority movements compared with the TSP implemented with MULTIBAND.

Furthermore, Figure 7*d* shows the number of stops for buses along an arterial. In general, BUSBAND and BUSBAND with TSP controls significantly outperform the TSP control strategies implemented with MULTIBAND in reducing the number of stops along the arterial. Taking the scenario of Demand 2 as an example, BUSBAND and BUSBAND with TSP control both show the number of stops as 2.90, indicating a reduction of 10.58% and 10.57%, respectively, when

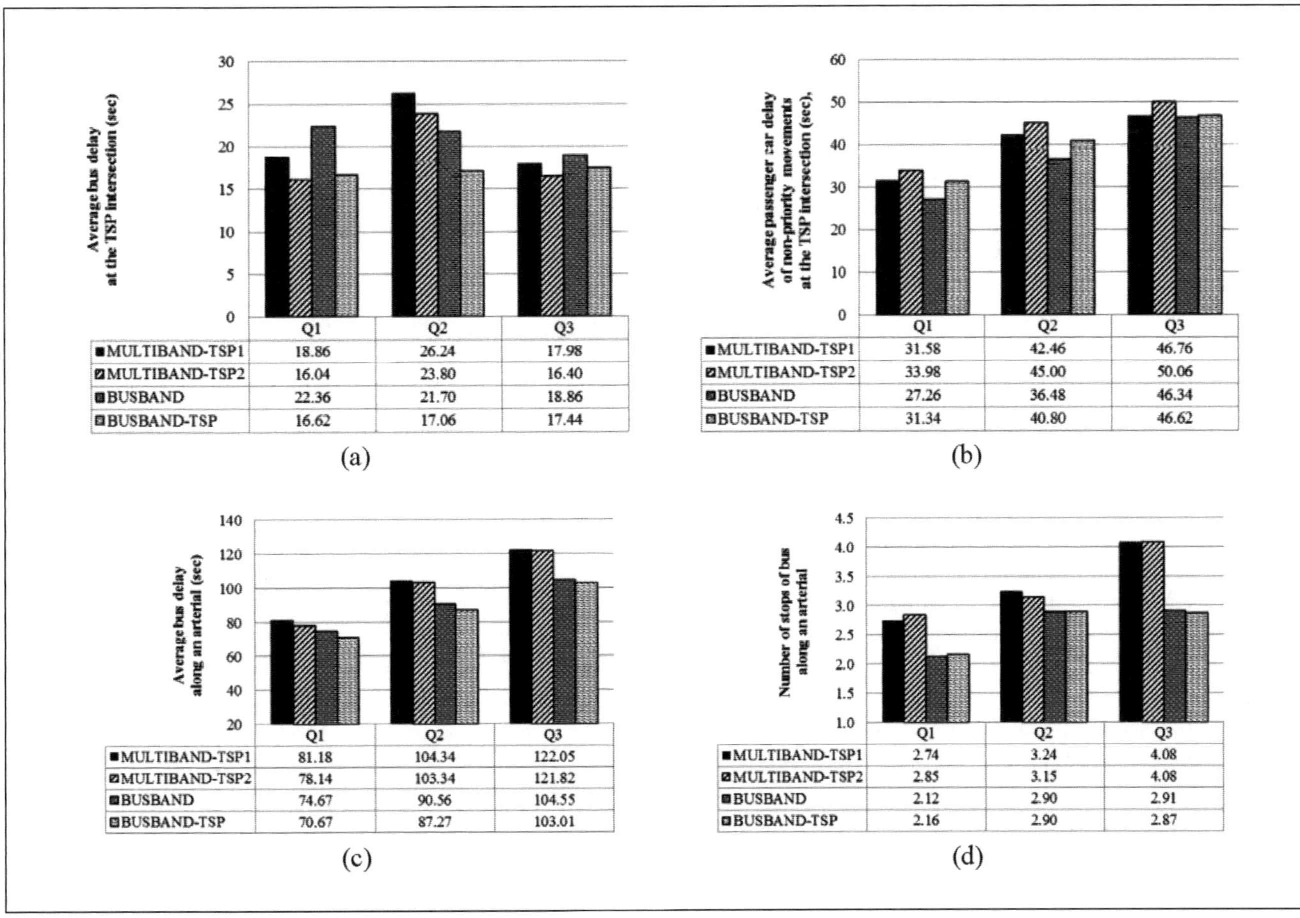

Figure 7. Delays at the target intersections and along an arterial (seconds) and number of stops of buses along the arterial: (*a*) Bus delays at the active TSP intersection; (*b*) Passenger cars: delays of non-priority movements at the active TSP intersection; (*c*) Bus delays along an arterial; (*d*) Number of stops of buses along an arterial.

compared with MULTIBAND with conditional TSP (3.24).

Total Person Delay along the Arterial. To evaluate the network's performance, average person delay is computed with the following equation:

$$d_p = \frac{\rho_c \cdot q_c \cdot d_c + \rho_b \cdot q_b \cdot d_b}{\rho_c \cdot q_c + \rho_b \cdot q_b} \tag{25}$$

where, d_p, d_c, and d_b denote the average person delay, average passenger car delay, and average bus delay, in seconds, respectively; ρ_c and ρ_b represent the loading factor of passenger cars and buses and are set at 1.5 and 30, respectively; and q_c and q_b are the flow rates of passenger cars and buses in vehicles per hour, respectively.

As shown in Figure 8*a*, BUSBAND and BUSBAND with TSP yield less average person delay compared with the results for MULTIBAND with conditional and unconditional TSP controls, especially under Demand 2 and Demand 3. Taking the scenario of Demand 2 as an example, BUSBAND with TSP decreases the average person delay by 10.41%, compared with MULTIBAND with conditional TSP.

Figure 8*b* shows the sensitivity of the average person delay with respect to various loading factors under the scenario of Demand 2. With an increase in the bus loading factor, BUSBAND and BUSBAND with TSP can reduce average person delays more. For example, if the loading factors of passenger car and bus are 1.2 and 40, respectively, BUSBAND and BUSBAND with TSP can decrease the average person delay by 11.2% and 12.0%, respectively, compared with MULTIBAND with conditional TSP. Such results evidence that the benefits of the proposed system become more pronounced for arterials with higher bus ridership.

Conclusions

This study has proposed a TSP system for an arterial with heavy transit flows which contains a bus-oriented base signal plan along an arterial and a rule-based,

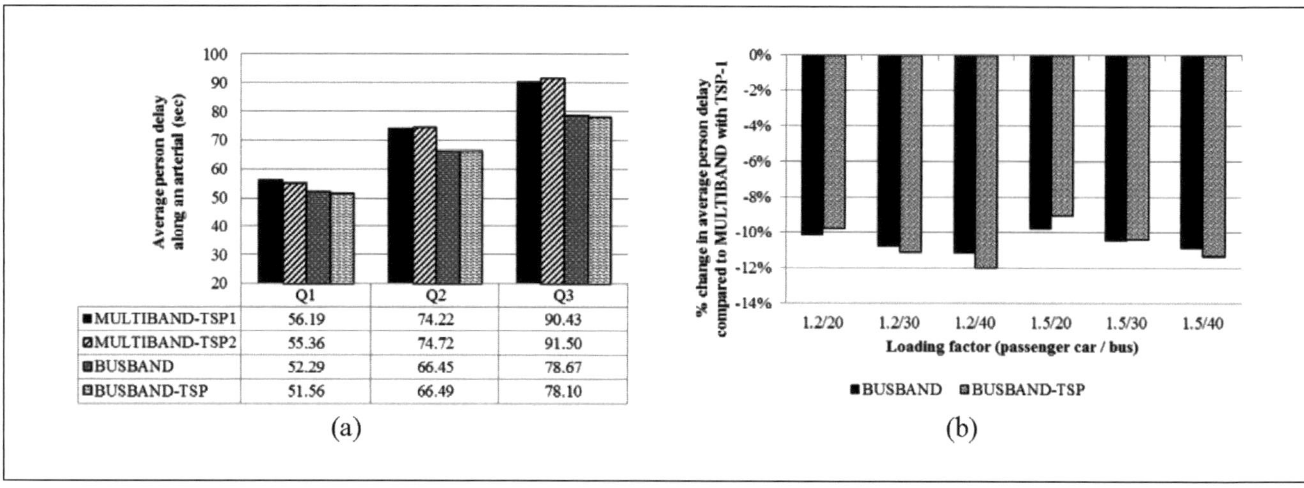

Figure 8. (*a*) Average person delays along an arterial (seconds) and (*b*) sensitivity of average person delays to various loading factors.

real-time TSP strategy. Its base signal plan is formulated with a two-stage model, where Stage 1 is designed to maximize the two-way bus progression bandwidth, considering the location of bus stops and the initial traffic queues at intersections, and Stage 2 functions to minimize the probability of activating a less favorable TSP control strategy in design of offsets. The rule-based TSP control functions to execute the priority to buses in real time while minimizing the negative impacts on non-priority movements. Its control rules account for the effectiveness of activations, the state of activation in the previous cycle, and the number of buses arriving at the target intersection during a red phase. The entire system features its potential to achieve the progression for bus flows with minimum investment on deploying real-time TSP at intersections.

The results from the extensive case study show that the proposed system can significantly improve the performance of buses with respect to various MOEs, including bus travel times and their variance, the number of stops along an arterial, and the negative impacts on the non-priority movements at TSP intersections, compared with the TSP controls with MULTIBAND. The results of experimental analyses show that the advantages of the proposed model would be more pronounced under the traffic scenarios of higher transit ridership and volume.

Considering the potential impacts of bus-based progression to general traffic, one of our on-going research issues is to develop an evaluation that will enable the responsible agencies to decide when to activate the arterial control with one of the following progression strategies: (a) bus-based signal progression; (b) auto-based signal progression; and (c) concurrent progression for both passengers and transit vehicles. In addition, further research to be undertaken includes (a) incorporating the bus turning ratios into and out of the arterial at each intersection into the design, (b) developing reliable models for estimating the physical length of traffic queues at the intersections, and (c) a method to determine at which intersections to deploy active TSP controls.

Author Contributions

The authors confirm contribution to the paper as follows: study conception and design: Hyeonmi Kim, Yao Cheng, Gang-Len Chang; data collection: Hyeonmi Kim; analysis and interpretation of results: Hyeonmi Kim, Yao Cheng, Gang-Len Chang; draft manuscript preparation: Yao Cheng, Hyeonmi Kim, Gang-Len Chang. All authors reviewed the results and approved the final version of the manuscript.

References

1. Ludwick, J. S. Simulation of an Unconditional Preemption Bus Priority System. *Transportation Research Record: Journal of the Transportation Research Board*, 1975. 536: 1–10.
2. Garrow, M., and R. B. Machemehl. *Development and Evaluation of Transit Signal Priority Strategies*. SWUTC/97/472840-00068-1. University of Texas at Austin, 1997.
3. Lin, Y., X. Yang, N. Zou, and M. Franz. Transit Signal Priority Control at Signalized Intersections: a Comprehensive Review. *Transportation Letters*, Vol. 7, No. 3, 2015, pp. 168–180.
4. Evans, H. K., and G. W. Skiles. Improving Public Transit through Bus Preemption of Traffic Signals. *Traffic Quarterly*, Vol. 24, No. 4, 1970, pp. 531–543.
5. *Bus Priority at Traffic Signals Using Selective Detection*. Technical Memorandum H2/77. HMSO, Department of Transport, London, 1977.
6. Moghimidarzi, S., P.G. Furth, and B. Cesme. Predictive–Tentative Transit Signal Priority with Self-Organizing Traffic Signal Control. *Transportation Research Record: Journal of the Transportation Research Board*, 2016. 2557: 77–85.

7. Furth, P., and T. H. Muller. Conditional Bus Priority at Signalized Intersections: Better Service with Less Traffic Disruption. *Transportation Research Record: Journal of the Transportation Research Board*, 2000. 1731: 23–30.
8. Skabardonis, A. Control Strategies for Transit Priority. *Transportation Research Record: Journal of the Transportation Research Board*, 2000. 1727: 20–26.
9. Hounsell, N. B., F. N. McLeod, K. Gardner, J. R. Head, and D. Cook. Headway-based Bus Priority in London Using AVL: First Results. *Proc. 10th IEE International Conference on Road Transportation Information and Control*, London, UK, 2000, pp. 218–222.
10. Lin, Y., X. Yang, G.-L. Chang, and N. Zou. Transit Priority Strategies for Multiple Routes under Headway-based Operations. *Transportation Research Record: Journal of the Transportation Research Board*, 2013. 2356: 34–43.
11. Head, L., D. Gettman, and Z. Wei. Decision Model for Priority Control of Traffic Signals. *Transportation Research Record: Journal of the Transportation Research Board*, 2006. 1978: 169–177.
12. Chang, G.-L., M. Vasudevan, and C.-C. Su. Modelling and Evaluation of Adaptive Bus-preemption Control with and without Automatic Vehicle Location Systems. *Transportation Research Part A: Policy and Practice*, Vol. 30, No. 4, 1996, pp. 251–268.
13. Hu, J., B. B. Park, and Y. J. Lee. Coordinated Transit Signal Priority Supporting Transit Progression under Connected Vehicle Technology. *Transportation Research Part C: Emerging Technologies*, Vol. 55, 2015, pp. 393–408.
14. Christofa, E., K. Ampountolas, and A. Skabardonis. Arterial Traffic Signal Optimization: A Person-based Approach. *Transportation Research Part C: Emerging Technologies*, Vol. 66, 2016, pp. 27–47.
15. Xu, M., Z. Ye, H. Sun, and W. Wang. Optimization Model for Transit Signal Priority under Conflicting Priority Requests. *Transportation Research Record: Journal of the Transportation Research Board*, 2016. 2539: 140–148.
16. Hu, J., B. B. Park, and Y. J. Lee. Transit Signal Priority Accommodating Conflicting Requests under Connected Vehicles Technology. *Transportation Research Part C: Emerging Technologies*, Vol. 69, 2016, pp. 173–192.
17. Ye, Z., and M. Xu. Decision Model for Resolving Conflicting Transit Signal Priority Requests. *IEEE Transactions on Intelligent Transportation Systems*, Vol. 18, No. 1, 2017, pp. 59–68.
18. Yu, Z., V. V. Gayah, and E. Christofa. Person-Based Optimization of Signal Timing: Accounting for Flexible Cycle Lengths and Uncertain Transit Vehicle Arrival Times. *Transportation Research Record: Journal of the Transportation Research Board*, 2017. 2620: 31–42.
19. Zeng, X., X. Sun, Y. Zhang, and L. Quadrifoglio. Person-Based Adaptive Priority Signal Control with Connected-Vehicle Information. *Transportation Research Record: Journal of the Transportation Research Board*, 2015. 2487: 78–87.
20. Zamanipour, M., K. L. Head, Y. Feng, and S. Khoshmagham. Efficient Priority Control Model for Multimodal Traffic Signals. *Transportation Research Record: Journal of the Transportation Research Board*, 2016. 2557: 86–99.
21. He, Q., K. L. Head, and J. Ding. Multi-modal Traffic Signal Control with Priority, Signal Actuation and Coordination. *Transportation Research Part C: Emerging Technologies*, Vol. 46, 2014, pp. 65–82.
22. Ding, J., M. Yang, W. Wang, C. Xu, and Y. Bao. Strategy for Multiobjective Transit Signal Priority with Prediction of Bus Dwell Time at Stops. *Transportation Research Record: Journal of the Transportation Research Board*, 2015. 2488: 10–19.
23. Delgado, F., J. C. Muñoz, R. Giesen, and N. H. Wilson. Integrated Real-Time Transit Signal Priority Control for High-Frequency Segregated Transit Services. *Transportation Research Record: Journal of the Transportation Research Board*, 2015. 2533: 28–38.
24. Zeng, X., Y. Zhang, K. N. Balke, and K. Yin. A Real-time Transit Signal Priority Control Model Considering Stochastic Bus Arrival Time. *IEEE Transactions on Intelligent Transportation Systems*, Vol. 15, No. 4, 2014, pp. 1657–1666.
25. Urbanik, I. I., R. W. Holder, and A. V. Fitzgerald. *Evaluation of Priority Techniques for High Occupancy Vehicles on Arterial Streets*. Research Report. 205-5, Texas Transportation Institute, Texas A & M University, 1977.
26. Lin, Y., X. Yang, N. Zou, and L. Jia. A New Passive Transit Signal Priority Control at Urban Arterials, *Journal of Northeastern University, Natural Science*, Vol. 34, No. 9, 2013. pp. 1134–1140. (in Chinese).
27. Lin, Y., X. Yang, and N. Zou. Passive Transit Signal Priority for High Transit Demand: Model Formulation and Strategy Selection. *Transportation Letters*, 2017, pp. 1–11.
28. Dai, G., H. Wang, and W. Wang. A Bandwidth Approach to Arterial Signal Optimisation with Bus Priority. *Transportmetrica A: Transport Science*, Vol. 11, No. 7, 2015, pp. 579–602.
29. Dai, G., H. Wang, and W. Wang. Signal Optimization and Coordination for Bus Progression Based on MAXBAND, *KSCE Journal of Civil Engineering*, Vol. 20, No. 2, 2016, pp. 890–898.
30. Cheng, Y., X. Yang, and G. -L. Chang. A Bus-Based Progression System for Arterials with Heavy Transit Flows. Presented at 94th Annual Meeting of the Transportation Research Board, Washington, D.C., 2015.
31. Bowen, G. T., R. D. Bretherton, J. R. Landles, and D. J. Cook. Active Bus Priority in SCOOT. *Proc., Seventh International Conference on Road Traffic Monitoring and Control*, London, UK, 1994. pp. 73–76.
32. Mirchandani, P., A. Knyazyan, L. Head, and W. Wu. An Approach Towards the Integration of Bus Priority, Traffic Adaptive Signal Control, and Bus Information/Scheduling Systems. *Computer-Aided Scheduling of Public Transport*, 2001, pp. 319–334.
33. Little, J. D. C., M. D. Kelson, and N. H. Gartner. *MAXBAND: A Versatile Program for Setting Signals on Arteries and Triangular Networks*. WP-1185-81, Massachusetts Institute of Technology, 1981.

34. Hounsell, N. B. Active Bus Priority at Traffic Signals. *IEE Colloquium on UK Developments in Road Traffic Signalling*, IET, 1988, pp. 4–1.
35. Gartner, N. H., S. F. Assman, F. Lasaga, and D. L. Hou. A Multi-band Approach to Arterial Traffic Signal Optimization. *Transportation Research Part B: Methodology*, Vol. 25, No. 1, 1991, pp. 55–74.

The Standing Committee on Traffic Signal Systems (AHB25) peer-reviewed this paper (18-03122).

Article

TRR
JOURNAL OF THE TRANSPORTATION RESEARCH BOARD

Transportation Research Record
2018, Vol. 2672(18) 15–26

DOI: 10.1177/0361198118773567
journals.sagepub.com/home/trr
SAGE

Peer-to-Peer Priority Signal Control Strategy in a Connected Vehicle Environment

Byungho Beak[1], Mehdi Zamanipour[2], K. Larry Head[1], and Blaine Leonard[3]

Abstract

This paper presents a methodology that enhances the priority signal control model in the multi-modal intelligent traffic signal system (MMITSS). To overcome the range limit of vehicle to infrastructure (V2I) and the intersection geometry message (MAP) distance limits, peer-to-peer intersection communications are utilized to send priority requests from adjacent intersections. Through integrated communication, the peer priority control strategy can create a signal plan for prioritized vehicles that considers longer term (headway) arrival times. Transit vehicles are considered in this study. The longer-term signal plan provides a flexible signal schedule that allows local phase actuation. The peer priority strategy is effective in reducing the number of stops and delay for priority eligible vehicles, while minimizing the negative impact on regular vehicles. To validate the strategy, a simulation experiment was designed to compare fully actuated control, coordination, and MMITSS priority control using two different VISSIM simulation networks (Arizona and Utah). The result shows that the peer-to-peer long term planning strategy can improve transit service reliability while limiting the adverse impact on other traffic.

Policies and strategies have been conceived to attract people to use public transportation by improving transit service in punctuality and regularity (*1*). Transit signal priority (TSP) has been one of the candidates that can improve transit service. TSP can reduce operation transit cost by reducing delays, travel time, having fewer stops, and requiring less vehicles to cover the same route.

There are three major categories of signal priority: passive TSP, active TSP, and adaptive TSP. Passive TSP tries to improve performance of TSP operation by using pre-determined signal timing plans (*2*). Stevanovic et al. presented a transit priority control strategy to adjust basic signal timing parameters, including cycle length, offset, splits, and phase sequences (*3*). The results showed that the adjustment of the signal settings can significantly improve transit operations. Estrada et al. developed an optimization tool to determine the optimal set of offsets that minimize travel times of transit (*4*). Passive TSP may be beneficial only if transit frequency is relatively high (*5*). Active TSP accounts for these limitation by detecting and responding to transit vehicles in real time through green extension, red truncation, and phase skipping. Active TSP strategies may sacrifice the service of other traffic movements when serving requests. Active priority may be conditional, based on lateness and occupancy criteria that determines when a request can be sent (or served) (*1, 6, 7*). To address the limitations of active and passive TSP, Ma and Yang integrated both active and passive strategies into one system, and showed that the integrated model can improve both variation and average bus delay (*5*). Adaptive TSP controls signal timing for transit vehicles with the goal of using phase time more effectively. Ling and Shalaby used a reinforcement learning method to find the optimal phase duration for transit vehicles taking into consideration transit headway and elapsed green time for the requested phase (*8*). Lee et al. developed a TSP operation model that selects the best plan among plans in a library according to a transit arrival prediction model (*9*).

Active and adaptive TSP requires a transit detection system to generate priority requests. The detection system can be based on several technologies such as loop detectors, video cameras, radar (UDOT primarily uses

[1]Department of Systems and Industrial Engineering, The University of Arizona, Tucson, AZ
[2]Turner-Fairbank Highway Research Center, Federal Highway Administration, McLean, VA
[3]Utah Dept of Transportation, Salt Lake City, UT

Corresponding Author:
Address correspondence to Byungho Beak: beak@email.arizona.edu

wavetronics detectors), infrared transmitters, and radio. Recently, TSP detection systems have begun to use the global positioning system (GPS). Koonce et al. (*10*) and Liao et al. (*11*) developed signal priority control models based on automated vehicle location systems (AVL) using GPS-based data. Through the emergence of connected vehicle technology, connected vehicle based signal priority control can provide significant benefit over traditional TSP strategies. Hu et al. developed a person-delay-based optimization approach for an intelligent TSP strategy using connected vehicle technology, assuming that the connected vehicle market penetration rate was 0% except for transit (*7*). He et al. (*12, 13*) and Zamanipour et al. (*14*) developed multi modal priority control strategies that consider requests from transit, as well as trucks and emergency vehicles.

Although TSP has been implemented for decades, challenges still exist. One of the shortcomings that has been observed in conventional TSP strategies is the difficulty in dealing with multiple conflicting priority requests. Most of the rule-based TSP strategies are based on if-then structures that assume a first-come, first-serve priority logic (*15–17*). They have difficulties when multiple conflicting priority requests are received at the same time. Head et al. presented an analytical model-based priority strategy based on a precedence graph structure (*18*). The paper showed that a rule-based first-come, first-served policy can lead to more delay than a model that considers multiple requests simultaneously. A similar result was found by Ma et al., who formulated a dynamic programming model for multiple priority requests by optimizing phase sequence (*19*). Hu et al. (*7*) and Zamanipour et al. (*14*) used a mixed integer linear programming model to minimize vehicle delay for multiple priority requests and the results showed that vehicle delay was significantly improved. Instead of a model-based approach, Zlatkovic et al. used a logic processor to resolve conflicting multiple requests and tested the model in a VISSIM simulation (*20*). The result showed that the first-come, first-served policy for conflicting TSP requests could result in worse performance than a policy that does not provide priority.

This paper presents an enhancement to the signal priority algorithm that is part of the multi-modal intelligent traffic signal system (MMITSS) project. MMITSS is a USDOT Cooperative Transportation System Pooled Fund project that focuses on connected vehicle technology (*21*). The limited range in the connected vehicle architecture means that the signal priority algorithm in MMITSS can only accommodate priority requests received in the near-term (e.g., 10–20 seconds). Through this near-term priority planning, the signal controller has a limited time window to adapt for the priority request, which may lead to priority request delay or insufficient delay reduction for a transit vehicle (*22*). Several studies have already revealed that the active priority policy based on such short notice can have adverse effects on other traffic (*1, 23*). As the time between transit vehicle detection and actual arrival time at the signal increases, the flexibility in signal adjustments are significantly increased (*24*).

A peer priority control logic is presented to provide a long-term priority plan using peer-to-peer (P2P) communications. P2P is based on the backhaul network between adjacent traffic signals. The basic structure of the mathematical optimization in this paper follows the signal priority control model in MMITSS (*14*). An improvement in the priority algorithm in (*14*) is created to integrate an implementation algorithm into the mathematical model. A platform is presented, which implements the peer priority control strategy based on the peer-to-peer priority request data from connected vehicles. Standard J2735 Dedicated Short Range Communications (DSRC) messages, including the signal request messages (SRMs), signal status messages (SSMs), basic safety messages (BSMs), signal phasing and timing messages (SPaTs), and MAP messages, defined by the Society of Automotive Engineers (SAE) are used to manage multiple peer priority requests in the framework.

Methodology of Peer-to-Peer Priority Signal Control

Peer Priority Control Framework

The peer priority control framework has been designed to address the limitation of the effective range of vehicle to infrastructure (V2I) communications (using DSRC) and the geographic extent of the MAP message. DSRC has a minimal range of 300 meters, but experience shows that signals can be received beyond 800 meters. However, a priority vehicle approaching an intersection will not be aware of the intersection until it receives a valid MAP message and is located within the extent of the MAP. The MAP message provides a description of the intersection geometry, including approaches, lanes, stop bar, cross walks, and so forth. The MAP description of each approach is based on GPS waypoints used to describe the lanes, because of the limitation in the MAP message size, the extent, that is the distance from the stop bar, is limited. Experience has shown that this is about 300 meters for relatively straight approaches and shorter distances for curved approaches. Once the priority requesting vehicle determines that it is inside the MAP extent, it can form a valid SRM that includes the in-lane (as defined in the MAP) and the time of desired service. Figure 1 illustrates the scenario when P2P priority control will be effective.

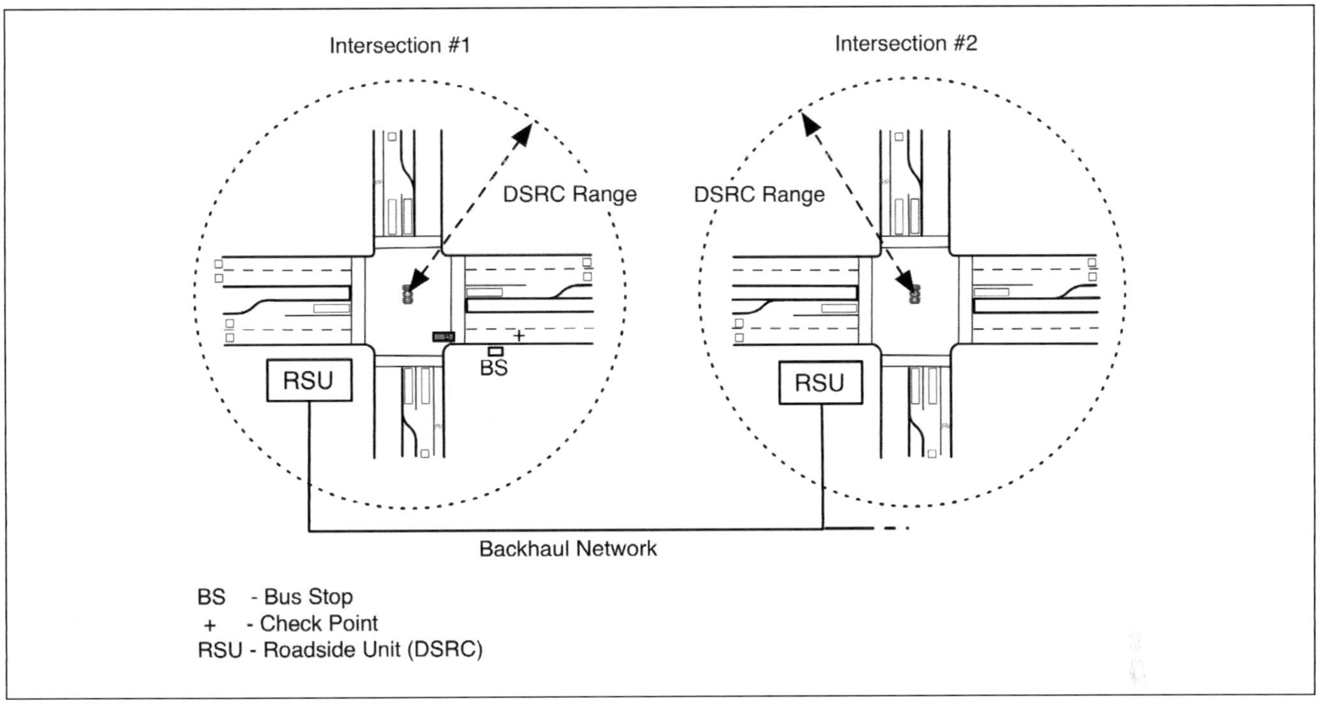

Figure 1. Illustration of peer priority request initiation.

When a prioritized vehicle passes a stop bar at an upstream intersection (intersection #1), the local, or near-term, priority request is canceled as the vehicle has been served. A peer priority request is initiated when the cancel message is received from the vehicle. The peer control component monitors the vehicle position until it passes one of the defined exit check points, which is the first GPS waypoint on an egress approach lane after the stop-bar if there is no far-side bus stop. If there is a far-side bus stop (shown as BS in Figure 1), a GPS waypoint after the bus stop, which is denoted as ' + ' symbol in Figure 1, becomes the check point. When the vehicle passes the exit check point, a peer priority request, including the requested signal phase, in-lane, and estimated arrival time to the stop bar at the downstream intersection, is sent to the designated intersection (intersection #2) through the backhaul network. The estimated arrival time is calculated based on an average travel time between the exit check point and the stop bar at the downstream intersection. In this paper, an average of free flow travel time for transit vehicles from the VISSIM simulation was used. The peer request makes it possible for the traffic signal controller at the downstream intersection to have a long term priority planning horizon, which can reduce delay and the number of stops for transit with more flexibility for other traffic. The system architecture of the P2P priority control system is shown in Figure 2.

The control system consists of several software components including the priority request generator, priority request server, priority solver, vehicle trajectory awareness, traffic control interface, and long term planning, all of which are implemented in both simulation and real traffic networks. All of the components reside at the roadside, either on a road side unit (RSU) or on a co-processor at each intersection. When a priority vehicle enters in the V2I range of the intersection, the signal request message (SRM) broadcast from the priority request generator on the on board unit (OBU) is received by the priority request server (PRS) through V2I communication. Once the PRS adds a new near-term priority request to the active requests table, a unified mixed integer linear Programming (MILP) model, which resides in the priority solver, creates an optimal near-term signal priority plan. The signal plan is sent to the traffic signal controller through the traffic controller interface, using the National Transportation Communications for ITS Protocol (NTCIP) commands, which include vehicle call, force-off, and hold (*25*).

When the prioritized vehicle passes the stop bar, the OBU broadcasts a cancel SRM message. The cancel SRM is received at the intersection and the PRS cancels the active signal request and checks to see if peer priority control is enabled for the downstream intersection. If it is enabled, the PRS sends a peer priority request to the long term planning (LTP) component. The LTP component

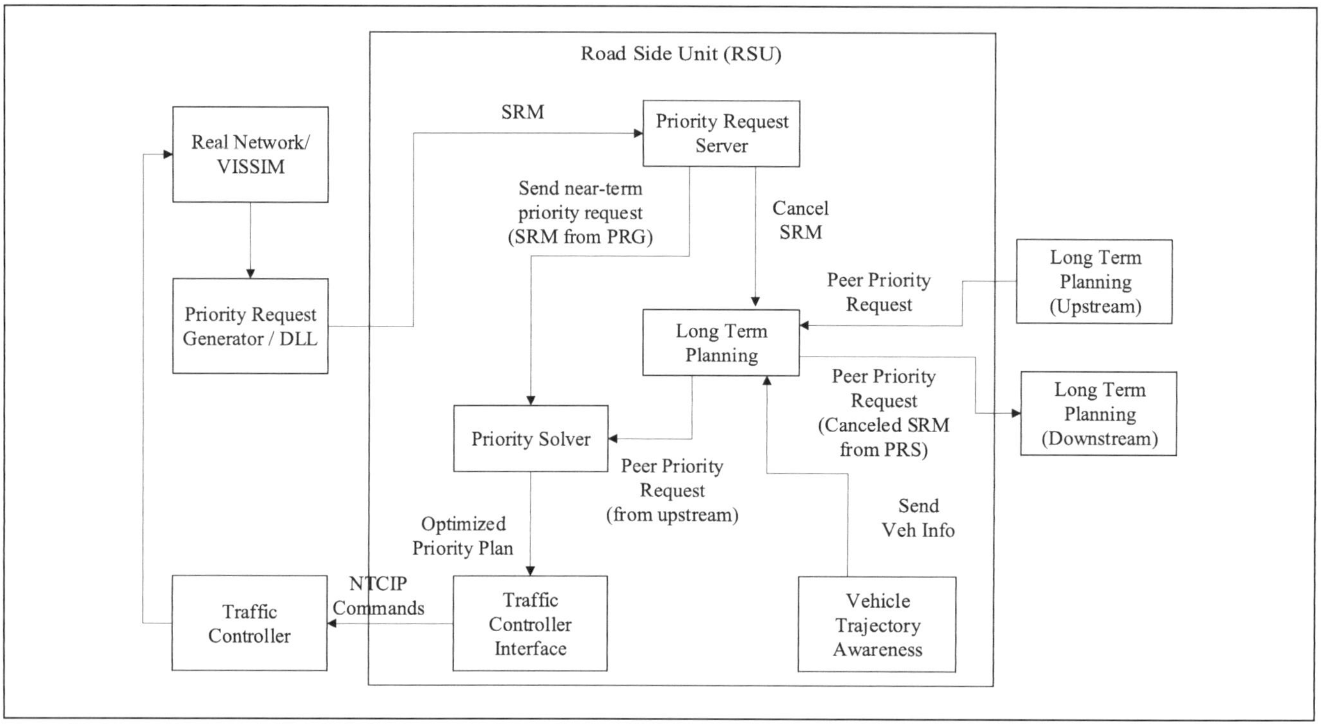

Figure 2. Architecture of peer-to-peer priority control system.

continues checking the vehicle position to see if it has passed one of the defined exit check points. When the vehicle passes the exit check point, the LTP sends a peer priority request to the designated downstream intersection. Once a LTP at a downstream intersection receives a peer priority request, it adds the request to the active peer request table and sends it to the priority solver. If there is no priority request in the V2I range of the downstream intersection, the priority solver solves the MILP only for the peer request and sends a feasible solution to the traffic controller interface. If there are priority requests that are currently being served in the V2I range of the downstream intersection, the peer priority request is solved along with them in the same optimization process.

A detailed procedure of the peer priority control algorithm can be found in Figure 3. Figure 3*a* demonstrates how a peer priority request is initiated at an upstream intersection, while Figure 3*b* shows how the peer request is processed once received at the downstream intersection.

Figure 4 shows two examples using a time-phase diagram, which explain the necessity of P2P priority control strategies (*26*). A priority signal plan can be visualized in the diagram, which is a useful tool to analyze the optimized plan, verify if the schedule is feasible, and evaluate priority delay given the signal plan and priority request arrival times.

A priority request (box) with an arrival time interval is shown in Figure 4, where the x-axis is the time and the left and right y-axes are showing phases in rings 1 and 2, respectively. The box with the letter 'L' represents a priority request that is out of the V2I range and is not currently observed at the intersection. The box with the letter 'N' represents a near-term priority request that has already been detected at the intersection. The V2I range is shown at 17 seconds. Two black dotted lines starting from the origin, which denotes current time and current phase, represent the shortest path (minimum green times for each phase) and longest path (maximum green times) over the planning horizon. A signal control plan must reside in the cone-shaped area between the two lines. The solid arrow shows a phase-serving path that the controller may possibly take within the feasible area based on actuation logic. This could be based on coordination plan splits. As shown in Figure 4*a*, the priority request may not be served in the current cycle because it is not observed by the local server (inside V2I range) before the signal completes the service phase (phase 2). Figure 4*b* shows that, even with a near-term priority request being served, the long term priority request may not be served after the near term request passes the intersection. This may result in a significant delay to the priority eligible vehicle.

Unified Mathematical Optimization Model

Zamanipour et al. developed a signal priority algorithm that includes two parts: a mathematical optimization

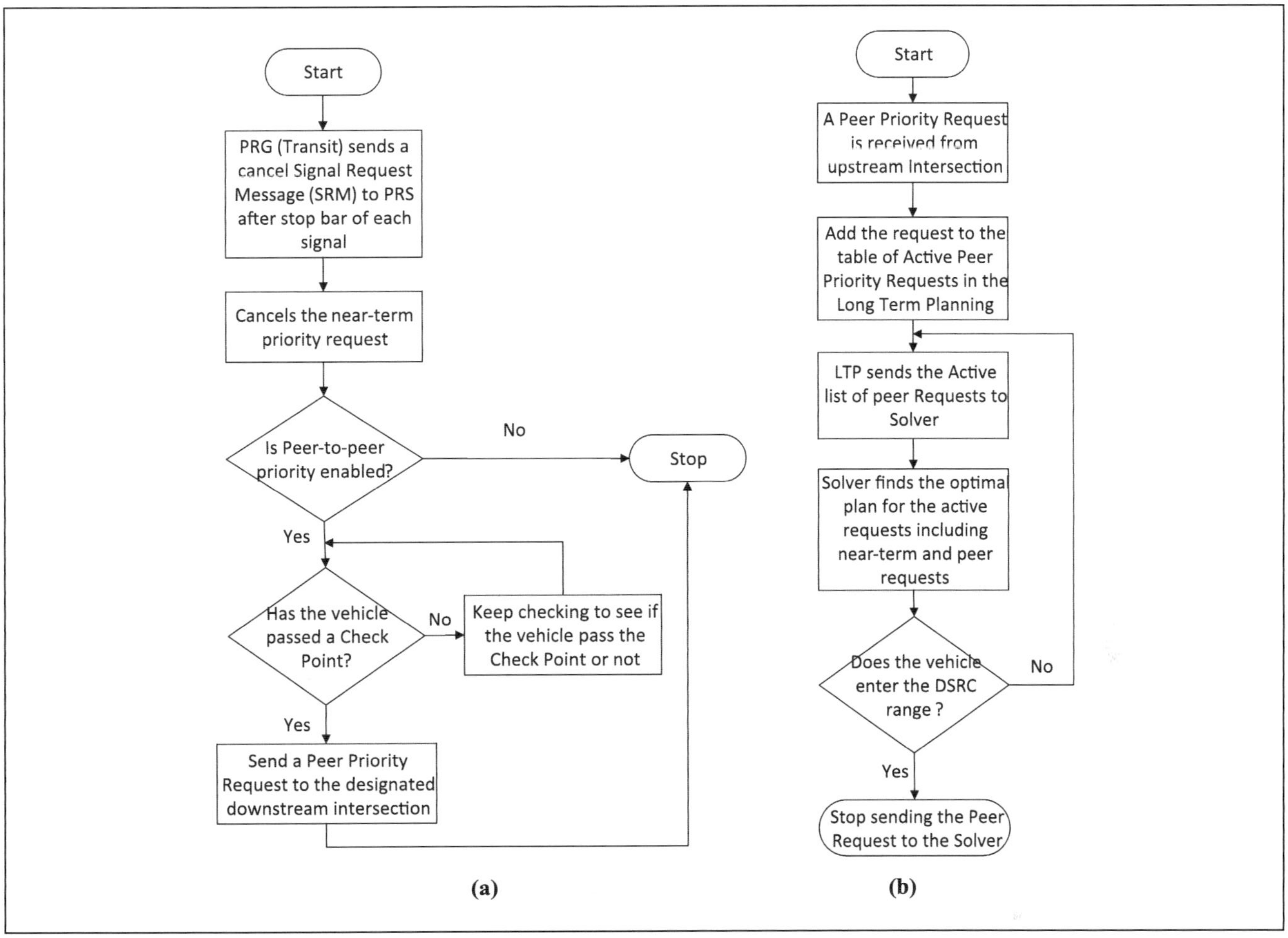

Figure 3. Flow chart of peer-to-peer priority control model: (*a*) upstream, (*b*) downstream.

model and a flexible implementation algorithm (*14*). Based on the standard North American National Electrical Manufacturing Association (NEMA) dual-ring structure, the mathematical optimization model was formulated as a MILP problem to minimize a weighted summation of multiple priority requests delay. To reduce the negative impacts on regular vehicles that may be caused by granting priority to priority-eligible vehicles, a flexible implementation algorithm was developed, which allows the non-priority phases to be actuated to cope with variation in traffic demand in real-time. This flexible optimized signal plan ensures that the signal controller can better serve non-prioritized vehicles on the phases that precede the requested phase, while minimizing the priority request delay. As shown in Figure 5, a feasible region is created by two piecewise lines starting from the origin to the priority request arrival time. The left dotted piecewise line represents a set of "HOLD" points for each phase, which denotes when the signal controller should hold a phase green. The right long-dashed piecewise line represents a set of "FORCE-OFF" points that provide less, or zero, delay to the priority request. Within the flexible region between the two lines, the controller is allowed to actuate the signal phase, which means that if there is no vehicle detection (call), the phase will gap out. Otherwise, the phase will extend until the "FORCE OFF" point.

The original flexible implementation algorithm was revised to directly address signal flexibility, considering two critical points as additional decision variables in the mathematical model. The revised model is:

$$Min_{t_{left},\, t_{right}} \sum_{m \in M} w^m d^m - \sum_{p \in P,\, k \in K} (t_{right}(k,p) - t_{left}(k,p))$$

s.t
Precedence Constraints
Phase Duration and Interval Constraints
Priority Request Delay Evaluation Constraints

In which d^m is the total priority request delay of mode m, w^m is the mode m weight determined by the multi-modal

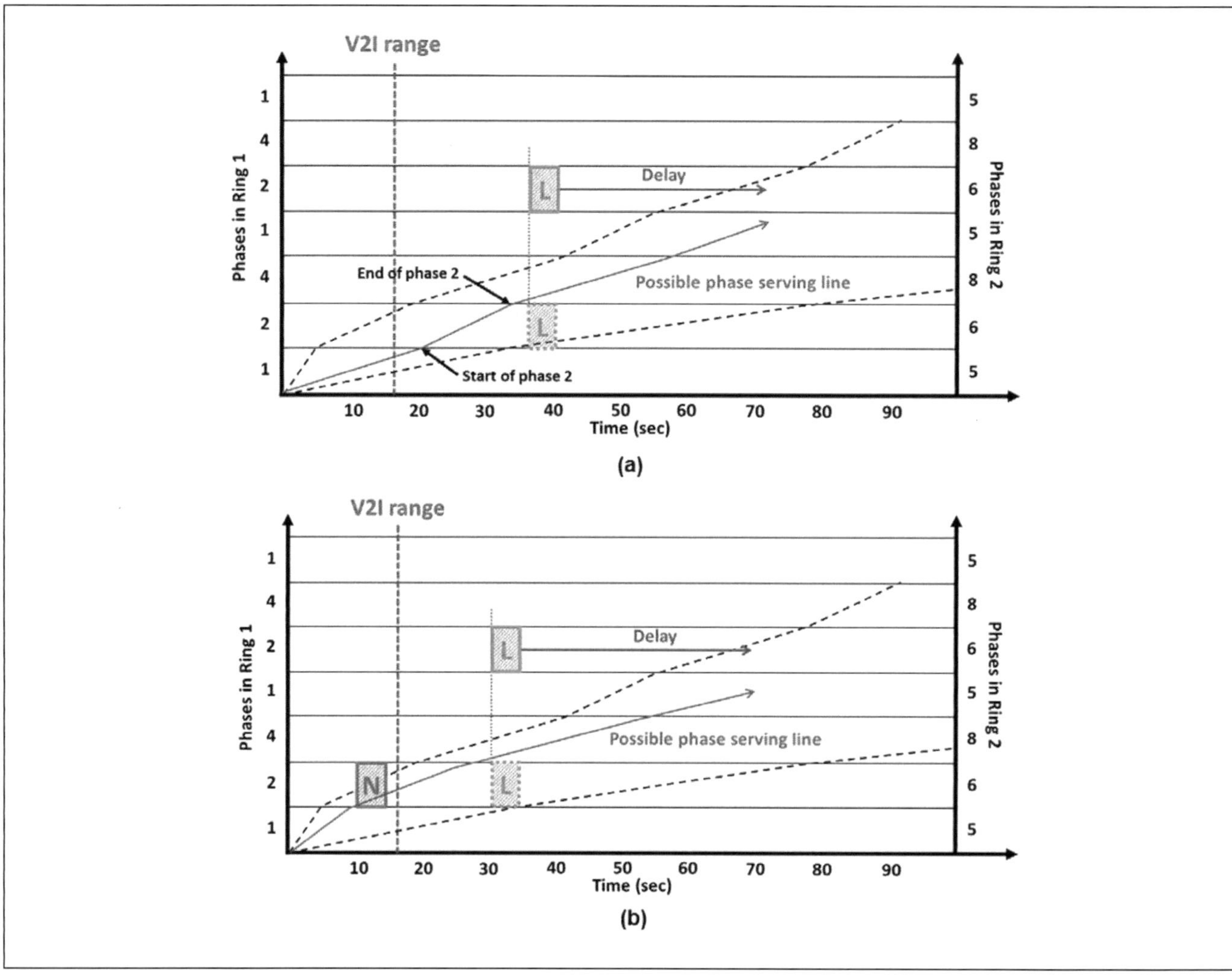

Figure 4. Two long term priority request scenarios: (*a*) without near-term request, (*b*) with near-term request.

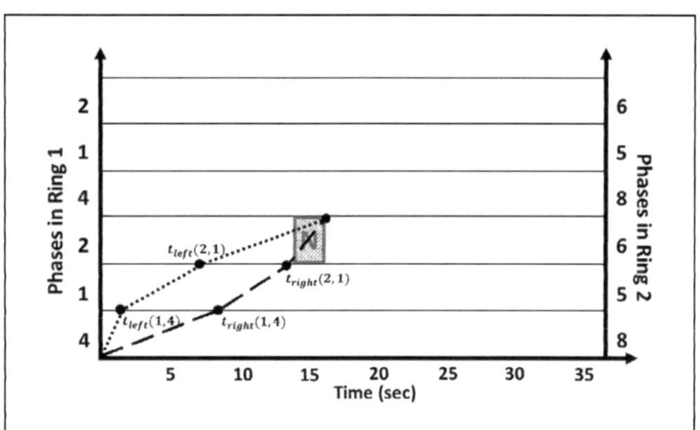

Figure 5. Flexible feasible signal plan in time-phase diagram.

policy. $t_{left}(k,p)$ is the left critical point of phase p in cycle k, while $t_{right}(k,p)$ is for the right critical point, as shown in Figure 5. To maximize the feasible region, a negative term is added to the objective function, which is the summation of time intervals between the "FORCE-OFF" point and "HOLD" point. One set of precedence constraints are considered for each ring to find the left and right critical points. The *Phase Duration and Interval Constraints* and *Priority Request Delay Evaluation Constraints* also hold for both left and right points independently. The other variables defined in (*14*) remain the same.

Figure 6 shows several priority plans that the peer priority control model creates for multiple priority requests, including both near-term and peer requests, based on the revised optimization model. Each signal plan in Figure 6 was drawn based on actual log data from the Priority Solver component. As shown in Figure 6*a*, the flexibility from the current phase to the requested phase is possible when there is only one peer request. The delay for the peer request will be zero, and the signal controller

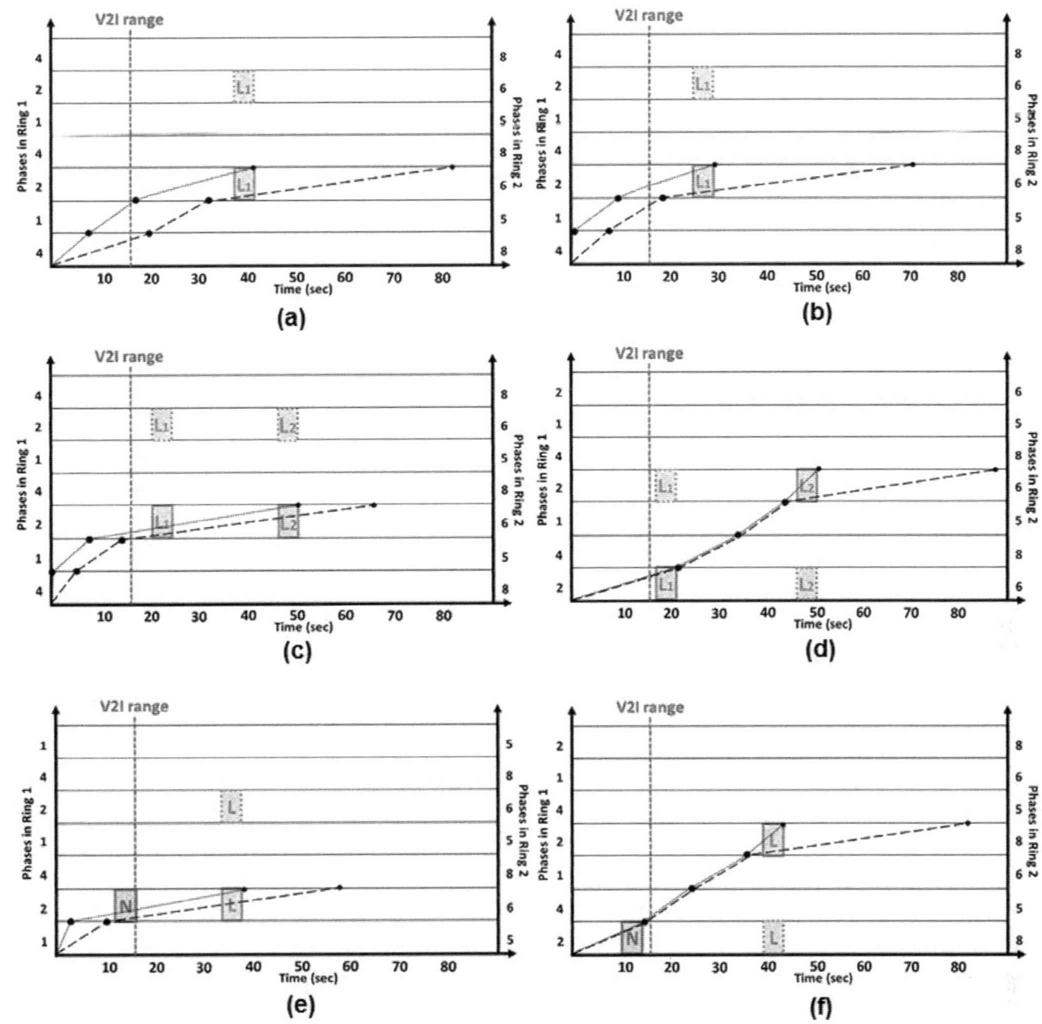

Figure 6. Time-phase diagram with peer priority requests and feasible plan.

has significant flexibility to serve for other traffic. Figure 6*b* shows the priority plan for the same request approximately 10 seconds later, which still maintains a large flexible region. Even with multiple peer requests, a flexible plan can exist as shown in Figure 6*c*. Sometimes, the flexibility can be limited to serve multiple requests as demonstrated in Figure 6*d*. According to the arrival interval of each peer request and traffic signal status, which includes current phases served and elapsed green times of the phases, peer requests can be served in one cycle (Figure 6*c*), or they can be served in different cycles (Figure 6*d*). Figure 6*e* and *f* illustrate the same case of Figure 6*c* and *d*, when the first peer requests enter the V2I range. After the near-term request (N) passes the stop-bar, the signal controller can hold the signal green for phase 2 because a peer request (L) exists that can be served within the maximum green time of phase 2 (Figure 6*e*). On the other hand, if the peer request cannot be served in the current cycle, the controller could force off the phase 2 right after the near-term request and follow the non-flexible line (a set of minimum green times for phases 4 and 1) up to the peer request (L) to have zero delay for the peer request (Figure 6*f*).

Experimental Design

The peer priority control strategy has been tested on two different VISSIM simulation networks: Daisy Mountain Drive in Anthem, Arizona and Redwood Road in Salt Lake City, Utah. As Figure 7 shows, the Daisy Mountain corridor has 6 intersections over 2.3 miles, while the Redwood corridor has 13 intersections over 5.12 miles. The main street in the Arizona network has three lanes in both directions, and the side streets have

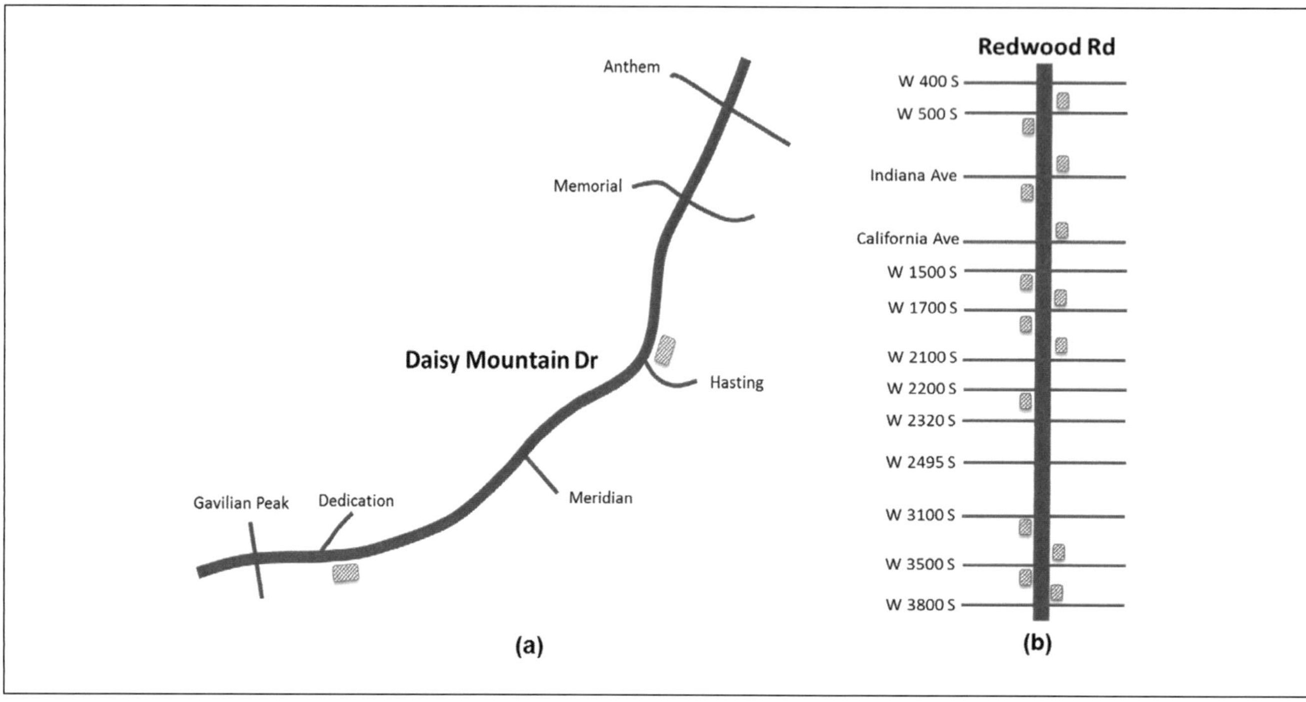

Figure 7. Simulation test bed layout: (*a*) Anthem, Arizona, (*b*) Salt Lake City, Utah.

two lanes, while the main street in the Utah network has two or three lanes, and the side streets have one or two lanes. The intersection spacing of both networks varies, although the Utah network has, on average, wider spacing ranging from 150 m to 1400 m than the Arizona network, which has a range of 300–900 m. An eastbound movement from Gavilian Peak to Anthem on Daisy Mountain Dr. was simulated as a bus route, while a southbound movement from W 400 S to W 3800 S on Redwood Rd. was selected as a bus route. The Utah Transit Authority (UTA) actually runs a bus service on Redwood Rd. at approximately 15 minute intervals, while there is no public transit service running in Anthem. For the test of the P2P priority control strategy, two far-side bus stops with bus bays and located within V2I range, shown as blue boxes in Figure 7*a*, were simulated in the Arizona network model. As shown in Figure 7*b*, the Utah network model has seven far-side bus stops without bus pullouts. The bus dwell time at each bus stop was distributed by normal distribution in the mean of 20–45 seconds with a standard deviation of 5 seconds for both simulation networks. The bus speed for both networks was simulated as 35 miles per hour, which was slightly lower than the actual speed limit of each road.

Both simulation models were run for two and half hours, which included a thirty minute warm-up period. The bus headway for the Arizona and Utah networks were set up as five minutes and ten minutes, respectively, so the experiments would include a sufficient number of buses. Both simulation experiments were replicated with ten different random seeds. Traffic demand on the main street and side streets in Anthem were 900 vehicles per hour (vph) and 200 to 400 vph, respectively. Traffic demand on the main street in Salt Lake City is 1200 vph, and side streets have a traffic volume ranging from 200 to 500 vph. An Econolite virtual controller (ASC/3), which can support the NTCIP interface, was used in the simulation.

Evaluation Results

Several measurements from VISSIM were evaluated including average vehicle delay, number of stops, and travel time. The peer priority control strategy was compared with three different signal control applications: fully actuated control (free mode), coordination control, and MMITSS priority control. The coordination parameters on the Arizona network, including phase splits, cycle length, and offsets, are determined by VISTRO (*27*) based on the vehicle demand, while the parameters for the Utah were provided by the Utah Department of Transportation.

Table 1 summarizes the performance of transit vehicles on a bus route for both networks for the average values of ten replications with different random seeds. The average vehicle delay represents the average delay of each transit vehicle that travels along the bus route.

Table 1. Performance Measure of Transit Vehicles

		Average			Standard deviation	
Simulation network	Measurement	Vehicle delay	Number of stop	Travel time	Vehicle delay	Travel time
Arizona						
Bus Route (eastbound)	Fully actuated control	79.67	1.32	334.74	27.32	28.00
	Coordination control	72.10	1.12	327.04	24.32	24.32
	Priority control	51.40	0.37	306.42	13.01	13.89
	Peer Priority control	41.15	0.07	296.00	8.44	10.34
Comparison	Peer VS Fully actuated	−48.35%	−94.94%	−11.57%	−69.12%	−63.07%
	Peer VS Coordination	−42.92%	−94.03%	−9.49%	−65.30%	−57.47%
	Peer VS Priority control	−19.95%	−81.82%	−3.40%	−35.14%	−25.52%
Utah						
Bus Route (eastbound)	Fully actuated control	298.42	5.15	998.80	50.39	49.53
	Coordination control	279.80	3.26	980.50	32.82	32.76
	Priority control	197.72	2.30	898.36	31.32	32.99
	Peer Priority control	163.99	1.31	864.75	23.68	24.44
Comparison	Peer VS Fully actuated	−45.05%	−74.60%	−13.42%	−53.01%	−50.66%
	Peer VS Coordination	−41.39%	−59.85%	−11.80%	−27.85%	−25.40%
	Peer VS Priority control	−17.06%	−42.91%	−3.74%	−24.41%	−25.92%

Note: VS = Versus.

Compared with fully actuated control in Arizona, the average vehicle delay, number of stops, and travel time from peer priority control were greatly improved by −48.35%, −94.94%, and −11.57%, respectively. Similar results were shown in the Utah network. When comparing the same measures with coordination control in both networks, peer control outperformed coordination by −41.39%~−42.92%, −59.85%~−94.03%, and −9.49%~−11.80%, respectively. The results show that the improvements were slightly reduced compared with fully actuated control. This is because signals were coordinated on the bus route, and buses benefit from the coordinated progression. However, as buses travel over a corridor, they cannot remain in the progression because of stopping and dwelling for passengers at bus stops. A comparison between peer priority control and MMITSS priority control also shows that all of the measures from the peer control were improved. In particular, the average number of stops for both networks were significantly decreased by −81.82% and −42.91%, respectively.

To show the reliability of the transit service under each signal control system, the standard deviation of travel time and vehicle delay were compared in Table 1. Compared with fully actuated control, the peer control model significantly improved the reliability of vehicle delay by −69.12% in Arizona and −53.01% in Utah. The reliability of travel time was also improved by −63.07% in Arizona and −50.66% in Utah. In comparison to coordination and MMITSS priority control, the results show that the reliability of the peer control outperformed in both networks as shown in Table 1.

Table 2 demonstrates how the peer priority control affects the performance of regular vehicles in the entire system. It was observed that in both networks the peer priority control does not have a negative impact on regular vehicles for the average vehicle delay, number of stop, and travel time, compared with fully actuated and MMITSS priority control. However, the average number of stops under coordination control in both networks was better than peer control, although vehicle delay and travel time from the peer control were better than under coordination. It turned out that the coordination plans from VISTRO and the UDOT were effective at minimizing the number of stops for regular vehicles, although it did not improve the number of stops for transit because the coordination cannot differentiate vehicle modes, as shown in Table 1.

A statistical analysis was performed to determine the significance of the simulation results. Using a paired-sample t-test at the confidence level of 95%, the mean value of ten replications were examined for each signal control comparison. The null hypothesis of the test was that the mean values of each signal control were

Table 2. Performance Measure (Average) of Regular Vehicle on Entire Network

Simulation network	Measurement	Average vehicle delay	Average number of stop	Average travel time
Arizona				
Entire network	Fully actuated control	40.68	1.57	158.22
	Coordination control	48.60	1.29	165.93
	Priority control	40.58	1.57	158.14
	Peer Priority control	40.45	1.55	157.96
Comparison	Peer VS Fully actuated	–0.57%	–1.03%	–0.16%
	Peer VS Coordination	–16.78%	20.07%	–4.80%
	Peer VS Priority control	–0.33%	–1.06%	–0.11%
Utah				
Entire network	Fully actuated control	45.59	1.57	135.59
	Coordination control	61.62	1.42	151.56
	Priority control	45.82	1.58	135.84
	Peer Priority control	45.97	1.58	136.09
Comparison	Peer VS Fully actuated	0.84%	0.24%	0.37%
	Peer VS Coordination	–25.24%	11.32%	–10.21%
	Peer VS Priority control	0.32%	0.03%	0.18%

Note: VS = Versus.

insignificant compared with peer priority control. According to p-values for all measures of transit, all of the null hypotheses are rejected at the 5% significance level, which indicates that the improvement of peer priority control on transit is significant. For the same measures of regular vehicles, the test fails to reject all of the null hypothesis except for the comparison of peer and coordination control, indicating that the mean values are insignificant. Accordingly, the results indicate that the peer priority control model does not have an adverse impact on the performance of regular vehicles.

Conclusions and Future Research

This paper presents a methodology that enhanced the MMITSS priority control strategy based on P2P and V2I communications in a connected vehicle environment. The peer priority control strategy makes it possible for a signal controller to have a flexible long-term plan for prioritized vehicles. They can benefit from the long-term plan within a secured flexible region and it can prevent the near-term priority actions from having a negative impact on other traffic by providing more flexibility for phase actuation. In addition, the strategy can be applied to all different modes of prioritized vehicles such as transit, freight, and emergency vehicles without consideration of bus stops.

In this paper, peer priority control was implemented without a consideration of coordination or pedestrians in the optimization. Future research will focus on an integration of peer priority-coordination signal control, which creates an optimal solution that considers multiple peer requests and periodic virtual coordination requests in the same optimization framework. Inclusion of pedestrians is a complex research topic that needs to be addressed, as pedestrians are transit customers. It is understood that pedestrian service intervals create significant constrains on the priority control, but it is believed that the longer optimization horizon of P2P priority will afford some additional benefits.

In addition, the estimation of arrival time for each peer request to the downstream intersection is critical. If the arrival time of the peer request is not accurate, the performance of the prioritized vehicles and other traffic could be negatively impacted. For example, if the arrival time to the downstream intersection was forecasted too early, it could force off competing phases too early, causing unnecessary delay or cycle failure for other traffic. On the other hand, if the arrival time was estimated too late, the peer priority may not take advantage of the long term planning. Thus, for future improvements, statistical methods could be used to estimate more accurately, instead of using the average travel time on the link.

Acknowledgments

This work has been supported by the UDOT MMITSS Transit Signal Priority Control Project from the Utah Department of Transportation.

Author Contributions

Study conception and design: Byungho Beak, Mehdi Zamanipour, Larry Head, Blaine Leonard; Data collection: Byungho Beak, Mehdi Zamanipour, Larry Head; Analysis and interpretation of results: Byungho Beak, Mehdi Zamanipour, Larry Head; Draft manuscript preparation: Byungho Beak, Mehdi Zamanipour, Larry Head, Blaine Leonard. All authors reviewed the results and approved the final version of the manuscript.

References

1. Furth, P., and T. H. Muller. Conditional Bus Priority at Signalized Intersections: Better Service with Less Traffic Disruption. *Transportation Research Record: Journal of the Transportation Research Board*, 2000. 1731: 23–30.
2. Sunkari, S. R., P. S. Beasley, T. Urbanik, and D. B. Fambro. Model to Evaluate the Impacts of Bus Priority on Signalized Intersections. *Transportation Research Record: Journal of the Transportation Research Board*, 1995. 1494: 117–123.
3. Stevanovic, J., A. Stevanovic, P. T. Martin, and T. Bauer. Stochastic Optimization of Traffic Control and Transit Priority Settings in VISSIM. *Transportation Research Part C: Emerging Technologies*, Vol. 16, No. 3, 2008, pp. 332–349.
4. Estrada, M., C. Trapote, M. Roca-Riu, and F. Robuste. Improving Bus Travel Times with Passive Traffic Signal Coordination. *Transportation Research Record: Journal of the Transportation Research Board*, 2009. 2111: 68–75.
5. Ma, W., and X. Yang. Design and Evaluation of an Adaptive Bus Signal Priority System Base on Wireless Sensor Network. *Proc., 11th International IEEE Conference on Intelligent Transportation Systems*. IEEE, 2008.
6. Skabardonis, A., and N. Geroliminis. Real-Time Monitoring and Control on Signalized Arterials. *Journal of Intelligent Transportation Systems*, Vol. 12, No. 2, 2008, pp. 64–74.
7. Hu, J., B. B. Park, and Y.-J. Lee. Coordinated Transit Signal Priority Supporting Transit Progression under Connected Vehicle Technology. *Transportation Research Part C: Emerging Technologies*, Vol. 55, 2015, pp. 393–408.
8. Ling, K., and A. Shalaby. Automated Transit Headway Control via Adaptive Signal Priority. *Journal of Advanced Transportation*, Vol. 38, No. 1, 2004, pp. 45–67.
9. Lee, J., A. Shalaby, J. Greenough, M. Bowie, and S. Hung. Advanced Transit Signal Priority Control with Online Microsimulation-Based Transit Prediction Model. *Transportation Research Record: Journal of the Transportation Research Board*, 2005. 1925: 185–194.
10. Koonce, P., J. Ringert, T. Urbanik, W. Rotich, and B. Kloos. Detection Range Setting Methodology for Signal Priority. *Journal of Public Transportation*, Vol. 5, No. 2, 2002, pp. 115–135.
11. Liao, C.-F., and G. Davis. Simulation Study of Bus Signal Priority Strategy: Taking Advantage of Global Positioning System, Automated Vehicle Location System, and Wireless Communications. *Transportation Research Record: Journal of the Transportation Research Board*, 2007. 2034: 82–91.
12. He, Q., K. Head, and J. Ding. Heuristic Algorithm for Priority Traffic Signal Control. *Transportation Research Record: Journal of the Transportation Research Board*, 2011. 2259: 1–7.
13. He, Q., K. L. Head, and J. Ding. Multi-Modal Traffic Signal Control with Priority, Signal Actuation and Coordination. *Transportation Research Part C: Emerging Technologies*, Vol. 46, 2014, pp. 65–82.
14. Zamanipour, M., K. L. Head, Y. Feng, and S. Khoshmagham. Efficient Priority Control Model for Multimodal Traffic Signals. *Transportation Research Record: Journal of the Transportation Research Board*, 2016. 2557: 86–99.
15. Conrad, M., F. Dion, and S. Yagar. Real-Time Traffic Signal Optimization with Transit Priority: Recent Advances in the Signal Priority Procedure for Optimization in Real-Time Model. *Transportation Research Record: Journal of the Transportation Research Board*, 1998. 1634: 100–109.
16. Balke, K., C. Dudek, and T. Urbanik II. Development and Evaluation of Intelligent Bus Priority Concept. *Transportation Research Record: Journal of the Transportation Research Board*, 2000. 1727: 12–19.
17. Dion, F., and B. Hellinga. A Rule-Based Real-Time Traffic Responsive Signal Control System with Transit Priority: Application to an Isolated Intersection. *Transportation Research Part B: Methodological*, Vol. 36, No. 4, 2002, pp. 325–343.
18. Head, L., D. Gettman, and Z. Wei. Decision Model for Priority Control of Traffic Signals. *Transportation Research Record: Journal of the Transportation Research Board*, 2006. 1978: 169–177.
19. Ma, W., Y. Liu, and X. Yang. A Dynamic Programming Approach for Optimal Signal Priority Control upon Multiple High-Frequency Bus Requests. *Journal of Intelligent Transportation Systems*, Vol. 17, No. 4, 2013, pp. 282–293.
20. Zlatkovic, M., A. Stevanovic, and P. Martin. Development and Evaluation of Algorithm for Resolution of Conflicting Transit Signal Priority Requests. *Transportation Research Record: Journal of the Transportation Research Board*, 2012. 2311: 167–175.
21. MMITSS Detail Design. *Final Report: Multi-Modal Intelligent Traffic Signal System – Phase II: System Development, Deployment and Field Test*. University of Arizona, University of California PATH Program, Savari Networks, Inc., Econolite, Center for Transportation Studies, California, 2016. http://www.cts.virginia.edu/cvpfs_research/.
22. Moghimidarzi, S., P. G. Furth, and B. Cesme. Predictive–Tentative Transit Signal Priority with Self-Organizing Traffic Signal Control. *Transportation Research Record: Journal of the Transportation Research Board*, 2016. 2557: 77–85.
23. Ma, W., X. Yang, and Y. Liu. Development and Evaluation of a Coordinated and Conditional Bus Priority Approach. *Transportation Research Record: Journal of the Transportation Research Board*, 2010. 2145: 49–58.
24. Duerr, P. Dynamic Right-of-Way for Transit Vehicles: Integrated Modeling Approach for Optimizing Signal Control on Mixed Traffic Arterials. *Transportation Research Record: Journal of the Transportation Research Board*, 2000. 1731: 31–39.
25. NTCIP 1202. *National Transportation Communications for ITS Protocol Object Definitions for Actuated Traffic Signal Controller (ASC) Units, v02.19*. American Association of State Highway and Transportation Officials (AASHTO),

Institute of Transportation Engineers (ITE), and National Electrical Manufacturers Association (NEMA), 2005.
26. He, Q., K. L. Head, and J. Ding. Heuristic Algorithm for Priority Traffic Signal Control. *Transportation Research Record: Journal of the Transportation Research Board*, Vol. 2259, 2011. 2259: 1–7.
27. PTV. *PTV VISTRO User Manual.* Karlsruhe, Germany, 2014.

The Standing Committee on Traffic Signal Systems (AHB25) peer-reviewed this paper (18-01483).

TRR
JOURNAL OF THE TRANSPORTATION RESEARCH BOARD

Article

Transportation Research Record
2018, Vol. 2672(18) 27–34

DOI: 10.1177/0361198118782796
journals.sagepub.com/home/trr
SAGE

Increasing Freeway Capacity by Efficiently Timing Its Nearby Arterial Traffic Signals

Xingan (David) Kan[1], Xiao-Yun Lu[2], and Alexander Skabardonis[1]

Abstract

Currently, arterial traffic signals adjacent to freeway on-ramps operate independent of freeway on-ramp metering. During peak periods, the traffic signals employ long signal cycles to maximize intersection capacity. This approach often results in long platoons of freeway-bound traffic advancing toward the on-ramp within a short period, which quickly fills up the on-ramp and causes spillback on the surface streets. Queue override is typically employed to prevent the queue spillback by suspending ramp metering or relaxing the metering rates. Unfortunately, this counteracts the objective of on-ramp metering, which is to prevent capacity drop at freeway merge bottlenecks. Significant benefits can be realized by integrating arterial signal timing with the adjacent ramp-metering system. A signal control strategy was developed and evaluated in this study. The algorithm takes available on-ramp storage and the freeway ramp-metering rate into account and dynamically adjusts the cycle length to prevent on-ramp queue spillback and activation of queue override. The proposed algorithm was tested through simulation and the results show that the proposed strategy reduces the freeway, arterial, and system-wide delay at a modest penalty for on-ramp bound traffic.

Under current practices, arterial traffic signals that are adjacent to freeway on-ramps operate independent of freeway ramp metering. During peak hours, arterial traffic signals respond to high demand by employing long signal cycles and green durations to maximize arterial capacity. The signals are also progressively coordinated along the major arterial that facilitates access to freeway on-ramps. This may lead to platoons of arterial traffic advancing to the metered freeway on-ramps. As a result, metering the large influx of on-ramp traffic requires sufficient space to store queued vehicles.

Unfortunately, physical constraints in road geometry and the surrounding land use do not allow for the construction of longer or wider on-ramps. This consequently leads to on-ramp oversaturation, and queue spillback can occur prior to the termination of the green phase. Due to the interference from queue spillback, a portion of the green phase cannot be used to serve traffic demand, resulting in a reduction in intersection capacity and large delays for the arterial traffic. Most of the operational ramp-metering systems employ a "queue override" feature that is intended to prevent the on-ramp queue from obstructing traffic flow along the adjacent surface streets. The override is triggered whenever a sensor placed at the entrance of the on-ramp detects a potential queue spillover of the on-ramp vehicles on the adjacent surface streets. This clears the on-ramp queue by temporarily turning off ramp metering or increasing the metering rate to the maximum allowable value (typically 900 vphpl).

The purpose of most commonly employed metering is to prevent capacity drop at freeway merge bottlenecks by restricting the flow of on-ramp traffic, as shown by many empirical studies (*1–3*). The activation of the queue override counteracts the intended benefit. As demonstrated by a recent empirical study, queue override reduces the effectiveness of the employed ramp-metering systems during the highest traffic demand, when ramp metering is most needed (*4*). This translates to approximately 10% reduction in the freeway bottleneck discharge flow (*4*). Therefore, it is important to control the arrival of arterial traffic to the freeway on-ramp. This

[1]Institute of Transportation Studies, Department of Civil and Environmental Engineering, University of California, Berkeley, CA
[2]Partners for Advanced Transportation Technology (PATH), Institute of Transportation Studies, Richmond, CA

Corresponding Author:
Address correspondence to Xingan (David) Kan: david.kan@berkeley.edu

would prevent on-ramp oversaturation and avoid queue override, thereby preserving the capacities of both the freeway bottleneck and the arterial intersection.

There has been very limited research in the area of signal timing for arterial intersections near metered freeway on-ramps during peak hours, and no generalizable control strategies have been implemented.

Tian et al. developed an algorithm for diamond interchanges that reduces green durations for movements with on-ramp access to prevent on-ramp queue spillback, but maintains the same cycle length (*5*). This approach may cause spillback of on-ramp demand onto the upstream arterial, especially under long cycle lengths, and temporary activation of queue override.

Recker et al. developed a system-wide optimization model for ramp metering and traffic signals from stochastic queuing theory, but the improvement observed after implementing the control strategy at a network of freeways and arterials was a result of using a more efficient ramp-metering control, rather than an improved signal timing plan (*6*). Moreover, the proposed approach requires solving non-linear optimization in real time, which is computationally intensive and not feasible in most situations.

Other research efforts focused only on control of isolated signalized intersections at or adjacent to freeway on-ramps. For example, Li and Tao proposed a signal optimization model for an arterial at an isolated freeway interchange using the cell transmission model but neglected ramp metering in their algorithm (*7*).

Su et al. developed a signal optimization model that takes the ramp meter rate and on-ramp queue length into account for an isolated diamond interchange (*8*). However, similar to the method by Tian et al. (*5*), the proposed algorithm simply reallocated green times without changing the cycle length, therefore it provided unnecessarily long green durations for the conflicting movements and disregarded the potential queue spillback into the upstream arterial intersection. Furthermore, the impact of queue override was not considered.

The objective of the study described in this paper was to develop and test a simple and readily implementable control strategy for an integrated operation of arterial signals and freeway ramp meters. Under this integrated operation, arterial traffic signal controllers would account for the freeway on-ramp metering rate and queue length to appropriately adjust the signal settings for intersections adjacent to the freeways, which would enable smaller platoons of arterial traffic to enter the freeway on-ramp and the green phase to terminate before on-ramp queue spillback occurs, allowing conflicting directions to clear the intersection. This approach would prevent queue override and potential impact on freeway capacity. This would also prevent the loss of intersection capacity caused by the presence of queue spillback during green phases. Finally, the approach would not be limited to specific road geometries or ramp-metering algorithms.

The rest of the paper is organized as follows: the next section discusses the control strategy in detail. The following section documents the approach for evaluation of the proposed control algorithm through simulation of a real-world freeway corridor, followed by a section detailing the simulation results. The final section summarizes the study findings and discusses the next steps in the ongoing research.

Proposed Signal Control Strategy

A simple and readily implementable control strategy is presented to resolve the current inefficient control of freeway ramp-metering and arterial traffic signals, which fail to recognize oversaturation of metered on-ramps by using long cycle lengths and providing progression to the heavier direction, as well as activating queue override. The control strategy is intended to manage the on-ramp queues at most freeway corridors with metered on-ramps and adjacent arterials primarily used to facilitate freeway access.

The proposed approach maintains the existing freeway ramp-metering algorithm in practice and is not limited to any specific freeway ramp-metering algorithm. For the adjacent arterial traffic signals, this approach recommends that the signal timing plans for arterials adjacent to freeway on-ramps should be developed similar to signal timings of oversaturated arterials. Recent research in this area (*9–11*) suggests that long cycle lengths and long green durations must be avoided and, instead, the cycle lengths and green durations should consider on-ramp queue storage space and should be designed to avoid queue spillback.

Consider a freeway on-ramp and the adjacent signalized intersection with three phases (Figure 1). The on-ramp is metered with a rate $r(t)$ and has queue storage capacity Q_r. In this example, signal phases 1 and 2 serve the traffic movements feeding the on-ramp, and the remaining phase does not have on-ramp access.

The queuing diagram in Figure 2 illustrates the upstream arrival pattern, the downstream departure rate, and the excess accumulation at the freeway on-ramp when ramp metering is active. When the vehicle arrivals $A(t)$ from phases 1 to 2 of the upstream signalized intersection are greater than the discharge flow $D(t)$, then we have excess accumulation $Q(t)$ on the on-ramp. When $Q(t)$ exceeds the storage capacity (in number of vehicles) Q_r, we have queue spillback that causes the activation of the queue override.

Since the arrival rates from phases 1 to 2 discharging at the saturation flow exceed the ramp-metering rate, the

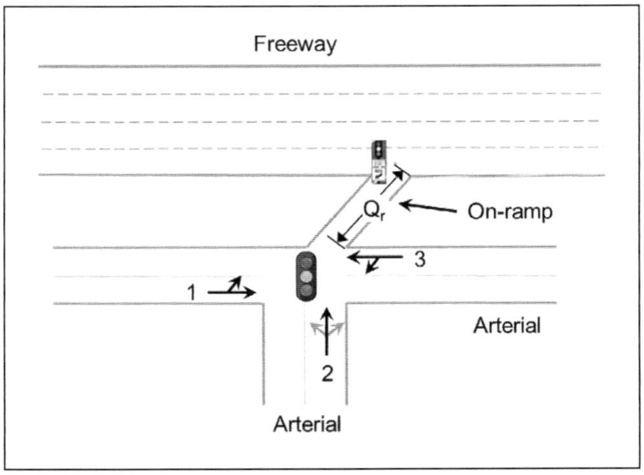

Figure 1. Example of a signalized intersection with on-ramp access.

corresponding phase green times must terminate before the excess accumulation $Q(t)$ reaches the on-ramp queue storage capacity Q_r. The remaining signal phase, which serves the intersection approach that does not feed the on-ramp, can be served earlier, and at that time the on-ramp queue dissipates. As a result, vehicles from phases 1 to 2 can enter the on-ramp in the next signal cycle.

Under this strategy, phases 1 and 2 receive shorter green times, which may cause spillback on the upstream intersection(s) of the arterial, as reported in the literature (*5*, *8*). Such a shortcoming can be remedied by maintaining the existing green distributions but using a shorter cycle length, as explained in the following sections.

The mathematical expressions for this control strategy are described as follows. First, the excess accumulation of the on-ramp must be estimated. Using the ramp meter rate $r(t)$ that is updated each time step t, the on-ramp excess accumulation $Q(t)$ at time step t can be determined based on the following process:

$$Q(0) = 0$$
$$Q(1) = Q(0) + A(1) - D(1)$$
$$Q(2) = Q(1) + A(2) - D(2)$$
$$\vdots$$
$$Q(t) = Q(t-1) + A(t) - D(t)$$

where $A(t)$ is the number of arrivals at the on-ramp during time step t, and $D(t)$ is the number of departures from ramp meter during time step t. The arrivals and departures can be measured by the loop detectors at the upstream and downstream ends of the freeway on-ramp, respectively. The on-ramp excess accumulation should be updated at the end of every cycle in order to perform real-time control.

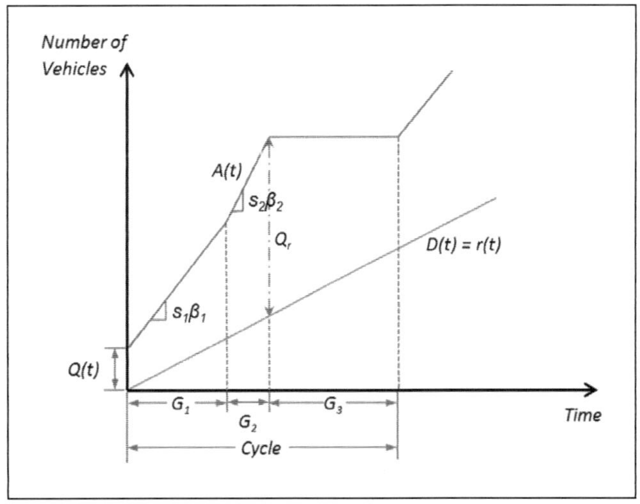

Figure 2. Queuing diagram of freeway on-ramp during a signal cycle (lost time not shown).

Equation 1 is developed using the queuing diagram in Figure 2, to ensure that the green time for phase 1 and phase 2 must terminate at or before the excess accumulation reaches the maximum on-ramp queue storage capacity Q_r therefore imposes an upper limit for the cycle length:

$$Q(t-1) + g_1 \cdot s_1 \cdot \beta_1 + g_2 \cdot s_2 \cdot \beta_2 - g_1 \cdot r(t) - g_2 \cdot r(t) - 2l \cdot r(t) \leq Q_r \quad (1)$$

where

g_1 = effective green time of Phase 1
g_2 = effective green time of Phase 2
s_1 = saturation flow of Phase 1
s_2 = saturation flow of Phase 2
β_1 = percentage of demand of Phase 1 that accesses the on-ramp
β_2 = percentage of demand of Phase 2 that accesses the on-ramp
$Q(t-1)$ = residual on-ramp queue from the previous cycle
$r(t)$ = ramp metering rate
l = lost time of each phase (s)

The effective green times can be expressed as functions of cycle length, based on a commonly practiced approach derived from an *M/D/1* queuing model (*12*). Thus, g_1 and g_2 are expressed as:

$$g_1 = \frac{y_1}{Y} \cdot (C - 3l) \quad (2)$$

$$g_2 = \frac{y_2}{Y} \cdot (C - 3l) \quad (3)$$

where

y_1 = ratio of arrival rate and saturation flow of Phase 1

y_2 = ratio of arrival rate and saturation flow of Phase 2
Y = sum of y's of the cycle
C = cycle length (s)

Substitute equations 2 and 3 into Equation 1, and Equation 1 can be expressed in terms of cycle length. Solving for cycle length in relation to the rest of the variables, the upper limit of cycle length is represented by the following:

$$C \leq \frac{[Q_r - Q(t-1) + r(t) \cdot 2l] \cdot Y + 3l \cdot \left[\sum_{i=1,2} s_i \beta_i y_i - \sum_{i=1,2} r(t) y_i\right]}{\left[\sum_{i=1,2} s_i \beta_i y_i - \sum_{i=1,2} r(t) y_i\right]} \quad (4)$$

The upper limit of the cycle length must be updated at the end of every cycle in order to perform real-time control that coordinates with freeway ramp metering.

Although the shorter cycle may introduce additional lost time, this ensures that all movements can clear the intersection at saturation flow rate and therefore prevent loss in intersection capacity due to queue spillback impeding the flow of traffic during green phases.

The above-selected cycle length does not provide for the maximum bandwidth of through traffic, which is appropriate for close to saturation arterial facilities that primarily provide access to multiple freeway on-ramps. The appropriate signal settings for additional intersections upstream of the one with immediate freeway on-ramp access, consist of shorter cycles and offsets that prevent long queues on the intersection approaches determined by methods presented in Lieberman et al. (*11*).

The proposed strategy is applicable to real-world multiphase signalized intersections by appropriately adjusting equations 1 to 4. Furthermore, the proposed strategy does not require any new surveillance technologies or infrastructural changes and can be accomplished using the existing infrastructure and detection capabilities. It only requires modification of signal and ramp meter controller settings (i.e., cycle length, queue override) and communication between the two controllers.

Evaluation

A real-world freeway corridor and its adjacent arterial streets were selected to evaluate the effectiveness of the proposed control strategy. A 3-mi section of freeway I-680 from Alum Rock Avenue to Berryessa Road in San Jose, California, shown in Figure 3 was selected. There are three recurrent bottlenecks on this stretch of the I-680, located near the on-ramps from Berryessa Road, McKee Road, and Alum Rock Avenue. At all three bottlenecks, high on-ramp demand from the westbound direction (diagonal on-ramp) of Berryessa Road, McKee Road, or Alum Rock Avenue, along with high on-ramp

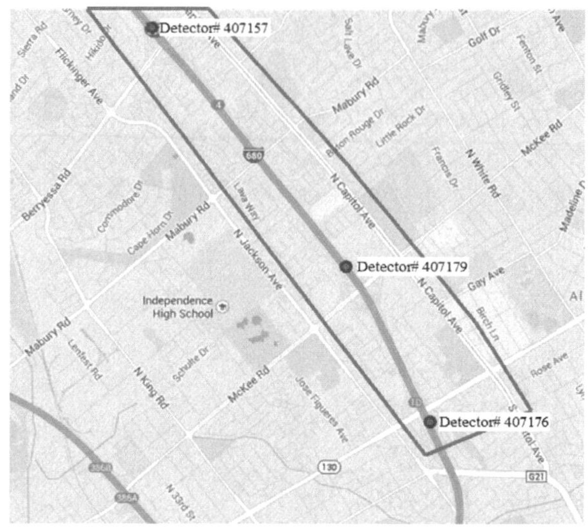

Figure 3. Map of study site and freeway detector locations.

demand from both directions of Capitol Avenue, result in high volumes of merging traffic entering the northbound freeway mainline during the morning peak (7:00–9:30 a.m.). On-ramp demand from the eastbound direction (loop on-ramp) of Berryessa Road, McKee Road, and Alum Rock Avenue is relatively low and does not contribute to peak-hour congestion.

This section of the freeway has four lanes in each direction, whereas the parallel arterial Capitol Avenue, as well as Alum Rock Avenue and Berryessa Road all have two lanes in each direction. McKee Road has three lanes in each direction. Typically, merging traffic from the arterial causes average speed to decrease to about 20 mph near the Alum Rock on-ramp, 30 mph near the McKee on-ramp, and 40 mph near the Berryessa on-ramp.

All of the on-ramps in this corridor are metered and the ramp meters operate under the local traffic responsive demand-capacity approach. The metering rates are assigned based on various thresholds of freeway mainline occupancies immediately upstream of the merging or weaving area. Under current practice, the ramp meters employ "queue override" in case queue spillback is detected by the loop detector at the on-ramp entrance. Queue override is typically activated from 7:30 to 8:00 a.m., though it sometimes lasts until 8:30 a.m. The queue override algorithm at this site increases the ramp-metering rate by 100 vphpl as soon as queue spillover is detected at the entrance of the on-ramp, and by another 100 vphpl if the queue spillover continues in the next 30-s cycle, until it reaches the maximum rate of 900 vphpl.

The signalized intersections of this corridor are traffic actuated and operate with time of day coordinated actuated timing plans. The existing common cycle length is relatively long (160 s) and the signal timing plan provides progression to the northbound direction, which is the one with higher demand. Light rail vehicles operate every 15 min in the median of Capitol Avenue and receive green extensions. Pedestrian volumes are relatively low and typically request actuation once every 15 min, when the light rail vehicles (in the median of Capitol Avenue) arrive at their stop. Pedestrian actuation requires a minimum cycle length of 120 s.

In order to evaluate the proposed signal control strategy, a microscopic simulation model was built in AIMSUN (*13*) using the most up to date road geometry, lane configurations, and speed limits of the freeway and the arterial at the selected site. Freeway and arterial data obtained from the morning peak (7:00–9:30 a.m.) of September 23, 2015 were used for the inputs in demand and turning percentages. During this morning peak, freeway ramp meters and arterial traffic signals were functioning properly and there were no incidents within the selected corridor.

For the freeway, 5-min interval loop detector data for flow were obtained from the Caltrans Performance Measurement System (*14*) and used as the demand input at the most upstream location of the simulation network and as the turning percentages at any applicable mainline-off-ramp split. Ramp-metering rates and algorithms were obtained from Caltrans District 4 and modeled in microscopic simulation via AIMSUN API (application programming interface).

For the arterial, video cameras were placed at the signalized intersections and arterial-on-ramp splits during the morning peak of September 23. Turning movement flows were recorded every 5 min, and they were used as the demand input at the entry points of the corridor (i.e., Southbound Capitol north of Berrryessa Road) and as turning percentages at the signalized intersections and the arterial-on-ramp splits. Field data suggest that the majority of the on-ramp bound traffic comes from the northbound left turns on Capitol Avenue and the through movements of westbound Alum Rock Avenue, McKee Road, and Berryessa Road. The signal timing plans for the coordinated actuated signals were provided by the city of San Jose.

The AIMSUN software used includes enhanced car-following and lane-changing algorithms recently developed and calibrated to accurately model the reduction of freeway discharge rate following the onset of congestion, known as the capacity drop (*4*, *15*). The microscopic simulation model was calibrated to existing conditions at the selected corridor prior to the evaluation of the proposed control strategy. The calibration results showed satisfactory agreement between the simulated and field observed freeway flow and speed and arterial flow (*4*, *16*). Twenty replications of the simulation model runs with different random number seeds were made using the existing demand, turning percentages, ramp-metering algorithm, and signal timing plans.

Simulation tests were conducted to determine the effectiveness of the proposed signal control strategy. First, the existing conditions, which employ long signal cycles and queue override at on-ramps, were simulated as the reference case for comparison. Next, the proposed signal control with shorter signal cycles, typically 100 s in length, was simulated via AIMSUN API, and queue override was not active during simulation. Additionally, pedestrian crossings were simulated to replicate real-world scenarios in which the cycle length must be at least 120 s due to long crossing times. Twenty replications with different random seeds were run in each of the scenarios described above, and paired *t*-tests were conducted to determine the statistical significance of changes in performance after introducing the proposed approach. The following performance metrics were used to analyze the simulation results:

- Freeway mainline: Total delay and vehicle miles traveled (VMT) of the entire peak period. Comparison of total delay is an indication of changes in speed and travel time on the freeway mainline and comparison of VMT implies any capacity changes at the freeway bottlenecks.
- Parallel arterial: Average delays of all turning movements corresponding to the parallel arterial (Capitol Avenue) at each signalized intersection over the entire peak period. The sum of the average delays at each signalized intersection will be presented.
- Cross streets: Average delays of all turning movements corresponding to the cross streets (Alum Rock Avenue, McKee Road, and Berryessa Road), across the entire peak period.
- Total system: Total delay of freeway and arterial during the entire peak.

Results

The comparison of the simulation results prior to and after implementing the proposed signal control is summarized in Table 1.

- Freeway mainline: The proposed control strategy prevented the activation of the queue override, which improved the freeway operating conditions. The total delay on the freeway mainline was reduced by 11.13%, and VMT was improved by 3.92%. Paired *t*-tests suggest that both improvements are statistically significant, as indicated by the numerical values in bold. The significant delay

Table 1. Performance Before and After Proposed Signal Control

	Before		After		% difference	
Freeway mainline						
	Total delay (vehicle hour)	Total distance traveled (vehicle mile)	Total delay (vehicle hour)	Vehicle miles traveled (VMT)	Change in total delay	Change in VMT
I-680 NB	833.41	43104.13	740.64	44792.95	**−11.13%**	**3.92%**
Arterial						
Average delay on main parallel arterial (min/vehicle)						
Capitol Ave. NB	8.63		10.51		**21.84%**	
Capitol Ave. SB	5.72		5.91		**3.33%**	
Average delay of cross street (s/vehicle)						
Alum Rock WB	48.05 (D)*		47.33 (D)*		−1.43%	
Alum Rock EB	37.27 (D)*		37.82 (D)*		1.47%	
McKee WB	56.76 (E)*		52.34 (D)*		**−7.79%**	
McKee EB	28.92 (C)*		16.51 (B)*		**−42.91%**	
Berryessa WB	47.27 (D)*		39.26 (D)*		**−16.73%**	
Berryessa EB	50.50 (D)*		37.55 (D)*		**−34.48%**	
Total system						
	Total delay (vehicle hour)		Total delay (vehicle hour)		Change in total delay	
Freeway and arterial	2881.37		2727.19		**−5.65%**	

Note: Bold % difference values = statistical significance; NB = northbound direction; SB = southbound direction; WB = westbound direction; EB = eastbound direction.
*Level of service.

reduction can be attributed to the increase in mainline speed and the capacity of the freeway bottlenecks. The higher freeway capacity is indicated by the increase in VMT once queue override is not active, and this conforms to the observations of the empirical study by Kan (*4*). The relatively small increase in VMT can be explained by the fact that queue override typically affects only 30 to 60 min of the 2.5 h morning peak, rather than the entire period.

- Parallel arterial (Capitol Avenue): Both directions of the parallel arterial experienced statistically significant change in average delay. The northbound direction, which consists of more than 60% left turn traffic intended to access the freeway on-ramps, experienced additional delay because it no longer benefits from the relaxed metering rate after queue override was avoided. However, the through movements of northbound Capitol Avenue, which do not feed the on-ramp, were not penalized. This could potentially encourage some drivers to avoid the left turn, continue on Capitol Avenue, and enter the freeway at an on-ramp further downstream to improve freeway congestion. This may require additional empirical observations to validate the potential change in route choice. On the other hand, the southbound direction of the parallel arterial remained mostly unaffected, due to the lower demand for on-ramp access. Lastly, the light rail vehicles experienced slight less (10%) delay as a result of the shorter cycle lengths.
- Cross streets: On the cross streets, delays were reduced because the proposed approach increased the intersection capacity by avoiding wasted green time caused by queue spillback and the new signal control adopted shorter cycles. There were statistically significant delay reductions at McKee Road and Berryessa Road but almost none at Alum Rock Avenue. This is because there is sufficient space for on-ramp queue storage near the Alum Rock Avenue on-ramp so the signal timing remained the same. The queue storage space was limited at the McKee Road and Berryessa Road on-ramps therefore the cycle lengths were reduced at the corresponding signalized intersections, which reduced the average delay on those cross streets. The westbound directions of the cross streets benefited much less than the eastbound directions because there is significant on-ramp bound arterial traffic in the westbound direction. As discussed earlier, the on-ramp bound traffic was penalized after avoiding the queue override's relaxed metering rate, thus negated some of the delay reduction experienced by those not intended

to access the on-ramp (in the other lanes of the same westbound approach).

- Total system: The proposed control strategy reduced the system-wide delay by more than 5%.

Conclusions

Currently, arterial traffic signals near freeway on-ramps operate independent of the ramp-metering rate and the presence of on-ramp queues. The arterial signal settings optimized to maximize arterial capacity and progression may send platoons of on-ramp bound vehicles toward the oversaturated on-ramps and cause queue spillbacks onto the adjacent arterial intersections. The conventional approach to mitigate queue spillbacks on the arterial is to employ queue override. However queue override counteracts the objective of on-ramp metering, which protects the freeway discharge rate and avoids congestion at merge bottlenecks. Empirical evidence suggests that the freeway bottleneck discharge rate drops by approximately 10% when queue override is activated.

However, nearby arterial traffic signals can help avoid queue spillback at freeway on-ramps if the signal control systems are integrated with those of the freeway ramp metering and if the traffic signals are timed effectively based on the current ramp-metering rate and on-ramp queue lengths. We recommend reducing the arterial signal cycle length and maintaining the same green time distribution such that only a short platoon feeds the adjacent freeway on-ramp every cycle. This could prevent queue spillback and subsequently avoid queue override by sending on-ramp bound traffic in smaller but more frequent platoons, without imposing significant penalties. This approach also increases the intersection capacity by avoiding wasted green time caused by queue spillback.

A section of the I-680 northbound freeway with Capitol Avenue as the parallel arterial in the city of San Jose, California, was selected as the test site to evaluate the proposed signal control. The a.m. peak was selected as the analysis period. Traffic operations at the site were simulated using the AIMSUN microscopic model after extensive data collection and model calibration. The simulation results show the proposed signal control strategy eliminated the queue spillback on the metered on-ramps that activated the queue override. This resulted in an 11.13% reduction in freeway mainline delay. Furthermore, the freeway capacity improved, as indicated by a 3.92% increase in VMT throughout the entire 2.5-h morning peak. Despite the penalty imposed on on-ramp bound traffic that no longer benefits from relaxed metering rate under queue override, the rest of the signalized approaches experienced significant delay savings.

Lastly, the proposed algorithm is simple and readily implementable at most freeway corridors with metered on-ramps and adjacent arterials primarily used to facilitate freeway access. The proposed approach is not limited to any specific freeway ramp-metering algorithm, and requires only software integrations between the on-ramp metering and arterial traffic signals. The implementation does not require any investment in new sensing or signal control technologies. Therefore, real-world field tests of the proposed signal control strategy are recommended in order to better estimate the effectiveness and potential benefits of the proposed control. The field test will be performed in the near future and will help us identify the practical constraints and issues that were not shown in the simulation. The real-world performance of the proposed signal control strategy will serve as a guideline for transportation agencies when designing arterial traffic signal timing plans near freeways. Other recommended future work includes tests for freeway corridors with different types of freeway-arterial interchange geometries and different ramp-metering algorithms.

Acknowledgments

This research is supported by the State of California Department of Transportation (Caltrans) under Interagency Agreement No. 65A0518. We thank the project technical panel for their guidance and support.

Author Contributions

The authors confirm contribution to the paper as follows: study conception and design: AS, XDK, X-yL; data collection: XDK; analysis and interpretation of results: AS, XDK, X-yL; draft manuscript preparation: XDK, AS. All authors reviewed the results and approved the final version of the manuscript.

References

1. Cassidy, M. J., and J. Rudjanakanoknad. Increasing the Capacity of an Isolated Merge by Metering Its On-Ramp. *Transportation Research Part B: Methodological*, Vol. 39, No. 10, 2005, pp. 896–913.
2. Zhang, L., and D. Levinson. Ramp Metering and Freeway Bottleneck Capacity. *Transportation Research Part A: Policy and Practice*, Vol. 44, No. 4, 2010, pp. 218–235.
3. Kim, K., and M. J. Cassidy. A Capacity-Increasing Mechanism in Freeway Traffic. *Transportation Research Part B: Methodological*, Vol. 46, No. 9, 2012, pp. 1260–1272.
4. Kan, X. *Increasing Freeway Capacity by Efficiently Timing Its Nearby Arterial Traffic Signals*. PhD dissertation. University of California, Berkeley, 2017.
5. Tian, Z., C. Messer, K. Balke, and T. Urbanik. Integration of Diamond Interchange and Ramp Metering Operations.

Transportation Research Record: Journal of the Transportation Research Record, 2005. 1925: 100–111.

6. Recker, W., X. Zheng, and L. Chu. *Development of an Adaptive Corridor Traffic Control Model*. California PATH Research Report, 2009.
7. Li, Z., and R. Tao. Integrate Control of Freeway Interchange Model Based on Enhanced Cell Transmission Model. *Transportation Research Record: Journal of the Transportation Research Record*, 2011. 2259: 179–191.
8. Su, D., X. Y. Lu, R. Horowitz, and Z. Wang. Coordinated Ramp Metering and Intersection Signal Control. *International Journal of Transportation Science and Technology*, Vol. 3, No. 2, 2014, pp. 179–191.
9. Denney, R. W., L. Head, and K. Spencer. *Signal Timing Under Saturated Conditions*. Federal Highway Administration Report No. FHWA-HOP-09-008, 2008.
10. Gettman, D., H. Liu, M. Abbas, and A. Skabardonis. *Operation of Traffic Signal Systems in Oversaturated Conditions*. National Cooperative Highway Research Program, 2012.
11. Lieberman, E. B., J. Chang, and E. S. Prassas. Formulation of Real-Time Control Policy for Oversaturated Arterials. *Transportation Research Record: Journal of the Transportation Research Record*, 2000. 1727: 77–78.
12. Newell, G. *Theory of Highway Traffic Signals*. Institute of Transportation Studies Research Report, UCB-ITS-RR-89-7, 1989.
13. TSS|Aimsun. http://www.aimsun.com/. Accessed March 19, 2017.
14. Caltrans PeMS. http://pems.dot.ca.gov/. Accessed March 31, 2016.
15. Lu, X., X. Kan, S. E. Shladover, D. Wei, and R. A. Ferlis. An Enhanced Microscopic Traffic Simulation Model for Application to Connected Automated Vehicles. Presented at 96th Annual Meeting of the Transportation Research Board, Washington, D.C., 2017.
16. Kan, D., X. Y. Lu, and A. Skabardonis. *Coordination of Freeway Ramp Metering and Arterial Traffic Signals-Phase IIA: Site Selection and Simulation Development*. PATH Research Report UCB-ITS-PRR-2016-02, 2016.

The Standing Committee on Traffic Signal Systems (AHB25) peer-reviewed this paper (18-05441).

The contents of this paper reflect the views of the authors, who are responsible for the facts and accuracy of the data presented here. The contents do not necessarily reflect the official views or policies of the State of California.

Article

TRR
JOURNAL OF THE TRANSPORTATION RESEARCH BOARD

Transportation Research Record
2018, Vol. 2672(18) 35–44

DOI: 10.1177/0361198118790860
journals.sagepub.com/home/trr
SAGE

Real-Time Detector-Free Adaptive Signal Control with Low Penetration of Connected Vehicles

Yiheng Feng[1], Jianfeng Zheng[2], and Henry X. Liu[1,2]

Abstract

Most of the existing connected vehicle (CV)-based traffic control models require a critical penetration rate. If the critical penetration rate cannot be reached, then data from traditional sources (e.g., loop detectors) need to be added to improve the performance. However, it can be expected that over the next 10 years or longer, the CV penetration will remain at a low level. This paper presents a real-time detector-free adaptive signal control with low penetration of CVs ($\leq$ 10%). A probabilistic delay estimation model is proposed, which only requires a few critical CV trajectories. An adaptive signal control algorithm based on dynamic programming is implemented utilizing estimated delay to calculate the performance function. If no CV is observed during one signal cycle, historical traffic volume is used to generate signal timing plans. The proposed model is evaluated at a real-world intersection in VISSIM with different demand levels and CV penetration rates. Results show that the new model outperforms well-tuned actuated control regarding delay reduction, in all scenarios under only 10% penetrate rate. The results also suggest that the accuracy of historical traffic volume plays an important role in the performance of the algorithm.

Driven by the rapid development of connected vehicle (CV) technologies, we are on the cusp of a new revolution in transportation safety and mobility on a scale not seen since the introduction of automobiles a century ago. To evaluate the CV technologies in real-world environments, U.S. Department of Transportation (USDOT) has initialized a number of deployment projects including Safety Pilot Model Deployment project in Ann Arbor, Michigan (*1*), CV pilot deployment projects (https://www.its.dot.gov/pilots/), and Smart City Challenge (https://www.transportation.gov/smartcity). Through these projects, thousands of vehicles and hundreds of intersections have/will be equipped with dedicated short-range communication (DSRC) devices, which enable vehicle-to-vehicle (V2V) and vehicle-to-infrastructure (V2I) communications to improve safety, mobility and sustainability.

Traffic signal control system, as one of the critical components of the urban transportation operations, can also benefit from the CV technology. Through V2I communications, the traffic control system receives vehicle trajectories from nearby CVs to make control decisions. Compared with traditional data from fixed-location infrastructure-based sensors, CV data provide much more information and have high potential in improving signal operations. A number CV-based signal control and performance estimation models have been proposed (*2–8*). However, results from existing studies have shown that minimum required penetration rates vary from different applications, but typically 20–30% penetration rate is necessary (*9*). If the critical penetration rate cannot be reached, then data from traditional sources (e.g., loop detectors) need to be added to improve the performance (*10*). Some studies intended to characterize individual vehicle behaviors through limited CV trajectories. For example, Goodall et al. estimated unequipped vehicle location (*5*), and Feng et al. inferred both location and speed of unequipped vehicles (*3*). Sun and Ban attempted to reconstruct the entire trajectory of unequipped vehicles (*11*). From a traffic signal control point of view, aggregated performance measures such as volume, queue length, travel time, and delay are sufficient to optimize traffic signals. Although these aggregated metrics can be

[1]University of Michigan Transportation Research Institute, Ann Arbor, MI
[2]Department of Civil and Environmental Engineering, University of Michigan, Ann Arbor, MI

Corresponding Author:
Address correspondence to Yiheng Feng: yhfeng@umich.edu

easily derived from individual vehicles,, it requires more information and therefore higher penetration rates. A systematic review of adaptive signal control with CVs can be found in Jing et al. (*12*).

Despite substantial efforts in investing and developing CV technologies in the past decade, over the next 10 years or longer the CV penetration rate is expected to remain at a low level. Therefore, optimizing traffic signals with low penetration rates of CVs is essential and will make an immediate impact on the state of the practice.

To the best of the authors' knowledge, there are only a few studies that have focused on a low-penetration environment. A study from Day and Bullock (*9*) proposed a proof-of-concept study to optimize signal offsets with limited CV market penetration. The penetration rates used in the paper were from 0.1% to 50%. However, instead of focusing on real-time implementation, their analysis periods were set to 3 h (offline) and 15 min (online). The selected analysis period may be sufficient for offset adjustment as offset may not change much over a few cycles. However, for real-time adaptive signal control, traffic conditions change significantly within 15 minutes. Moreover, the data used in this study were sampled from loop detectors, and does not represent real CV trajectory. Recent work done by Zheng and Liu utilized aggregated CV trajectory data to estimate traffic volumes (*13*). The model was formulated as a maximum likelihood estimation problem and solved by expectation maximization algorithm. The overall penetration rates varied from 3% to 12% at different approaches and time of day, and the mean absolute percentage error (MAPE) of the estimated volume was about 10%. However, this method cannot be directly implemented for real-time signal operations because the trajectory data need to be aggregated over days.

This paper extends the study by Zheng and Liu by combining both historical and real-time trajectory data to perform detector-free adaptive signal control. A simple probabilistic model is applied to estimate cycle-by-cycle vehicle arrival times and delays based on estimated average historical volume and a limited number of observed critical CV trajectories. Then a dynamic programming (DP)-based adaptive signal control algorithm is applied to generate the optimal signal plan, using estimated vehicle delay as the objective function. The proposed model is tested in software-in-the-loop (SIL) simulation with various low penetration rates (10%, 5%, 2%, and 0%) and demand levels at a real-world intersection. Results are compared with well-tuned actuated signal control.

The rest of this paper is organized as follows. The next section describes the methodologies for CV trajectory-based delay estimation and adaptive signal control models. Simulation results and discussions are then presented, and conclusions and areas for further research are provided in the final section.

Methodology

Before presenting the method, Figure 1 shows the CV trajectories in one lane under 10% penetration rate with a demand level of 700 veh/h/ln, which represent the raw data used in this paper. The figure shows that some CVs passed the intersection without stop while others stopped in the queue for the red signal. Some of the vehicle trajectories are only partial because of lane changes. Note

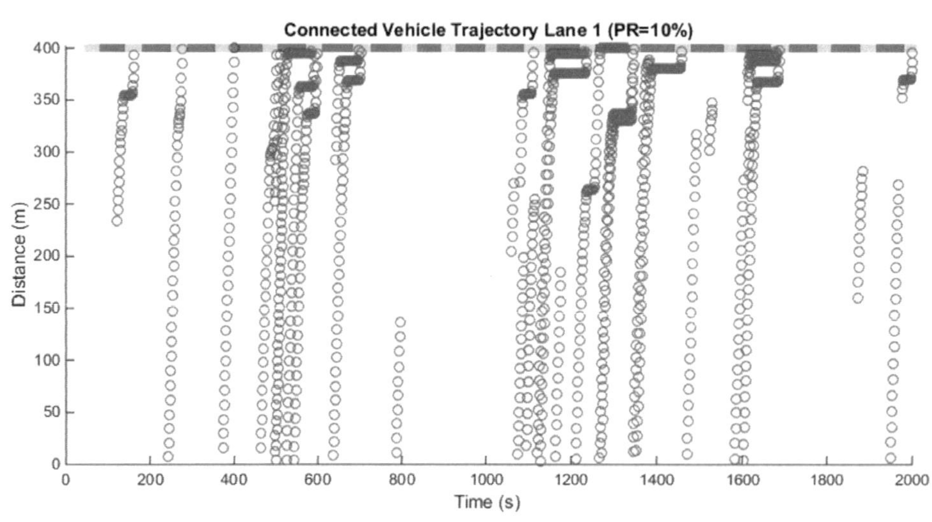

Figure 1. Illustration of CV trajectories under 10% penetration rate.

that during most of the cycles only one or two CVs were observed, and during some cycles there was no CV at all.

Vehicle Delay Estimation

The core idea of using limited trajectories to estimate delay is to utilize critical CV information. Critical CVs are defined as the last stopped CV and the first non-stopped CV. The last stopped CV provides a lower boundary of queue length, whereas the first non-stopped CV provides an upper boundary because the queue has to be fully discharged before the arrival of the non-stopped CV. For those cycles that do not have any observed CV, an average hourly volume is used to generate vehicle arrival and departure times for delay estimation. The hourly volume can be estimated from the aggregation of historical CV trajectory data (*13*). It was assumed the vehicle arrivals follow Poisson process with mean arrival rate λ. The cumulative number of arrivals during time interval t is expressed as $N(t) \sim \text{Poisson}(\lambda t)$.

Four cases are identified according to the existence of observed CVs as shown in Figure 2.

Case 1: No Observed CV. If no CV is observed during the entire cycle (Figure 2*a*), the only information that can be utilized is the average volume estimated by historical data. Given cycle length C, the total number of vehicles arrive within the cycle $n = \lambda C$, which is the mean of the Poisson distribution. Total vehicle delay D is the summation of delay from each vehicle:

$$D = \sum_{i=1}^{round(n)} D_i \qquad (1)$$

Delay of each vehicle is expressed by the following equation:

$$D_i = t_i^d - t_i^e - t_f \qquad (2)$$

where t_i^d is the departure time of vehicle i, t_i^e is the entrance time of vehicle i, and t_f is the free flow travel time from entrance location d_e to the intersection. Free flow travel time is assumed to be the same for all vehicles. The entrance location is defined based on the communication range (e.g., 300 m).

The departure time of each vehicle is calculated based on the entrance time, free flow speed, signal timing, and vehicle position in the queue:

$$t_i^d = \max(t_g + t_{sl} + i^* s_f, t_i^e + t_f) \qquad (3)$$

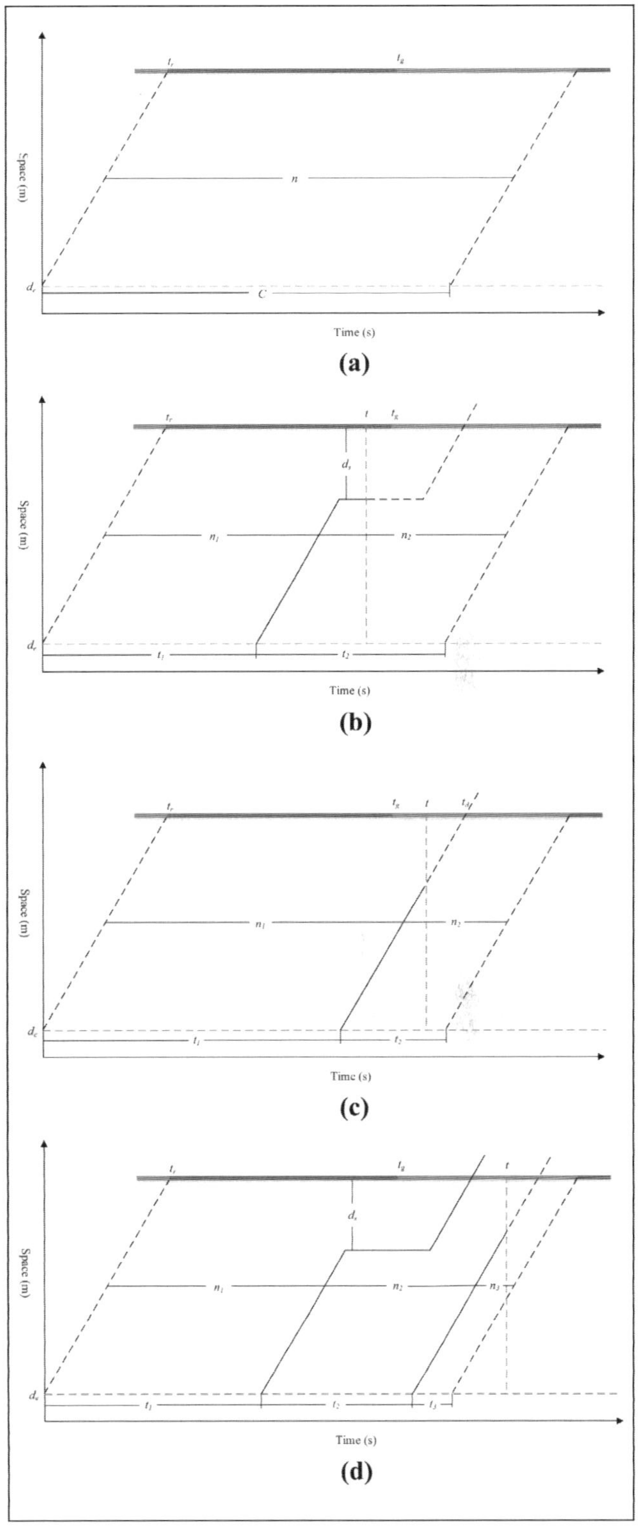

Figure 2. Four scenarios based on critical CV trajectory: (*a*) no CV; (*b*) only stopped CV; (*c*) only non-stopped CV; (*d*) both stopped and non-stopped CV.

where t_g is the green start time, t_{sl} is the start-up lost time, and s_f is the saturation headway (e.g., 2 s/veh). We do not consider lane change so that the ith vehicle approaching the intersection is also the ith vehicle in the queue.

Given the cumulative number of arrivals n during interval C, the event occurrence time (entry time of each individual vehicle) $t_1^e, \ldots t_n^e$ have the same distribution as the order statistics corresponding to n independent random variables uniformly distributed within the interval $(0,C)$ (*14*). The entrance time of each vehicle can be expressed as:

$$t_i^e = t_r - t_f + CU_{(i)} \tag{4}$$

where t_r is red start time and $U_{(i)}$ is i^{th} order statistics from a standard uniform distribution. It can be proved that i^{th} order statistics of the standard uniform distribution follows a beta distribution $U_{(i)} \sim B(i, n+1-i)$ (*14*). The mean of the beta distribution is used to present arrival times of each vehicle:

$$U_{(i)} = i/(n+1) \tag{5}$$

Combining Equations 1–5, total delay can be calculated.

Case 2: Only Stopped CV. If only stopped CVs are observed during a cycle, then the cycle time is divided into two intervals (Figure 2*b*). The first interval is the time period from the entry time of the first stopped vehicle to the entry time of the stopped CV (t_1), and the second interval is the time period after the entry time of the stopped CV until the last vehicle that passes during the green time (t_2), with $t_1 + t_2 = C$.

All vehicles that enter during t_1 are stopped vehicles, as the stopped CV provides a lower boundary of the queue. Based on the location of the stopped CV, the total number of stopped vehicle n_1 can be estimated, assuming vehicle length is uniform and known:

$$n_1 = round\left(\frac{d_s}{l}\right) \tag{6}$$

where d_s is the stopping distance of the CV and l is the average vehicle length.

The number of vehicles that enter during t_2 is estimated based on the average arrival rate because no more CV information is available. That is $n_2 = \lambda t_2$. The departure time of each vehicle during t_1 and t_2 is calculated using Equations 7 and 8, respectively. Vehicles arrive during t_1 must depart at saturation flow rate because they are all in the queue. Vehicles arrive during t_2 may pass the intersection without stop if they arrive after the queue is fully discharged:

$$t_i^d = t_g + t_{sl} + i^*s_f \quad i \le n_1 \tag{7}$$

$$t_i^d = \max(t_g + t_{sl} + i^*s_f, t_i^e + t_f) \quad n_1 < i \le n_1 + n_2 \tag{8}$$

The entry time and delay calculation are similar to Case 1.

Case 3: Only Non-Stopped CV. If only non-stopped CVs are observed during the cycle, then the cycle time is also divided into two intervals (Figure 2*c*). The first interval is the time period from the entry time of the first stopped vehicle to the entry time of the non-stopped CV (t_1), and the second interval is the time period after the entry time of the non-stopped CV until the last vehicle that passes during the green time (t_2), with $t_1 + t_2 = C$.

The non-stopped CV provides an upper boundary of the queue. Unlike stopped CV, it only gives the maximum possible number of vehicles entered during t_1, because the queue can be cleared before the arrival of the non-stopped CV. The maximum possible number of vehicles is calculated as:

$$n_{\max} = (t_d - t_g - t_{sl})/s_f \tag{9}$$

where t_d is the departure time of the non-stopped CV. The number of actual vehicles arrived during t_1 can be calculated as the summation of conditional probabilities:

$$n_1 = \sum_{i=1}^{round(n_{\max})} i^*P(X = i|X \le round(n_{\max})) \tag{10}$$

where $X \sim$ Poisson(λt_1). The total estimated delay of vehicles entered during t_1 is the summation of total delays of all possible numbers of entered vehicles multiplied by the corresponding probability:

$$D_1 = \sum_{j=1}^{round(n_{\max})} \left\{ P(X = j|X \le round(n_{\max})) \sum_{i=1}^{j} D_{i,j} \right\} \tag{11}$$

where $D_{i,j}$ is the delay of vehicle i given total j vehicles enter during t_1. The calculation of $D_{i,j}$ for each vehicle is similar to Equations 3–5. The number of vehicles that enter during t_2 is estimated based on the average arrival rate as in Case 2. Note that since the queue is already fully discharged before the non-stopped CV, vehicles that enter after the CV do not cause any delay.

Case 4: Both Stopped and Non-Stopped CV. In this case, both the lower boundary and the upper boundary of vehicle queue are provided by the stopped CV and non-stopped CV, respectively (Figure 2*d*). Therefore, the cycle time is divided into three intervals. The first interval is the time period from the entry time of first stopped vehicle to the entry time of the non-stopped CV (t_1). The second interval is the time period from the entry time of the stopped CV to the entry time of the non-stopped CV (t_2), and the third interval is after the entry time of the non-stopped CV until the last vehicle that passes during the green period (t_3), with $t_1 + t_2 + t_3 = C$. It is easy to see that delay estimation of the three intervals are included in previous three cases. To avoid redundancy, the detailed calculation is skipped.

If multiple stopped and non-stopped CVs are observed within one cycle, only the last stopped CV and the first non-stopped CV are utilized because they represent the critical information.

Adaptive Signal Optimization

The adaptive control algorithm is adapted from previous research by Feng et al. (*3*). The algorithm generates an optimal signal phase sequence and duration using a two-level optimization model. The model is based on DP and can apply different objective functions, including total delay minimization and total queue length minimization. In this paper, only total delay minimization is chosen as the objective.

The algorithm uses an arrival table as the input to the optimization model. The arrival table is a two-dimensional matrix with time and phase. The value of each cell is the number of vehicles that will arrive at the stop-bar at time point t requesting phase p. It is generated based on CV trajectory data at the time of executing the signal optimization. The original model adds all queuing vehicles to the first line of the arrival table, which does not consider accumulative delay generated by already stopped vehicles. Delay of all vehicles is calculated from the time point when the signal optimization is conducted. In the proposed delay estimation model, entry times of each individual vehicle are generated so that the arrival time of each vehicle at the stop-bar can be calculated. As a result, the accumulative delay of each queuing vehicle can be obtained. A new arrival table is constructed to incorporate the delay from vehicles that already stopped before the planning time.

Simulation Results and Discussion

To test the proposed models, a SIL simulation framework is designed and implemented with VISSIM microscopic simulation software. The SIL simulation architecture is shown in Figure 3.

CVs in VISSIM simulation network generate basic safety messages (BSMs) at a frequency of 10 Hz and broadcast to the Data Processor module. This module also requests Signal Phasing and Timing (SPaT) data from the Econolite ASC/3 virtual controller. Processed CV trajectory and signal information are then sent to the delay estimation module. This module generates the arrival table and sends to the adaptive control algorithm, which is responsible for producing optimal signal timing plan with the objective to minimize total vehicle delay. The optimal signal plan will be converted into a series of control commands by the Signal Controller Interface module and control virtual signal controllers in VISSIM.

A real-world intersection at Huron Pkwy and Plymouth Rd in Ann Arbor, Michigan is modeled in VISSIM 9. The intersection geometry and signal phasing are shown in Figure 4.

To evaluate the proposed delay estimation model, the VISSIM model is run for 1 h, and all vehicle trajectories are recorded and served as the ground truth. The traffic signals are under actuated control so that the cycle lengths and phase splits change over time. Figure 5

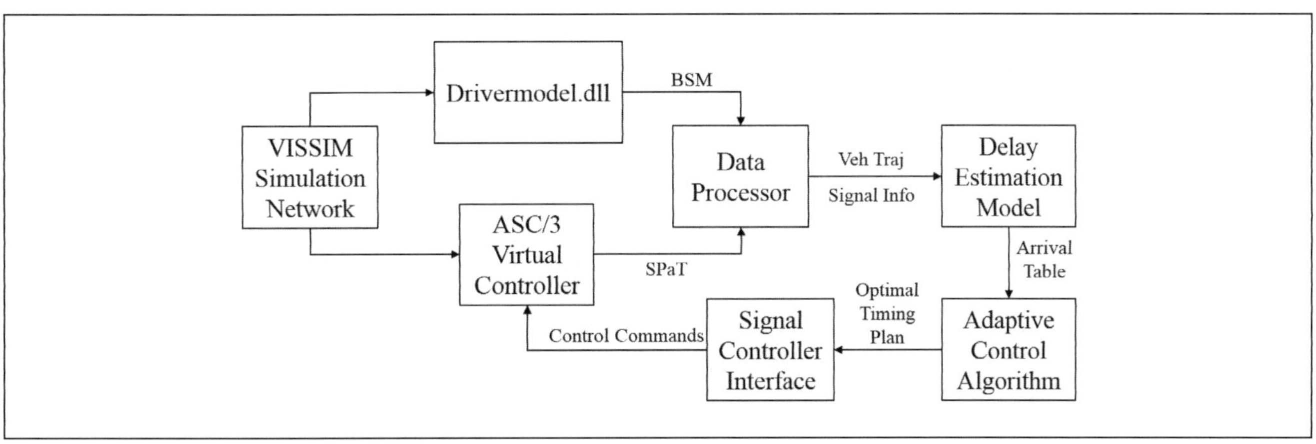

Figure 3. Software-in-the-loop simulation architecture.

Figure 4. Geometry and signal phasing at Huron Pkwy & Plymouth Rd intersection.

shows the comparison of the estimated total vehicle delay and the actual vehicle delay of Phase 6 by lane with 10% penetration rate. There are total 31 full cycles operated within 1 h.

To further quantify the accuracy, the MAPE was calculated for the estimated delay using the following equation:

$$MAPE = \frac{1}{N}\sum_{i=1}^{N}\frac{|D_i^e - D_i^t|}{D_i^t} \quad (12)$$

where N is the total number of cycles, D_i^e is the estimated vehicle delay of cycle i, and D_i^t is the actual vehicle delay of cycle i.

Under 10% penetration rate, the MAPE for Lane 1 and Lane 2 are 18.99% and 14.56%, respectively. If two lanes are combined together, the MAPE for Phase 6 is 14.30%. The model was also tested under 0% penetration rate, under which only hourly volume is used to generate vehicle arrivals (i.e., always in Case 1 because of no observed CV). The MAPE for Lane 1 and Lane 2 are 32.60% and 28.65%, respectively. If two lanes are combined together, the MAPE values for Phase 6 is 30.49%. The result indicates that if only hourly volume is used as input for the delay estimation model, the estimated delays in each cycle significantly differ from the actual delays. From Figure 5*c*, it can be seen that the vehicle delay of each cycle varied from less than 500 veh·s to over 2000 veh·s. Estimation using only 10% CV's data can reduce the MAPE significantly, from more than 30% to less than 15% percent. This suggests that just a few critical CV trajectories are needed to improve the vehicle delay estimation to a relatively accurate level.

As the delay estimation algorithm generates individual vehicle arrival times, an arrival table can be easily constructed and served as the input to the adaptive control algorithm. Two scenarios with two different demand levels and four penetration rates are evaluated. Scenario 1 assumes that the estimated hourly volume of each phase (or average arrival rate λ) is accurate. Scenario 2 assumes the estimated hourly volume of each phase has 10% error, which is more realistic based on field data (*13*). Scenario 2 adds 10% of demand on phase one to four and deducts 10% of demand on phase five to eight. The objective of such adjustment is to maximize the disturbance on the signal timing. Two demand levels are considered as medium (critical v/c ratio 0.82) and congested (critical v/c ratio 0.93) traffic conditions. Four penetration rates under evaluation are: 10%, 5%, 2%, and 0%. Under 0% penetration rate, the adaptive control basically becomes a fixed-time signal plan, which is generated by the hourly volume (always Case 1 in delay estimation algorithm). The traffic demands used in each scenario are summarized in Table 1. Note that the estimated hourly volume with 10% error is only used in the delay estimation model. The vehicle inputs in the VISSIM model are the same for the two scenarios, which is the actual hourly volume.

A total duration of 3900 s is executed in VISSIM simulation for each scenario, each demand level, and each penetration rate, with 300 s of warm-up period and 3600 s of data collection time. To capture the stochastic demand pattern, each simulation run is repeated with

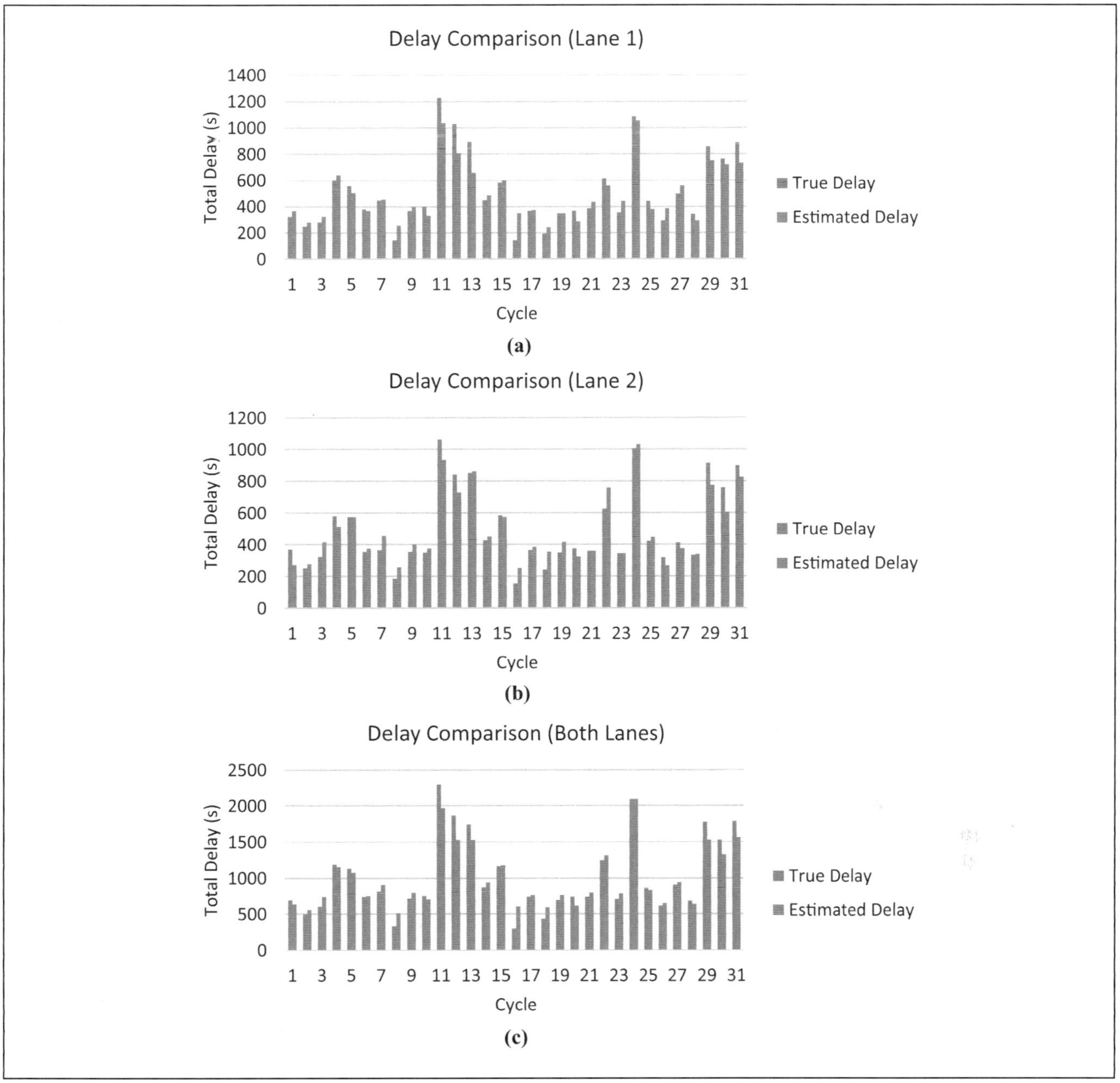

Figure 5. Estimated vehicle delay under 10% penetration rate: (*a*) lane 1; (*b*) lane 2; (*c*) combination of both lanes.

Table 1. Traffic Demands of Each Phase under Two Scenarios and Two Demand Levels

Unit: veh/h/ln	P1	P2	P3	P4	P5	P6	P7	P8
Medium demand (Scenario 1)	187	675	133	450	150	656	150	333
Medium demand (Scenario 2)	206	742	146	495	135	591	135	300
Congested demand (Scenario 1)	212	765	167	525	170	744	175	417
Congested demand (Scenario 2)	233	841	182	577	153	670	157	375

five random seeds. The results are compared with a well-tuned fully actuated control, in which the minimum green time, maximum green time, yellow interval, and all red clearance interval are set to be the same as in the adaptive control algorithm. The unit extension time of the actuated control is set to 1.6 s, which is obtained by

Table 2. Total Vehicle Delay in Seconds under Medium Demand Level

Random seed	1	2	3	4	5	Average (SD)	Delay reduction
Scenario 1: accurate hourly volume estimation							
10% PR	143336	152534	135818	151338	137554	144311 (7674)	5.23%
5% PR	148165	157135	141530	158741	149372	150988 (7034)	0.84%
2% PR	168963	190877	152779	178334	168224	171835 (14046)	−12.84%
Actuated	145736	162606	150933	158352	143770	152279 (8070)	N/A
Scenario 2: 10% hourly volume estimation error							
10% PR	144404	155736	143002	155517	149726	149677 (5983)	1.71%
5% PR	157791	168744	146392	159259	151568	156750 (8447)	−2.94%
2% PR	164093	182495	145614	170820	164004	165405 (13386)	−8.62%
Actuated	145736	162606	150933	158352	143770	152279 (8070)	N/A

Note: SD = standard deviation.

Table 3. Total Vehicle Delay in Seconds under Congested Demand Level

Random seed	1	2	3	4	5	Average (SD)	Delay reduction
Scenario 1: accurate hourly volume estimation							
10% PR	227684	248169	222959	260393	231441	238129 (15656)	16.33%
5% PR	240871	258387	222687	260856	231085	242777 (16692)	14.70%
2% PR	259532	281069	240524	280446	242579	260830 (19631)	8.35%
0% PR	327241	367273	288306	344282	261268	317674 (42731)	−11.62%
Actuated	256728	305282	279268	330017	251736	284606 (33074)	N/A
Scenario 2: 10% hourly volume estimation error							
10% PR	252124	282365	243068	279463	258485	263101 (17189)	7.56%
5% PR	267432	283013.	242671	271912	249347	262875 (16577)	7.64%
2% PR	270629	339032	254176	317639	281232	292541 (34897)	−2.79%
0% PR	346828	380832	356243	442983	313010	367979 (48470)	−29.29%
Actuated	256728	305282	279268	330017	251736	284606 (33074)	N/A

Note: SD = standard deviation.

the recommendations from Signal Timing Manual (*15*). Tables 2 and 3 show the delay comparison under two demand levels.

The following observations are made by analyzing the results:

1. When the penetration rate is 10%, the proposed model outperforms well-tuned actuated control in all cases. The total vehicle delay is decreased by 16.33% under congested demand level with accurate volume estimation. Under medium demand

level with 10% volume estimation error, the vehicle delay is still reduced by 1.71%. As the penetration rate decreases, the total delay tends to increase.

2. The hourly volume estimated from historical data has a significant impact on the performance. Under same demand level and same penetration rate, the results with 10% volume estimation error are all worse than no error in volume estimation. When the algorithms are executed under low penetration rates, it is common that no CV is observed within the entire cycle. Then the hourly volume serves as the only data for determining the phase duration.
3. Besides penetration rate, the absolute number of observed CVs is also crucial to the performance of the algorithm. This explains why the algorithm performs better under congested demand level than medium demand level with the same penetration rate. Under congested demand level with accurate volume estimation, even 2% penetration has a delay reduction of 8.35%. However, under medium demand level with accurate volume estimation, model performance with 5% penetration rate is almost the same as actuated control.
4. Vehicle delays with 10% and 5% penetration rates under congested demand level are similar, in both scenarios. This indicates that a few critical vehicle trajectories are enough to make an accurate estimation of vehicle delay. Higher penetration rates only bring marginal benefits.
5. When the algorithm is executed under 0% penetration rate, the adaptive control becomes a fixed-time control. Because no CV trajectory is available, the control decision is made only based estimated hourly volume, which is a set value. The results under such conditions are significantly worse than other cases, which supports a well-accepted argument that fixed-time control cannot accommodate short-time demand fluctuation, even if the average volume is accurate. Moreover, under congested demand level, the intersection under fixed-time control may enter oversaturated condition as a result of demand fluctuation, and the delay increases significantly. On the other hand, actuated and adaptive control can handle the demand fluctuation better and prevent the intersection entering the oversaturated condition.

Conclusion and Further Research

This paper presents a detector-free real-time adaptive signal control model in a low CV penetration environment. Critical CVs were defined, which referred to the last stopped CV in the queue and the first non-stopped CV that passed the intersection. They provided the lower and upper boundaries of queue length, respectively. Based on critical CV information, a simple delay estimation model was proposed. Then the model was integrated with an adaptive control algorithm to generate optimal signal plans with the objective of minimizing vehicle delay. Microscopic simulation results showed the proposed model worked well under 10% penetration rate in all scenarios. Compared with well-tuned actuated control, the total delay reduction can reach as much as 16.3%.

The new model has two significant advantages. First, it does not require any data from infrastructure-based sensors, which usually have considerable high installation and maintenance costs. Second, it only needs at most 10% CV penetration rate, so that it can be implemented at an early stage of CV deployment. For example, the Ann Arbor Connected Vehicle Test Environment (AACVTE) project is targeting to equip up to 5,000 vehicles in the next few years, which accounts for about 5% of total vehicles in Ann Arbor metro area (https://www.its.dot.gov/research_archives/safety/aacvte.htm). The proposed model has great potential to be implemented at real-world intersections in the near future.

One direction for further research is to extend the current model to a corridor level, where the vehicle arrivals may not be Poisson distributed, and signal coordination needs to be considered. One of the difficulties lies in the determination of platoon size and speed on coordinated phases to dynamically update offset and split. In addition, the current model relies on the estimated average volume from historical data as the first step. It would be very interesting to develop an integrated platform that combines the volume estimation algorithm, as in Zheng and Liu (*13*), and the real-time adaptive signal control together so that the estimated volumes can be updated dynamically when new CV trajectories become available.

Acknowledgments

This research was partially funded by the U.S. Department of Energy (EERE Award No. DEEE0007212) and USDOT Center for Connected and Automated Vehicles (CCAT).

Author Contributions

The authors confirm contribution to the paper as follows: study conception and design: Yiheng Feng, Jianfeng Zheng, Henry Liu; data collection: Yiheng Feng; analysis and interpretation of results: Yiheng Feng, Jianfeng Zheng; draft manuscript preparation: Yiheng Feng, Jianfeng Zheng, Henry Liu. All authors reviewed the results and approved the final version of the manuscript.

References

1. Bezzina, D., and J. Sayer. *Safety Pilot Model Deployment Test Conductor Team Report.* Publication DOT HS 812

171. National Highway Traffic Safety Administration, Washington, D.C., 2015.
2. Argote-Cabañero, J., E. Christofa, and A. Skabardonis. Connected Vehicle Penetration Rate for Estimation of Arterial Measures of Effectiveness. *Transportation Research Part C: Emerging Technologies*, Vol. 60, 2015, pp. 146–160.
3. Feng, Y., K. L. Head, S. Khoshmagham, and M. Zamanipour. A Real-Time Adaptive Signal Control in a Connected Vehicle Environment. *Transportation Research Part C: Emerging Technologies*, Vol. 55, 2015, pp. 460–473. https://doi.org/10.1016/j.trc.2015.01.007.
4. Goodall, N., B. Smith, and B. Park. Traffic Signal Control with Connected Vehicles. *Transportation Research Record: Journal of the Transportation Research Board*, 2013. 2381: 65–72.
5. Goodall, N. J., B. Park, and B. L. Smith. Microscopic Estimation of Arterial Vehicle Positions in a Low-Penetration-Rate Connected Vehicle Environment. *Journal of Transportation Engineering*, Vol. 140, No. 10, 2014, p. 04014047. https://doi.org/10.1061/(ASCE)TE.1943-5436.0000716.
6. He, Q., K. L. Head, and J. Ding. PAMSCOD: Platoon-Based Arterial Multi-Modal Signal Control with Online Data. *Transportation Research Part C: Emerging Technologies*, Vol. 20, No. 1, 2012, pp. 164–184. https://doi.org/10.1016/j.trc.2011.05.007.
7. Lee, J., B. Park, and I. Yun. Cumulative Travel-Time Responsive Real-Time Intersection Control Algorithm in the Connected Vehicle Environment. *Journal of Transportation Engineering*, Vol. 139, No. 10, 2013, pp. 1020–1029. https://doi.org/10.1061/(ASCE)TE.1943-5436.0000587.
8. Priemer, C., and B. Friedrich. A Decentralized Adaptive Traffic Signal Control Using V2I Communication Data. Presented at the 12th International IEEE Conference on Intelligent Transportation Systems, ITSC '09, St. Louis, MO, 2009.
9. Day, C. M., and D. M. Bullock. Detector-Free Signal Offset Optimization with Limited Connected Vehicle Market Penetration. *Transportation Research Record: Journal of the Transportation Research Board*, 2016. 2558: 54–65.
10. Feng, Y., M. Zamanipour, K. L. Head, and S. Khoshmagham. Connected Vehicle–Based Adaptive Signal Control and Applications. *Transportation Research Record: Journal of the Transportation Research Board*, 2016. 2558: 11–19.
11. Sun, Z., and X. J. Ban. Vehicle Trajectory Reconstruction for Signalized Intersections Using Mobile Traffic Sensors. *Transportation Research Part C: Emerging Technologies*, Vol. 36, 2013, pp. 268–283. https://doi.org/10.1016/j.trc.2013.09.002.
12. Jing, P., H. Huang, and L. Chen. An Adaptive Traffic Signal Control in a Connected Vehicle Environment: A Systematic Review. *Information*, Vol. 8, No. 3, 2017, p. 101. https://doi.org/10.3390/info8030101.
13. Zheng, J., and H. X. Liu. Estimating Traffic Volumes for Signalized Intersections Using Connected Vehicle Data. *Transportation Research Part C: Emerging Technologies*, Vol. 79, 2017, pp. 347–362. https://doi.org/10.1016/j.trc.2017.03.007.
14. Ross, S. *Introduction to Probability Models*. Elsevier Science & Technology, Oxford, UK, 2009.
15. Urbanik, T., A. Tanaka, B. Lozner, E. Lindstrom, K. Lee, S. Quayle, S. Beaird, S. Tsoi, P. Ryus, D. Gettman, S. Sunkari, K. Balke, and D. Bullock. *NCHRP Report 812: Signal Timing Manual Second Edition*. Transportation Research Board of the National Academies, Washington, D.C., 2015.

The Standing Committee on Traffic Signal Systems (AHB25) peer-reviewed this paper (18-04795).

The views presented in this paper are those of the authors alone.

Article

TRR
JOURNAL OF THE TRANSPORTATION RESEARCH BOARD

Transportation Research Record
2018, Vol. 2672(18) 45–57

DOI: 10.1177/0361198118774184
journals.sagepub.com/home/trr
SAGE

Signal Priority Request Delay Modeling and Mitigation for Emergency Vehicles in Connected Vehicle Environment

Jiangchen Li[1], Chen Qiu[1], Liqun Peng[1], and Tony Z. Qiu[1,2]

Abstract

Connected vehicle-based signal priority control is widely regarded as an advanced method for improving travel efficiency of an emergency vehicle when passing through intersections. However, the vehicle-to-everything (V2X) communication delay is a critical factor affecting the performance of signal request and has rarely been considered in existing studies. This paper conducted a comprehensive delay analysis of the preemption signal request and its influence on practical preemption control for the emergency vehicle. First of all, a general end-to-end delay decomposition model is formulated to analyze significant delay uncertainties from different sources. Then, a compensated distance strategy is adopted for cooperative preemption control to ensure the reliability of preemption control and minimize impacts on performance caused by communication delay. Based on the analysis of field data and numerical results, the proposed model is able to reveal characteristics of communication delay for multimodal traffic signal control with priority. The proposed communication delay compensation strategy shows clear benefits in improving the performance of signal preemption control priority for an emergency vehicle at intersections and therefore has potential to enhance V2X applications in a connected vehicle environment.

Signal priority strategy is essential for urban transportation as it allocates the right of the road to different vehicles based on reasonable priority ranking. Among all signal priority strategies, emergency vehicle preemption is the simplest and most important one, implemented for approaching emergency vehicles (e.g., fire trucks) to pass through a signalized intersection as quickly as possible under safety guarantee via the provision of a preempted green signal. The provided preemptions ensure no conflicting vehicles will enter the target intersection, and guarantees the emergency vehicle will minimize its response time by maintaining a relatively high speed (*1*, *2*). Thus, emergency vehicle response time is an essential measure of effectiveness for an implemented emergency service system (*3*). Moreover, statistics have shown a strong relationship between the response time of emergency vehicles and the mortality rate, with 10% of deaths occurring just a few minutes after accidents (*3*). Consequently, decreases in emergency vehicle response time will effectively reduce the mortality rate within the first golden hour (*3*). Therefore, it is imperative to effectively increase performance of the emergency vehicle systems by reducing the response time (i.e., response delay).

To minimize emergency vehicle response time, which means reducing delays caused by surrounding traffic congestion and an approaching intersection's signal controls, various preemption control methods are proposed to identify and improve the performance of preemption control (*2*). From an identification perspective, the identified impacts include emergency vehicle demands (*4*), data sources (*1*, *5*), and service area coverage (*6*). All these investigated parameters are found to have great influence on the performance of preemption systems. On the other hand, many studies have proposed solutions for improving the performance of the preemption control for single and route-based intersections. These include optimal control and coordination strategies (*2*, *7*, *8*), route reliability-based strategy (*9*), and exiting preemption strategies (*10*, *11*). However, few of these methods are designed and validated with a solid real-time basis.

Therefore, the real-time basis of emergency vehicle preemption systems will not be achieved without the availability of intelligent transportation system (ITS)

[1]Department of Civil and Environmental Engineering, University of Alberta, Edmonton, AB, Canada
[2]Intelligent Transportation Systems Research Center, Wuhan University of Technology, Wuhan, Hubei, China

Corresponding Author:
Address correspondence to Tony Z. Qiu: zhijunqiu@ualberta.ca

technologies. Examples of these include the OPTICOM system, the STROBECOM system, and the MIRT system (*2*). Emergency vehicle detection in these ITSs includes inductive loop detectors, GPS-based systems, magnet-based technologies, light and infrared systems, acoustic systems, video detection, radio-based systems, and so forth (*12*, *13*). These detection systems provide real-time information to traffic signal controllers and actuate various signal priority control approaches. Recently, the latest advances in emerging connected vehicle (CV) technology and systems using dedicated short range communication (DSRC) have presented the traffic system with another opportunity for improving emergency response services, where intelligence can be used to provide cooperative services (*14*). CV technology enables vehicles to communicate with each other and share information with infrastructure and other road users, for example, pedestrians and cyclists. Three major forms of communication in the CV environment are vehicle-to-vehicle (V2V) communication, vehicle-to-infrastructure (V2I) communication, and vehicle-to-everything (V2X) communication. Information shared through CV technology is real time and features low latency, and includes position, speed, heading, signal phase and timing (SPaT), surrounding geometric description (map), and so forth (*14*). Based on these characteristics, several studies have utilized V2I and/or V2V communication to evaluate and improve emergency vehicle response time (*3*, *15–21*). Wang et al. (*16*) and Weinert and Düring (*18*) presented and implemented prototype systems for preemption signal control based on V2I communication, and the test results showed the high feasibility of the CV-based system. In addition, from Noori's work in 2013 (*15*), a preemption method is proposed based on the real-time distance between the emergency vehicle and the signalized intersection via V2I communication. From Jordan and Cetin's work in 2015 (*19*), a new strategy is developed to identify vehicle queue lengths and to preempt traffic signals earlier based on V2V communication. Furthermore, Noori et al. in 2016 (*3*) proposed another strategy considering both the distance and the predicted queue length from traffic volume to better serve the intersection and decrease the response time. For V2V communication, one study (*13*) proposed a path clearance method of providing the emergency vehicle with specific instruction via alert message using V2V communication. Recently, a deployment of a multimodal intelligent traffic signal system that covers an area of several urban blocks by Head et al. was studied for multimode traffic signal priority requests, and the results show that there were effective delay benefits for various applications (*14*, *20*, *22*). These studies lack the communication delay analysis of the multi-priority request in V2X communication.

Although the previous studies have verified several prototype systems and evaluated some strategies, few of them have systematically focused on the comprehensive delay analytics of the preemption signal request and its influence on practical preemption control. To address that, this paper focuses on the delay analytics in a real-time CV-based environment.

First, a general end-to-end delay decomposition model is formulated to analyze significant delay uncertainties. Compared with previous studies (*3*, *15–21*), this proposed model focuses on communication delay analytics of preemption signal requests and provides quantitative correlations based on potential significant traffic variables.

Then, a compensated distance-based cooperative preemption control is adopted to ensure the reliability of preemption control to minimize impacts of communication delay on system performance. A compensated distance is calculated so that the request message is guaranteed to be received and triggered in preemption control.

Third, field tests and comprehensive analysis of a complex condition are conducted to verify the proposed model and control method. The results show the proposed model and compensated distance-based preemption control are efficient.

Problem Definition

An emergency vehicle may encounter a red light at a signalized intersection and thus experience stop delay, as shown in Figure 1*a*, which increases the response time dramatically (*3*). To mitigate stop delay, a CV-based preemption system is applied as a real-time and thus efficient strategy (*3*, *15–21*). However, applying CV-based strategies introduces additional communication delay, which has critical impacts on the performance of the solution (*16*, *22*).

As illustrated in Figure 1*b*, the CV-based system includes two major components. One is an on-board equipment (OBE) unit in an emergency vehicle that acts as a signal priority request dissemination station. Another is a roadside equipment (RSE) unit connected to an intersection signal controller that behaves as a request receiver to trigger preemption and return an acknowledgment.

In such a CV-based system, communication delay is divided into two parts: a stochastic waiting time t_{sw} and a packet transmission time t_{trans} (*23*). The t_{sw} refers to a time interval from the beginning of a message request first accessing a wireless medium to the moment the message has been successfully transmitted. This t_{sw} is significantly influenced when multiple priority requests from different traffic modes and objects exist (*23–25*). The t_{trans} represents the transmission period for the request message packet, which is determined by the size of the

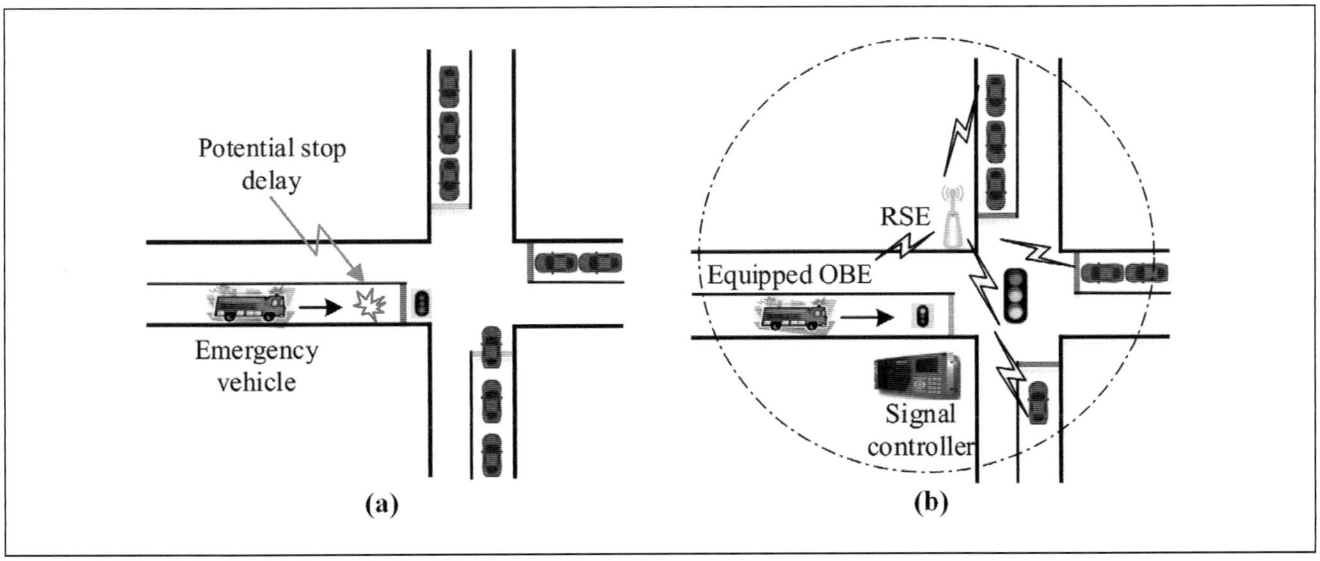

Figure 1. Delay illustration for emergency vehicle's response time: (*a*) typical stop delay for emergency vehicle (*3*); (*b*) a representative CV-based preemption system with communication delay (*21*).

SAE J2735 packets and corresponding wireless access in vehicular environments (WAVE) message (*20*). The larger the message size is, the larger the transmission time will be, which in the end increases the communication delay (*25*).

Therefore, impacts of stochastic waiting time and transmission time should be assessed and minimized in priority request communication applications.

Methodology

A methodology including signal priority request delay modeling and mitigation is proposed.

Probabilistic End-to-End Delay Decomposition Model

A probabilistic analysis model is formulated (*25*) to analyze the aforementioned uncertainties' impacts on signal request communication. This analysis provides a more detailed method to understand the delay decomposition compared with previous studies in designing a CV-oriented signal control (*3*, *15–21*).

As demonstrated in Figure 2*a* (*14*), a two-step process for a successful request message communication between a single OBE in the emergency vehicle and the RSE on the roadside of the intersection is presented, in which step one is request message broadcasting and step two is acknowledgment message broadcasting. Later, a preemption signal control is used as an example to represent any priority request applications. Figure 2*b* illustrates packet transmissions with a contention-based carrier sense multiple access/collision avoidance (CSMA/CA) scheme (*22*, *23*), in which transmissions of multiple request messages happen simultaneously during either one or two steps. For each step, the total delay includes stochastic waiting time and deterministic transmission time. Without loss of generality, this analysis can be applied to any CV-based application based on DSRC technology.

Stochastic Waiting Time. Stochastic waiting time is generated in the media access control layer and controlled by the contention-based CSMA/CA scheme.

First, assume n vehicles are within the RSE's communication (*14*). The number of vehicles simultaneously transmitting request messages (denoted as M) is calculated as follows:

$$n = \rho * CR * LN \quad (1a)$$

$$M = n * P_{smt} \quad (1b)$$

where ρ is the average vehicle density per lane per kilometer; CR is the RSE's communication coverage; LN is the total lane numbers; and P_{smt} is the probability of vehicles simultaneously sending requests. Among M vehicles, one is the target emergency vehicle and the remaining (M–1) vehicles are other traffic modes with priority requests (e.g., transit, freight). If there are multiple emergency vehicles sending preemption requests, when one is chosen as the target vehicle, the others are treated as normal vehicles. Assume that one vehicle broadcasts only one request packet at one time.

In Figure 2*b*, the scheme is used to avoid collisions among packets simultaneously transmitted from multiple vehicles (*22*, *23*). At the beginning, each packet from each vehicle (the emergency vehicle, vehicle 1, and vehicle 2 in

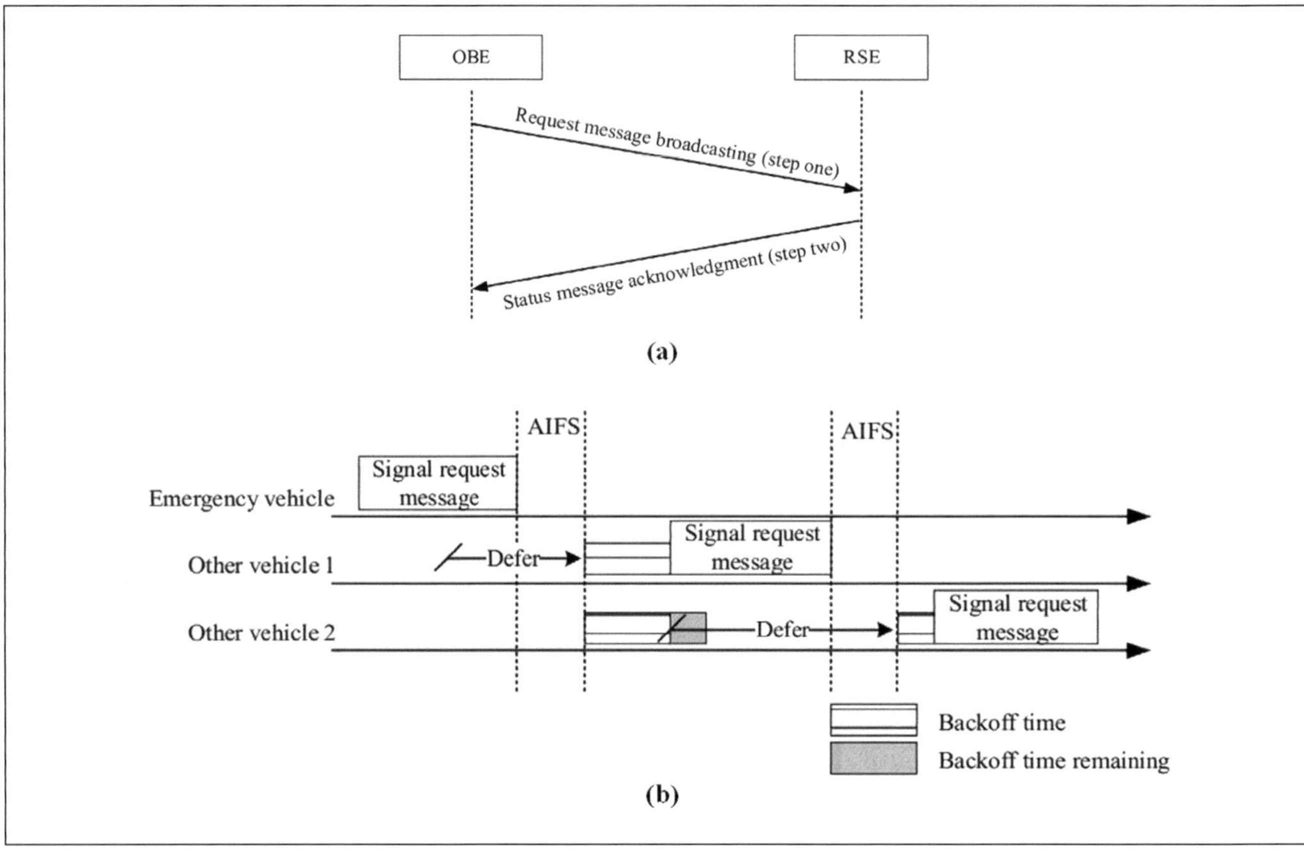

Figure 2. (*a*) two-step process for a successful request message communication between a OBE and RSE, which is abstracted from Figure 1*b*; (*b*) illustration of packet transmissions for multiple signal request messages with a contention-based scheme.

Figure 2*b*) randomly draws a backoff counter from a pre-defined section $[CW^{min}, CW^{max}]$, where CW^{max} and CW^{min} are the size of the maximum and minimum contention windows, respectively. Backoff counter multiplied by a time slot is a backoff time, which is marked in Figure 2*b*. The drawn counter value decides the transmission order of each packet. Then, each packet has to defer a while and pass through the backoff stage before being transmitted if the wireless medium is busy after a fixed waiting time (marked as AIFS process in Figure 2). In the last phase, the packet with zero remaining contention windows is transmitted via the wireless medium to the destination. Here, the time interval in this process is the stochastic waiting time. This process and the following analysis can be applied to any CV-based application based on a contention-based CSMA/CA scheme.

Now, consider a general scenario where the packet broadcasted from the target emergency vehicle is transmitted to the RSE after m vehicles have transmitted their packets. According to the scheme, the stochastic waiting time for this scenario (denoted as $t_{sw}(m,n_d)$) is calculated as below:

$$t_{sw}(m,n_d) = \underbrace{m(T_{AIFS} + t_{trans}) + \sum_{i=0}^{m} n_d^i T_d}_{previous} + \underbrace{T_{AIFS} + n_d T_d}_{current} \tag{2}$$

where, $t_{sw}(m,n_d)$ is the stochastic waiting time; T_{AIFS} denotes the arbitration inter-frame spacing (AIFS) time used for checking the busy status of the wireless medium; t_{trans} denotes the identical average transmission time for all previous m packets; m is the total number of previous packets that have been transmitted and $0 \leq m \leq (M-1)$; T_d denotes the length of a time slot; and n_d and n_d^i denote the minimum size of the randomly drawn contention windows of the current packet and the ith packet ($0 \leq i \leq m$), respectively.

Because the backoff mechanism follows a uniform distributed drawn strategy (*23*), the value n_d is uniformly drawn from the section $[CW^{min}, CW^{max}]$. The point probability of the event X_{n_d} that draws a certain value (i.e., n_d) as remaining contention windows for the current emergency vehicle is shown as follows:

$$P_{X_{n_d}}(X_{n_d} = n_d) = \left(\frac{1}{1 + CW^{max} - CW^{min}}\right) \quad (3a)$$

where $CW^{min} \leq n_d \leq CW^{max}$.

Accordingly, the probability of an event that m vehicles' messages are transmitted before the transmission of the emergency vehicle's request message is equal to a conditional probability of event X_m, given that the event X_{n_d} has occurred. The event X_m expressed for all possible sizes n_d^m of drawn contention windows for the m vehicles is less than n_d (i.e., $CW^{min} \leq n_d^i < n_d$). Thus, the conditional probability of X_m given event X_{n_d} is expressed below:

$$P_{X_m|X_{n_d}}(X_m = n_d^m < n_d | X_{n_d} = n_d) = \left(\frac{1 + n_d - CW^{min}}{1 + CW^{max} - CW^{min}}\right)^m \quad (3b)$$

Thus, the joint probability of obtaining $t_{sw}(m, n_d)$ is shown as follows:

$$P(t_{sw}(m, n_d)) = P_{X_{n_d}} P_{X_m|X_{n_d}} = \left(\frac{1}{1 + CW^{max} - CW^{min}}\right)\left(\frac{1 + n_d - CW^{min}}{1 + CW^{max} - CW^{min}}\right)^m \quad (3c)$$

Accordingly, the average waiting time $\hat{t}_{sw}$ of the emergency vehicle transmitting the priority message is calculated as follows:

$$\hat{t}_{sw} = E(t_{sw}(m, n_d)) = \sum_{m=0}^{M-1} \sum_{n_d = CW^{min}}^{CW^{max}} P(t_{sw}(m, n_d)) \bullet t_{sw}(m, n_d)$$
$$= \sum_{m=0}^{M-1} \left(\frac{1 + n_d - CW^{min}}{1 + CW^{max} - CW^{min}}\right)^m \sum_{n_d = CW^{min}}^{CW^{max}} \frac{1}{1 + CW^{max} - CW^{min}} \left(m(T_{AIFS} + t_{trans}) + \sum_{i=0}^{m} n_d^i T_d + T_{AIFS} + n_d T_d\right) \quad (4)$$

where Equation (2) and 3(c) are substituted.

Furthermore, several additional factors (*22*), such as channel fading and hidden terminals, may have minor impacts on the aforementioned request message broadcasts, but are negligible.

Deterministic Transmission Time. The t_{trans} depends on both the data rate and the message size, which is derived as follows:

$$t_{trans} = N / R_{DataRate} \quad (5)$$

where, N is message size; and $R_{DataRate}$ is data rate. The sizes of J2735 packets and corresponding WAVE messages vary from application to application.

Total End-to-End Communication Delay. Considering the **S**ignal priority **R**equest **M**essage (SRM) broadcast (i.e., step one, marked as SRM for corresponding variables) and **ack**nowledgment receipt (step two, marked as ACK for corresponding variables), the total stochastic end-to-end delay $t_{tot}(m^{SRM}, n_d^{SRM}, m^{ACK}, n_d^{ACK})$ with probability $P(t_{tot}(m^{SRM}, n_d^{SRM}, m^{ACK}, n_d^{ACK}))$ is expressed as follows, respectively:

$$t_{tot}(m^{SRM}, n_d^{SRM}, m^{ACK}, n_d^{ACK}) = t_{sw}^{SRM}(m^{SRM}, n_d^{SRM}) + t_{sw}^{ACK}(m^{ACK}, n_d^{ACK}) + t_{trans}^{SRM} + t_{trans}^{ACK} \quad (6a)$$

$$P(t_{tot}(m^{SRM}, n_d^{SRM}, m^{ACK}, n_d^{ACK})) = P(t_{sw}^{SRM}(m^{SRM}, n_d^{SRM}))P(t_{sw}^{ACK}(m^{ACK}, n_d^{ACK})) \quad (6b)$$

Accordingly, the average total end-to-end delay $\hat{t}_{tot}$ is as follows:

$$\hat{t}_{tot} = \hat{t}_{sw}^{SRM} + \hat{t}_{sw}^{ACK} + t_{trans}^{SRM} + t_{trans}^{ACK} \quad (6c)$$

where, $\hat{t}_{sw}^{SRM}$ and $\hat{t}_{sw}^{ACK}$ are the average wait times for broadcasting and acknowledgment, respectively; and t_{trans}^{SRM} and t_{trans}^{ACK} are the transmission times for broadcast and acknowledgment, respectively.

Reliability Estimation of Successful Communication

The probability of successful communication is a critical factor (*25*) that significantly affects the performance of various CV-based applications. Thus, an enhanced T-window reliability (*22*) (denoted as R_{pre}) for preemption control is proposed. The R_{pre} is defined as a probability $P_{app}(x, T_a)$ of successfully receiving one acknowledgment packet at distance x, within a given time interval T_a and is calculated by Method 1.

Minimum Compensated Distance Estimation

To minimize effects caused by the probability of successful communications, the R_{pre} should be as large as possible before vehicle passes a predefined triggering position (denoted as X_0). Thus, the request message should be sent out before reaching X_0, where the position sending the request message in advance is defined as advanced position X_a. The distance between X_a and X_0 is defined as a compensated distance S_{comp}. Therefore, how to find an optimal S_{comp} is the purpose. Thus, a compensated distance estimation is proposed as follows:

Method 1: Estimation of the Reliability R_{pre} for Preemption Control	
Input: all $(m^{SRM}, n_d^{SRM}, m^{ACK}, n_d^{ACK})$, Equation (6*a*) and (6*b*), and T_a	
Ouput: R_{pre} and $P_{app}(x, T_a)$	
Step 1:	Calculate $t_{tot}(m^{SRM}, n_d^{SRM}, m^{ACK}, n_d^{ACK})$ for each $(m^{SRM}, n_d^{SRM}, m^{ACK}, n_d^{ACK})$ by Equation (6a)
Step 2:	Calculate the corresponding $P(t_{tot}(m^{SRM}, n_d^{SRM}, m^{ACK}, n_d^{ACK}))$ of the $t_{tot}(m^{SRM}, n_d^{SRM}, m^{ACK}, n_d^{ACK})$ for each $(m^{SRM}, n_d^{SRM}, m^{ACK}, n_d^{ACK})$ by Equation (6*b*)
Step 3:	For each $(m^{SRM}, n_d^{SRM}, m^{ACK}, n_d^{ACK})$, we get one mapped pair of result $(t_{tot}(m^{SRM}, n_d^{SRM}, m^{ACK}, n_d^{ACK}), P(t_{tot}(m^{SRM}, n_d^{SRM}, m^{ACK}, n_d^{ACK})))$
Step 4:	Find all corresponding probabilities $P(t_{tot}(m^{SRM}, n_d^{SRM}, m^{ACK}, n_d^{ACK}))$ under the condition $t_{tot}(m^{SRM}, n_d^{SRM}, m^{ACK}, n_d^{ACK}) < T_a$ for the same combination
Step 5:	Adding up all found probabilities $P(t_{tot}(m^{SRM}, n_d^{SRM}, m^{ACK}, n_d^{ACK}))$ and the summed value is the needed $P_{app}(x, T_a)$
Step 6:	$R_{pre} = P_{app}(x, T_a)$

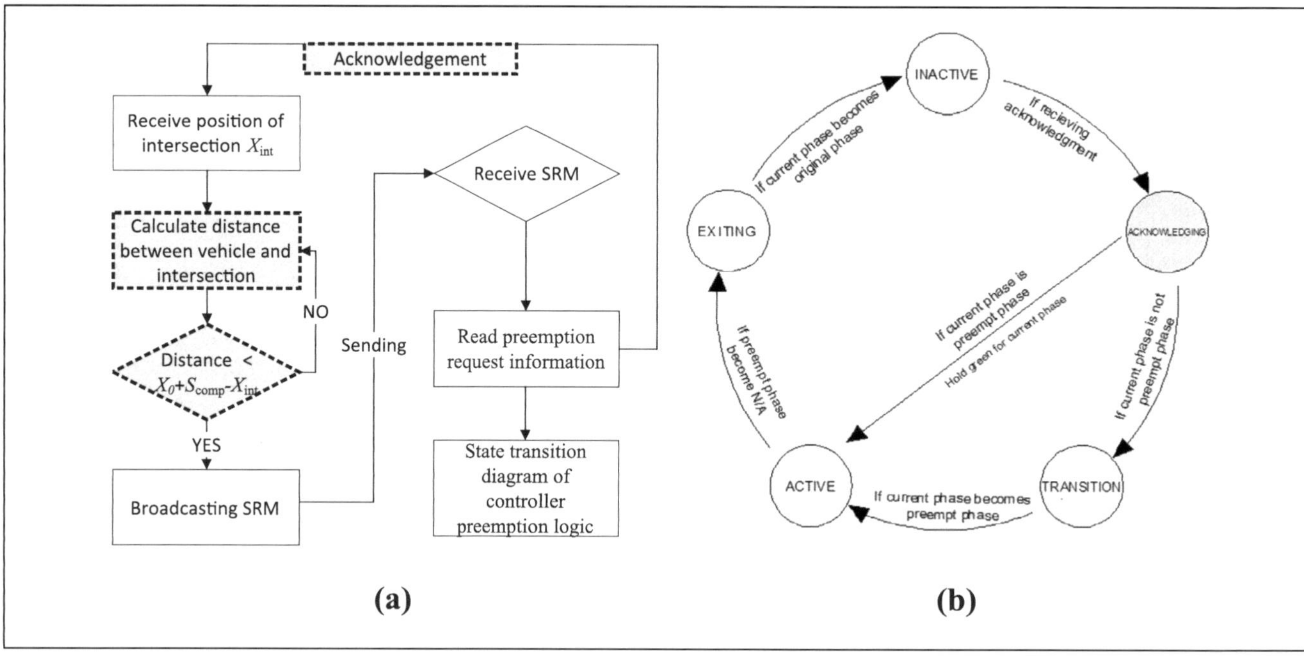

Figure 3. Preemption control algorithms: (*a*) a compensated distance-based cooperative preemption control; (*b*) state transition diagram of controller preemption logic in RSE with compensated distance-based state.

where, S_{comp}^{min} is the compensated distance to ensure the reliability of preemption control no less than the P_{app}^{given}, when the target vehicle passes X_0.

Compensated Distance-Based Cooperative Preemption Control

After finding S_{comp}^{min}, a compensated distance-based cooperative preemption control is proposed and shown in Figure 3.

In Figure 3*a*, the S_{comp} eliminates impacts of end-to-end delays. The OBE receives the position of the intersection (denoted as X_{int}) and broadcasts the SRM when approaching one intersection. The preemption is triggered to broadcast when the distance to the intersection is smaller than the modified predefined threshold (i.e., *Distance* $< X_0 + S_{comp} - X_{int}$ in Figure 3*a*). When the SRM is received, RSE obtains the preemption information; returns an acknowledgment; and triggers the controller to actuate the preemption control.

Figure 3*b* shows the designed state transition diagram to give a green signal to target the preempt phase as soon as possible. The preempt state is set to the *INACTIVE* state initially without any preemption request. When receiving a preemption request and acknowledgment from the RSE, the controller switches to the

ACKNOWLEDGE state and compares the current phase number with the requested preempt phase number. If the current phase is the requested preemption phase, the controller holds the green and switches to the *ACTIVE* state. Otherwise, the controller will force off the current phase, omit every next phase until the preemption phase, and switch to the *TRANSITION* state. The safety consideration is satisfied since the phase is not forced off within the minimum green interval. When all other phases are omitted, and the current phase matches the preempt phase, the controller switches state from *TRANSITION* to *ACTIVE* and holds green for preemption phase. When the detecting RSE stops sending a preemption request, the controller state is switched to *EXITING*. When the current phase matches the original forced-off phase, the state switches back to *INACTIVE* state and remains on standby for the next preemption.

Evaluations and Analysis

Field Implementation

The intersection of 87 Avenue and 116 St, Edmonton, was chosen as a field test location. Figure 4, *a* and *b* illustrate geometrical and phase information, and setup of devices. An Econolite ASC/3 traffic signal controller was installed at this intersection; and a RSE was connected to the ASC/3 controller through Ethernet cable. During testing, an authorized emergency vehicle equipped with OBE broadcasted SRM after passing an ingress and stopped broadcasting when it received an acknowledgment. Once SRM was received, the RSE activated the controller to actuate preemption control and returned an acknowledgment (embedded in a SPaT message). Both SRM and SPaT were transmitted by WAVE short message protocol (WSMP) via control channel 178, and were set to always stay on the same channel. However, it was generally expected to be actively switching between control channels (CCH) and service channels (SCH) to accommodate the needs of messages generated by all vehicles nearby in the future.

To present messages, a demo app in an on-board smartphone was designed with an interface showing real-time SPaT and other CV parameters for the ACTIVE-AURORA project. The smartphone was connected to OBE through WiFi. SPaT of both green and red intervals were presented in the interface shown in Figure 4, *c–d*, with signal color in the center background and remaining time counting down in the center figure. When preempted, the controller held the preempt phase instead of phase changing and the app presented a green interval with "0" counting figure as distinguished from regular SPaT. Both the controller and smartphone interface for preemption green are shown in Figure 4, *e* and *f*. Figure 4, *g* and *h* give data frame structures of the implemented SRM and SPaT messages evaluated in the field (*23*, *25*).

During field tests, the end-to-end delay and the distance x between the emergency vehicle and the RSE were recorded for quantitative analysis.

Field Test Results

Figure 5 shows quantitative results (marked as dotted and starred curves for two different filed tests and two OBE, respectively) of scenarios with single and double priority requests. As illustrated in Figure 5, *a* and *e*, end-to-end delay distribution is restricted to a section for different distances intuitively, which reveals delay does not change too much for both scenarios.

Figure 5, *b* and *c*, and *f* and *g*, depict a point and cumulative probability with respect to different delays for two field tests and two OBE, respectively. Consistent with results in Figure 5, *a* and *e*, delay is concentrated on a fixed section, which is 5–35 ms in a probabilistic perspective. And from Figure 5, *c* and *g*, the reliability of the delay under 30 ms and 35 ms is larger than 0.99, respectively.

Accordingly, reliability R_{pre} for different S_{comp} is given Figure 5*d* and *h*. The S_{comp}^{min} of a single request for 72 km/h, 90 km/h, and 108 km/h are 0.5 m, 0.6 m, and 0.7 m for 99% reliability, respectively, while S_{comp}^{min} of two requests are 0.7 m, 0.88 m, and 1.1 m, respectively. Also comparing results from Figure 5, *a* to *d* with results from Figure 5, *e* to *f*, delays are almost the same when there are few requests.

In summary, delays always exist, are not negligible, and follow an evenly distributed pattern over different distances, as seen from Figure 5, *a–c* and *e–g*, wherever the vehicle is located within communication coverage of RSE. The reason is the stochastic waiting time follows a uniform distribution and has a maximum contention window in a single request condition via Method 1. Then, because delays are restricted to a certain area that cannot be negligible, a compensated distance is needed to improve the emergency vehicle's reliability before it passes the predefined ingress. The required minimum S_{comp} for a given reliability increases with the emergency vehicle's speed.

Thus, it is essential for operators using the proposed methodology to improve the reliability in the few requests condition, considering negative impacts of the communication delay.

Comprehensive Analysis of Complex Conditions

Though field results are good in the few requests condition, the field implementation still has limitations in the multiple signal priority requests condition considering

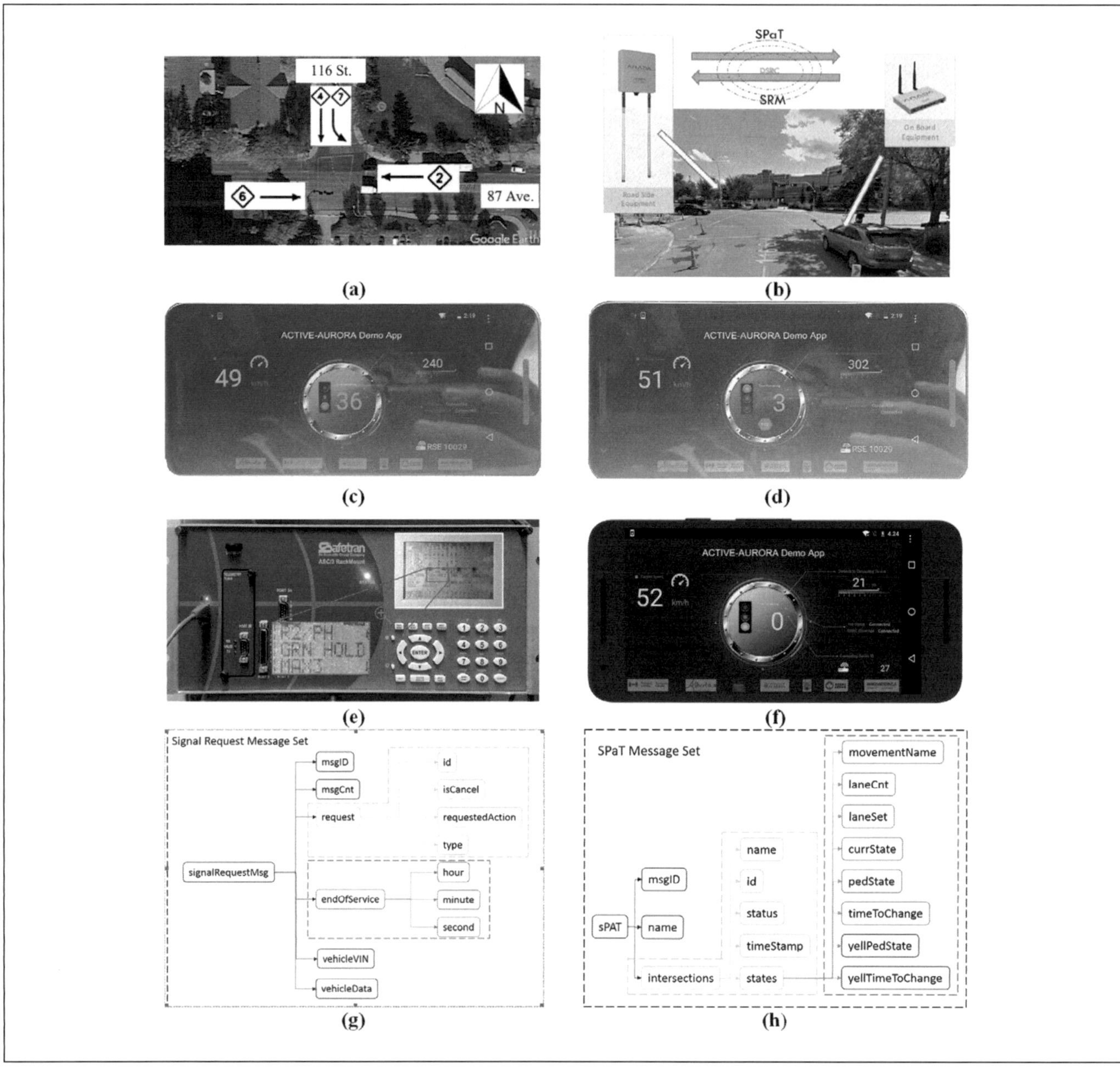

Figure 4. Field implementations and results: (*a*) phase and geometric layout of target intersection; (*b*) physical architecture of field setup; (*c, d*) real-time SPaT status (green and red phase); (e) green hold status in controller interface; (*f*) green hold status in smartphone interface (screenshot); (*g–h*) partial selected data frame's structure of implemented SRM and SPaT message.

the CV penetration rate is low under the current stage. Thus, a comprehensive analysis of a complex scenario with multiple active signal priority requests was conducted by numerical simulations using Matlab. The basic contention parameters and slot time are set according to definitions in the standard draft (*23*, *24*). For commonly used control or service channels (e.g., 178), the operation is on a 10 MHz-wide channel. Basic aCWmin is 15 and aCWmax is 1023, respectively. Time slot (T_d) is 13 us. The data rate is 27 Mbps. Different priorities of messages are assigned to different access categories (ACs), where there are four types of ACs (i.e., AC0, AC1, AC2, and AC3) in total. These types have different channel access settings including different *CWmin*, *CWmax*, and *AIFS*, where the settings of AC0 are used as the default set in further simulations. Figure 6 demonstrates quantitative results for the complex scenario.

Figure 6, *a* and *b* shows results of delay and associated probability in step one for different sending indexes

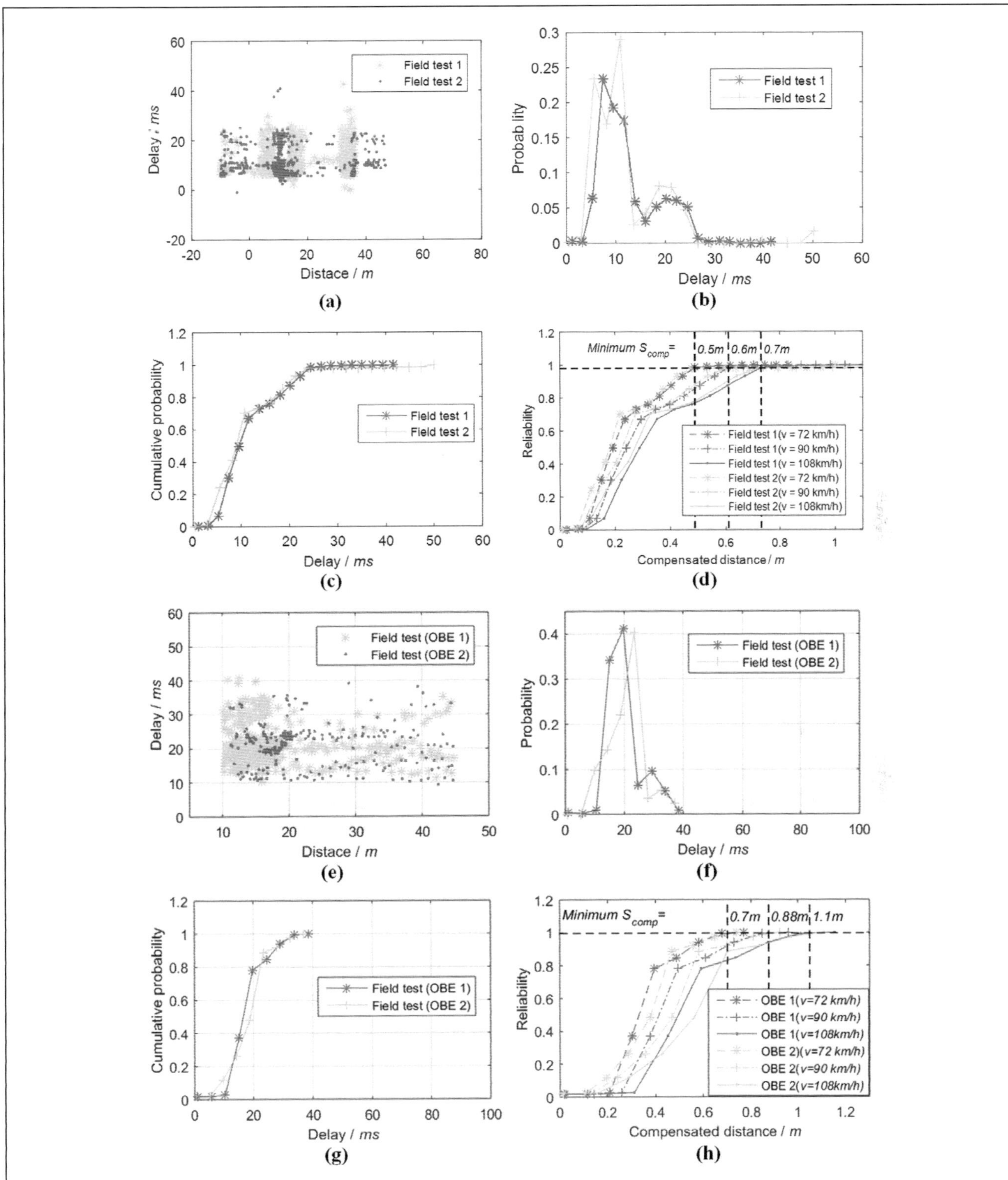

Figure 5. Quantitative analysis of field tests: (*a*) end-to-end delays for different distances; (*b*) probabilities for different end-to-end delays; (*c*) cumulative probabilties for different end-to-end delays; (*d*) reliabilities R_{pre} for different S_{comp} and the minimum S_{comp} (v = 72 km/h; 90 km/h and 108 km/h); (e) end-to-end delays for different distances (two OBE); (*f*) probabilities for different end-to-end delays (two OBE); (*g*) cumulative probabilties for different end-to-end delays (two OBE); (*h*) reliabilities R_{pre} for different S_{comp} and the minimum S_{comp} (v = 72 km/h, 90 km/h and 108 km/h) (two OBE).

m and sizes n_d (shown in Equation (2)), where the total number of vehicles simultaneously sending requests is 20. From Figure 6*a*, it is notable that the delay increases when the sending index and contention window size increase, and reaches a significant quantity (i.e., hundred-ms level). Therefore, the delay in the complex scenario is not ignored because of reaching a hundred-ms level and increases significantly.

Then, probabilities and cumulative probabilities for different delays calculated in Method 1 are shown in Figure 6*c*. Accordingly, results of reliability and minimum S_{comp} are shown in Figure 6*d* by Method 2. From

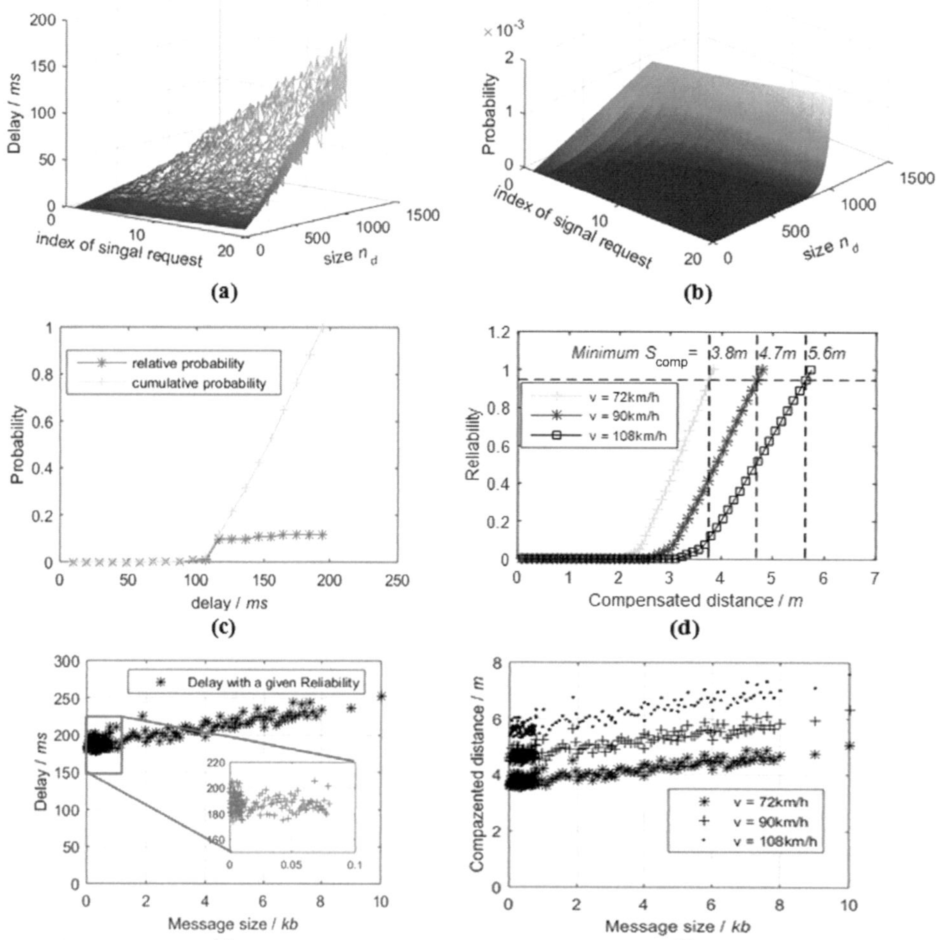

AC	CWmin	CWmax	AIFS
0	$aCWmin$	$aCWmax$	9
1	$\frac{aCWmin+1}{2}-1$	$aCWmin$	6
2	$\frac{aCWmin+1}{4}-1$	$\frac{aCWmin+1}{2}-1$	3
3	$\frac{aCWmin+1}{4}-1$	$\frac{aCWmin+1}{2}-1$	2

(g)

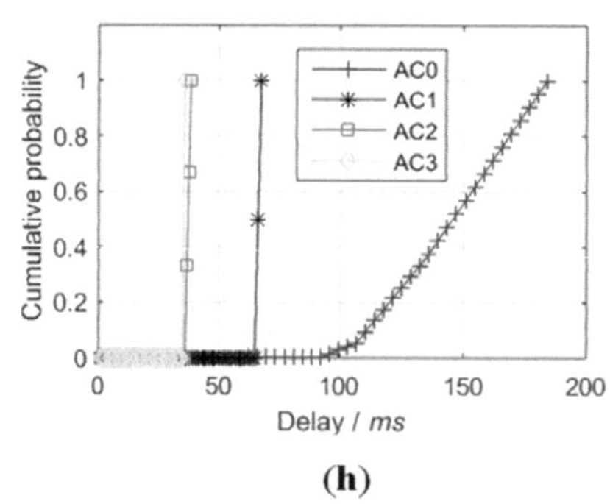

Figure 6. (Continued)

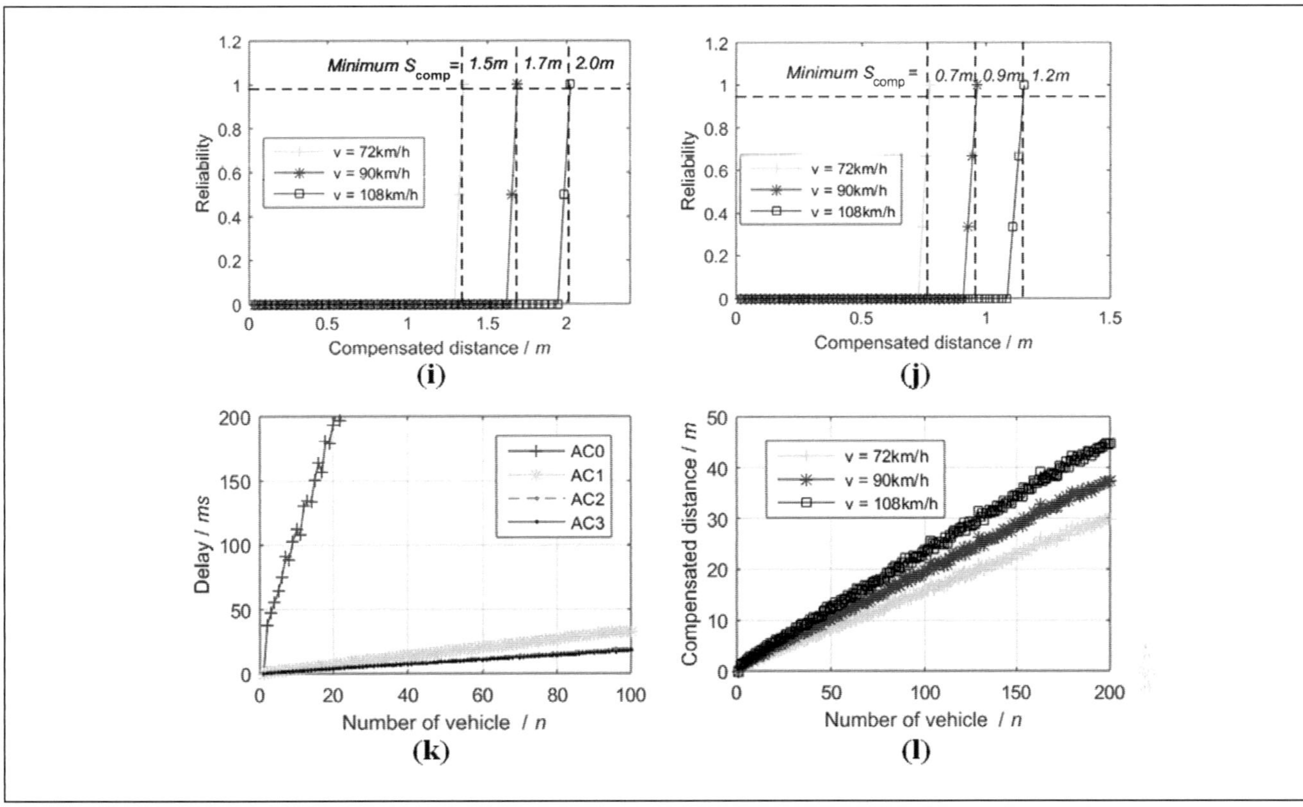

Figure 6. Simulation results: (*a*) different stochastic waiting time for one-way (i.e., step one in Figure 2*a*); (*b*) point probability of different stochastic waiting time for one-way (i.e.; step one in Figure 2*a*); (*c*) probabilistic end-to-end delay; (*d*) reliability and the minimum S_{comp} for different speeds; (*e*) end-to-end delay for different message sizes (default); (*f*) minimum S_{comp} for different message sizes (default); (*g*) different channel access settings for different priorities of messages (ACs); (*h*) probabilistic end-to-end delay for different priorities of messages (different ACs); (*i*; *j*) minimum S_{comp} for different speeds (AC1 and AC2/AC3); (*k*) end-to-end delay for different numbers of vehicles (different ACs); (*l*) minimum S_{comp} for different numbers of vehicles (default).

Figure 6*d*, the minimum S_{comp} increases and reaches 3.8 m, 4.7 m, and 5.6 m when speed is 72 km/h, 90 km/h, and 108 km/h, respectively. Thus, it is noted that S_{comp} is required for the complex scenario.

Next, Figure 6, *e* and *f* gives the end-to-end delay and corresponding minimum S_{comp} for different message sizes in default priority, respectively. In Figure 6*e*, the delay increases significantly when message size is larger than a few kilobits (kb), whereas it does not change too much when the size is as small as hundreds of bits. The minimum S_{comp} in Figure 6*f* shows similar trends.

Furthermore, effects of the ACs determining the priority are shown in Figure 6, *g*–*j*. Figure 6*g* gives different settings for four ACs, where each AC has a set of CW_{min}, CW_{max}, and *AIFS*. Figure 6*h* shows different AC's end-to-end delays, where the highest priority AC3 has the lowest delay reaching around 35 ms and is with a sharp increase. Figure 6, *i* and *j* demonstrate the corresponding S_{comp} for AC1, AC2, and AC3 priority, where the highest priority AC3 has the smallest minimum S_{comp} reaching around 0.7 m as predicted. Thus, results suggest messages with the highest priority are a proper choice during transmission.

Next, Figure 6, *k* and *l* gives the end-to-end delays and corresponding minimum S_{comp} for different numbers of vehicles, respectively. In both figures, the delay and the minimum S_{comp} increase when the number of vehicles rises. In Figure 6, *k* and *l*, the delay of the AC0 priority is significant, being as large as several seconds and requiring a large minimum S_{comp} to mitigate the negative influences. Instead, the highest priority AC3 has the smallest delay from Figure 6*k*. But the delay still reaches around 50 ms when there are hundreds of vehicles, and will be even larger considering multiple applications running in one vehicle. Therefore, S_{comp} is necessary for the complex scenario with multiple applications and vehicles in urban areas.

In summary, based on analysis for both single requests in the field and multiple requests via numerical simulations, delay always exists and is not negligible when the

Method 2: Estimation of Compensated Distance S_{comp} and S_{comp}^{min}

Input: all t_{tot} and $P(t_{tot})$ via Equation (6*a*) and (6*b*) with respect to m^{SRM}, n_d^{SRM}, m^{ACK} and n_d^{ACK}, given reliability P_{pre}^{given}, time step T_d and vehicle speed V
Output: S_{comp}, reliability $R_{pre}(S_{comp})$, a minimum time interval T_{a_min}, and corresponding minimum compensated distance S_{comp}^{min}
Step 1: Initial: $T_{a\_min} = T_d$
Step 2: For each combination, one mapped pair of results (t_{tot},$P(t_{tot})$) is obtained according to Method 1
Step 3: do
Step 4: Find all corresponding $P(t_{tot}(m^{SRM}, n_d^{SRM}, m^{ACK}, n_d^{ACK}))$ under the condition $t_{tot}(m^{SRM}, n_d^{SRM}, m^{ACK}, n_d^{ACK}) < T_{a\_min}$ for the same combination
Step 5: Sum all probabilities $P(t_{tot}(m^{SRM}, n_d^{SRM}, m^{ACK}, n_d^{ACK}))$, and the summarized value is denoted as $EP_{app}(x,T_a)$
Step 6: Using kinetic laws, following equations are obtained,
$S_{comp} = T_{a\_min}*V$
$R_{pre}(S_{comp}) = EP_{app}(x,T_a)$
Step 7: if $EP_{app}(x,T_a) < P_{app}^{given}$, $T_{a\_min} = T_{a\_min} + T_d$ and go to Step 4
Step 8: elseT_{a_min} is the needed interval where $EP_{app}(x,T_a) \geq P_{app}^{given}$
Step 9: End: The S_{comp}^{min} is obtained as follows:
$S_{comp}^{min} = T_{a\_min} \cdot V$

emergency vehicle is within the communication coverage of a RSE. Using the compensated distance strategy will enable a preemption signal request with better reliability in the single request condition. Furthermore, the delay will greatly increase in the multiple active requests condition, reaching the hundred-ms level quantity, resulting in the compensated distance strategy being necessary for an emergency vehicle maintaining a high speed to achieve a more reliable preemption control with less trip response time. Also, larger message size, lower message priority type, and denser vehicle density in urban areas apparently increase the delay and make the compensated distance strategy more necessary.

Conclusion

This paper focuses on the comprehensive delay analytics of the preemption signal request and its mitigation measure of preemption control for the emergency vehicle. In detail, an end-to-end delay decomposition model and a compensated distance-based cooperative preemption control are proposed for delay modeling and mitigation, respectively. Both the end-to-end delay decomposition model and the compensated distance-based cooperative preemption control are validated in efficient ways.

The field and numerical results show communication delay is a critical factor affecting the reliability of emergency response time and should be compensated. Communication delay is not negligible and always exists whenever the OBE on an emergency vehicle communicates with RSE for both single and multiple request conditions. Also, the communication delay cannot be ignored when the compensated distance reaches the hundreds of milliseconds level, and will increase significantly in the multiple active requests condition in an urban area. Thus, the minimum compensated distance is essential for a high-speed emergency vehicle.

In future, more complex scenarios, such as field tests validating the multiple active signal priority requests as well as a signalized intersection with congested traffic flow, need to be conducted to analyze the effects of communication delay for current and other V2X applications. For large CV deployments, two alternatives can be used for tackling the existing traffic signal preemption systems in local areas. One is to abandon them, and another is to make them compatible with the CV-based system, creating full utilization of two systems.

Acknowledgments

The infrastructure for this research was supported by Transport Canada, under the Asia-Pacific Gateway and Corridor Initiative Transportation Infrastructure Fund (APGCI TIF), as well as Alberta Transportation and the City of Edmonton.

Author Contributions

The authors confirm contribution to the paper as follows. Study conception and design: Jiangchen Li; data collection: Jiangchen Li, Dr. Liqun Peng; analysis and interpretation of results: Jiangchen Li, Chen Qiu, Dr. Liqun Peng; draft manuscript preparation: Jiangchen Li, Chen Qiu, Dr. Liqun Peng and Dr. Tony Qiu. All authors reviewed the results and approved the final version of the manuscript.

References

1. Chou, C.-S., and A. P. Nichols. Characterizing Emergency Vehicle Preemption Operation with High-Resolution Traffic Signal Event Data. *Transportation Research Record: Journal of the Transportation Research Board*, 2016. 2557: 108–121.

2. Qin, X., and A. M. Khan. Control Strategies of Traffic Signal Timing Transition for Emergency Vehicle Preemption. *Transportation Research Part C: Emerging Technologies*, Vol. 25, 2012, pp. 1–17.
3. Noori, H., L. Fu, and S. Shiravi. Connected-Vehicle-Based Traffic Signal Control Strategy for Emergency Vehicle Preemption. Presented at 95th Annual Meeting of the Transportation Research Board, Washington, D.C., 2016.
4. Nelson, E., and D. Bullock. Impact of Emergency Vehicle Preemption on Signalized Corridor Operation: An Evaluation. *Transportation Research Record: Journal of the Transportation Research Board*, 2000. 1727: 1–11.
5. Ten, H., Y. Qi, J. C. Falcocchio, K.-B. Kim, R. Patel, and E. Athanailos. Simulation Testing of Adaptive Control, Bus Priority and Emergency Vehicle Preemption in New York City. Presented at 82nd Annual Meeting of the Transportation Research Board, Washington, D.C., 2003.
6. Yang, S., M. Hamedi, and A. Haghani. Online Dispatching and Routing Model for Emergency Vehicles with Area Coverage Constraints. *Transportation Research Record: Journal of the Transportation Research Board*, 2005. 1923: 1–8.
7. Kang, W., G. Xiong, Y. Lv, X. Dong, F. Zhu, and Q. Kong. Traffic Signal Coordination for Emergency Vehicles. *Proc., 17th IEEE International Conference on Intelligent Transportation Systems*, IEEE, Qingdao, China, 2014, pp. 157–161.
8. Toy, C., K. Leung, L. Alvarez, and R. Horowitz. Emergency Vehicle Maneuvers and Control Laws for Automated Highway Systems. *IEEE Transactions on Intelligent Transportation Systems*, Vol. 3, No. 2, 2002, pp. 109–119.
9. Zhang, Z., Q. He, J. Gou, and X. Li. Performance Measure of Travel Time Reliability of Emergency Vehicles in an Urban Region. Presented at 94th Annual Meeting of the Transportation Research Board, Washington D.C., 2015.
10. Obenberger, J., and J. Collura. Methodology to Assess Traffic Signal Transition Strategies for Exit Preemption Control. *Transportation Research Record: Journal of the Transportation Research Board*, 2007. 2035: 158–168.
11. Obenberger, J., and J. Collura. Transition Strategies to Exit Preemption Control: State-of-the-Practice Assessment. *Transportation Research Record: Journal of the Transportation Research Board*, 2001. 1748: 72–79.
12. Huang, Y. S., Y. S. Weng, and M. Zhou. Design of Traffic Safety Control Systems for Emergency Vehicle Preemption Using Timed Petri Nets. *IEEE Transactions on Intelligent Transportation Systems*, Vol. 16, No. 4, 2015, pp. 2113–2120.
13. Qi, L., M. Zhou, and W. Luan. Emergency Traffic-Light Control System Design for Intersections Subject to Accidents. *IEEE Transactions on Intelligent Transportation Systems*, Vol. 17, No. 1, 2016, pp. 170–183.
14. He, Q., K. L. Head, and J. Ding. Multi-Modal Traffic Signal Control with Priority, Signal Actuation and Coordination. *Transportation Research Part C: Emerging Technologies*, Vol. 46, 2014, pp. 65–82.
15. Noori, H. Modeling the Impact of Vanet-Enabled Traffic Lights Control on the Response Time of Emergency Vehicles in Realistic Large-Scale Urban Area. *Proc., IEEE International Conference on Communications Workshops*, Budapest, Hungary, IEEE, New York, NY, 2013, pp. 526–531.
16. Wang, Y., Z. Wu, X. Yang, and L. Huang. Design and Implementation of an Emergency Vehicle Signal Preemption System Based on Cooperative Vehicle-Infrastructure Technology. *Advances in Mechanical Engineering*, Vol. 5, 2013, p. 834976.
17. Jordan, C., M. Cetin, and R. Robinson. Path Clearance for Emergency Vehicles Through the Use of Vehicle-to-Vehicle Communication. *Transportation Research Record: Journal of the Transportation Research Board*, 2013. 2381: 45–53.
18. Weinert, F., and M. Düring. Development and Assessment of Cooperative V2X Applications for Emergency Vehicles in an Urban Environment Enabled by Behavioral Models. In M. Behrisch, and M. Weber (eds.), *Modeling Mobility with Open Data*, Springer, Berlin, Germany, 2014, pp. 125–153.
19. Jordan, C. A., and M. Cetin. Signal Preemption Strategy for Emergency Vehicles Using Vehicle to Infrastructure Communication. Presented at 94th Annual Meeting of the Transportation Research Board, Washington, D.C., 2015.
20. Head, L. *Multi-Modal Intelligent Traffic Signal System – Phase II: System Development, Deployment and Field Test.* Final Report. University of California PATH Program, Savari Networks, and Econolite, Anthem, AZ, 2016, pp. 1–50.
21. Ma, X., J. Zhang, and T. Wu. Reliability Analysis of One-Hop Safety-Critical Broadcast Services in VANETs. *IEEE Transactions on Vehicular Technology*, Vol. 60, No. 8, 2011, pp. 3933–3946.
22. Ding, J., Q. He, L. Head, F. Saleem, and W. Wu. Development and Testing of Priority Control System in Connected Vehicle Environment. Presented at 92nd Annual Meeting of the Transportation Research Board, Washington, D.C., 2013.
23. Hartenstein, H., and K. Laberteaux. (eds.). *VANET: Vehicular Applications and Inter-Networking Technologies.* Wiley, Chichester, UK, 2010.
24. Lu, N., N. Cheng, N. Zhang, X. Shen, and J. W. Mark. Connected Vehicles: Solutions and Challenges. *IEEE Internet of Things Journal*, Vol. 1, No. 4, 2014, pp. 289–299.
25. Park, H., A. Miloslavov, J. Lee, M. Veeraraghavan, B. Park, and B. L. Smith. Integrated Traffic-Communication Simulation Evaluation Environment for IntelliDrive Applications Using SAE J2735 Message Sets. *Transportation Research Record: Journal of the Transportation Research Board*, 2011. 2243: 117–126.

The Standing Committee on Traffic Signal Systems (AHB25) peer-reviewed this paper (18-01843).

Article

TRR
JOURNAL OF THE TRANSPORTATION RESEARCH BOARD

Transportation Research Record
2018, Vol. 2672(18) 58–70

DOI: 10.1177/0361198118793007
journals.sagepub.com/home/trr
SAGE

OD-NETBAND: An Approach for Origin–Destination Based Network Progression Band Optimization

Tugba Arsava[1], Yuanchang Xie[2], and Nathan Gartner[2]

Abstract

Traditional progression band optimization methods are focused on providing uninterrupted flow along arterial streets. For arterials with significant traffic streams joining and leaving from side streets, these approaches often generate poor traffic signal control performance. To address this deficiency, an origin–destination (OD) information based progression band optimization model, OD-BAND, was formulated to coordinate signals for arterials with major side-street traffic streams. This paper aims to extend the OD-BAND model further to address the OD based traffic signal coordination problem in multi-arterial grid networks. The extended model is able to create separate progression bands for each major OD stream in the network. In this expanded model, individual arterials are connected with loop constraints to ensure that offsets derived via different paths for a particular intersection are equal. The new OD-NETBAND model is formulated as a mixed integer linear program that maximizes the sum of each major OD stream's progression bandwidth. It can optimize simultaneously cycle length, offsets, and phase sequences for the entire network. Performance of the new model is evaluated with AIMSUN microscopic simulation and is compared with MAXBAND-86 and Synchro results.

Coordinating a group of neighboring traffic signals can result in less delay and fewer stops for drivers, reduced fuel consumption and emissions, uniform and improved travel speed, decreases in accidents and red light violations, and increase in the use of arterials instead of parallel minor streets. Given these potential benefits, many studies have been conducted to continue improving existing traffic signal coordination methods or explore new ideas. These methods can be largely classified into two categories based on their optimization objectives: delay minimization and bandwidth maximization. Delay is typically defined as the difference between expected and actual travel times, whereas bandwidth is the width of a progression band measured in seconds, representing the available green time interval for a traffic stream to travel multiple consecutive intersections without stopping. Although bandwidth maximization methods do not explicitly include delay functions in their objectives, both research and practice have demonstrated that these methods work well due to the intricate connection between progression band and delay (*1*).

Little (*2*) initially introduced a mixed integer linear program that generates progression bands for arterial through traffic. This model was further enhanced by Little et al. (*3*) and is widely known as MAXBAND, which generates uniform inbound and outbound progression bands along an arterial and maximizes them by calculating cycle time, offsets, and phase sequence patterns. MAXBAND was later extended by Gartner et al. (*4, 5*) as MULTIBAND, which creates symmetrical progression bands along an arterial with varying bandwidths. In MULTIBAND, the objective function of MAXBAND was reformulated to maximize the sum of progression bandwidths along individual arterial segments weighted by the directional link volumes. Recently, Zhang et al. (*6*) formulated an AM-BAND model by relaxing the symmetrical progression band requirement in MULTIBAND. The AM-BAND model allows arterial progression bands to be asymmetrical with respect to the progression line in MULTIBAND, adding more flexibility in the signal control space.

The arterial models mentioned above have also been extended to handle network problems. Little (*2*) has

[1]Department of Civil Engineering and Technology, Wentworth Institute of Technology, Boston, MA
[2]Department of Civil and Environmental Engineering, University of Massachusetts Lowell, Lowell, MA

Corresponding Author:
Address correspondence to Yuanchang Xie: yuanchang_xie@uml.edu

already pointed the way to formulate the network synchronization problem as a mixed integer linear programming problem by considering the network loop constraints. Gartner (*7*) described a scheme to formulate a fundamental set of loop constraints in a given network. These basic formulations were later implemented in computer programs using the MAXBAND model (*8*, *9*) as well as the MULTIBAND model (*10*, *11*). Heuristics were developed to improve the computational efficiency of these models for implementation in large-scale networks (*12*, *13*). Other important issues related to arterial signal coordination problems have also been extensively studied, including lead-lag phasing sequence optimization (*14*), signal coordination under oversaturated conditions (*15*), and formulating bandwidth maximization as an unconstrained nonlinear program considering relative offset and vehicle arrival functions (*16*).

All the aforementioned models use as input the total link volumes, that is, they consider only the turning traffic volumes at the individual intersections without knowing from where they originate and where they are eventually destined to go. These models aim to maximize progression bands along arterial streets and ignore traffic streams coming in from, and going out to, the side streets. Recently, an OD-BAND model (*17*) was proposed that takes OD and side-street traffic information into consideration. Arsava et al. (*18*) further enhanced the OD-BAND model by adopting a revised objective function that considers the number of road segments that a traffic stream or progression band traverses. They also generalized the original model to consider all possible major OD streams and to include left-turn phase sequence optimization as well. In the new OD-BAND model, progression bands are generated based on traffic OD data and for each major OD stream a separate progression band is created. With this approach, major traffic streams flowing in and out of the arterial can also be properly modeled and the network performance improved.

In this paper, the basic OD-BAND model is reformulated and extended for network progression band optimization and the new model is named OD-NETBAND. The new model has the capability of optimizing cycle time, phase sequences, and offsets while creating a separate progression band for each selected major OD stream. OD-NETBAND is also formulated as a mixed integer linear program and solved by the CPLEX Optimization Studio. The effectiveness of this model is evaluated using AIMSUN simulation and is compared with the performances of well-known network progression optimization models/tools such as MAXBAND-86 and Synchro.

Methodology

In this paper, OD-NETBAND is described based on a three-by-three grid network with nine intersections and eight centroids (i.e., origins and destinations) as in Figure 1. Two case studies are conducted and presented. Case I considers seven major OD streams and an additional OD stream is included in Case II. In both case studies, all lane widths are set to 3.5 meters with 200 meters used as the reserved lane visibility distance in AIMSUN. In Figure 1, the eight small circles with blue dots in the center represent centroids, where vehicles enter and exit the network. The centroids are named with two numbers. The first number is for column index and the second one is for row index. Each intersection (i.e., node in Figure 1) is controlled by a pre-timed signal plan. The OD streams and their selected paths considered in this study are presented in Table 1 and Figure 2, respectively. In Figure 2, the colored arrows represent the paths of major OD streams. The paths for OD streams considered in both Cases I and II are represented by orange arrows; and the path for the extra OD stream considered in Case II only is represented by purple arrows.

OD-BAND Model

Similar to OD-BAND, OD-NETBAND is formulated as a mixed integer linear program. Its objective is to maximize the sum of progression bandwidths created for major OD streams in the network. For ease of understanding of OD-NETBAND, OD-BAND is described first. Additional loop constraints are introduced into the basic model to account for the underlying network structure. The following variables are used in OD-BAND:

m_i = loop integer

$b(\bar{b})$ = outbound (inbound) arterial progression through bandwidth (cycles)

$a_i(\bar{a}_i)$ = weight for outbound (inbound) arterial progression band at section i

$y_{N_s,N_e}(\bar{y}_{N_s,N_e})$ = outbound (inbound) progression cross-bandwidth (cycles) for major OD

$a_{(o,d)}(i)(\bar{a}_{(o,d)}(i)\ )$ = weight for $y_{N_s,N_e}(\bar{y}_{N_s,N_e})$ progression cross-bands at section i

C_1, C_2 = lower and upper limits on cycle length

$R_i(\bar{R}_i)$ = outbound (inbound) common red time to allow side street movement at intersection N_i (cycles)

$L_i(\bar{L}_i)$ = time allocated for outbound (inbound) left turn green at intersection N_i (cycles)

$r_i(\bar{r}_i)$ = outbound (inbound) red time for arterial direction at intersection N_i (cycles), $((r_i = R_i + \bar{L}_i)$ $(\bar{r}_i = \bar{R}_i + L_i))$

$w_i(\bar{w}_i)$ = outbound (inbound) interference (cycles), measured from the right (left) end of outbound (inbound) red to the left (right) boundary of outbound (inbound) arterial progression band at intersection N_i (cycles)

$\bar{w}_i$‘ = inbound interference (cycles), measured from right end of inbound red to the left boundary of inbound arterial progression band at intersection N_i (cycles)

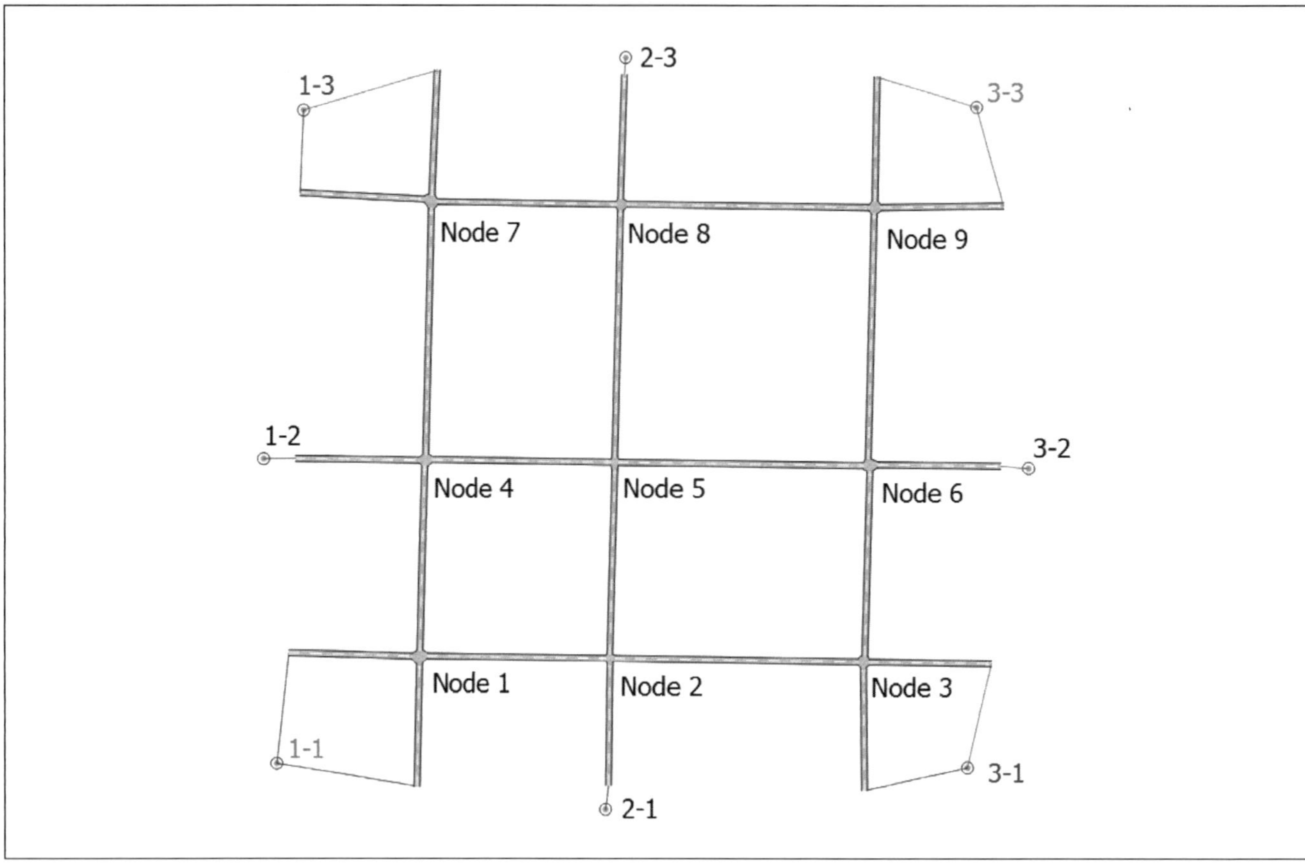

Figure 1. Case study network.

Table 1. OD Streams Considered in Case Studies

Origin	Destination	Case I	Case II
1-1	1-3	500	500
1-1	3-1	500	500
1-1	3-3	800	800
2-1	2-3	500	500
3-1	3-3	500	500
3-1	2-3	–	800
1-2	3-2	500	500
1-3	3-3	500	500
Total demand (vehicles/hour)		3800	4600

$x_{N_s,N_e,i}(\bar{x}_{N_s,N_e,i})$ = interference (cycles) measured from the left end of outbound (inbound) red to the beginning of $y_{o,d}(\bar{y}_{o,d})$ band at intersection N_i (cycles). Subscripts indicate the corresponding progression band information. (origin, destination, current intersection, respectively)

$z = 1/C =$ signal frequency (cycles/second)

k = ratios of inbound volume to outbound volume of arterial progression band

$k_{o,d}(j)(\bar{k}_{o,d}(j))$ = ratios of outbound (inbound) traffic volume turned into main street from side street to outbound (inbound) arterial through traffic volume

$\phi_{i,i+1}(\bar{\phi}_{i,i+1})$ = internode offsets (cycles) at intersection N_i (cycles)

Δ_i = internode offset, time difference between center of r_i and nearest center of $\bar{r}_i$ at intersection N_i (cycles)

δ_i, $(\bar{\delta}_i)$ = zero/one variables that indicate left turn patterns for outbound (inbound) along the arterial at intersection N_i (cycles)

$t_{i,i+1}(\bar{t}_{i,i+1})$ = travel time from N_i to N_{i+1} outbound (N_{i+1} to N_i inbound) (cycles)

$d_{i,i+1}(\bar{d}_{i,i+1})$ = distance between N_i and N_{i+1} outbound (inbound) (feet)

$G_i(\bar{G}_i)$ = outbound (inbound) green time for through traffic at S_i (cycles)

$e_i, f_i(\bar{e}_i, \bar{f}_i)$ = lower/upper limits on outbound (inbound) speed (feet/second) at section i

$g_i, h_i(\bar{g}_i, \bar{h}_i)$ = lower/upper limits on change in outbound (inbound) speed (feet/second) at section i

$V(\bar{V})$ = outbound (inbound) arterial through volume

$V_{(o,d)}(\bar{V}_{(o,d)})$ = outbound (inbound) traffic volume of specific OD pair

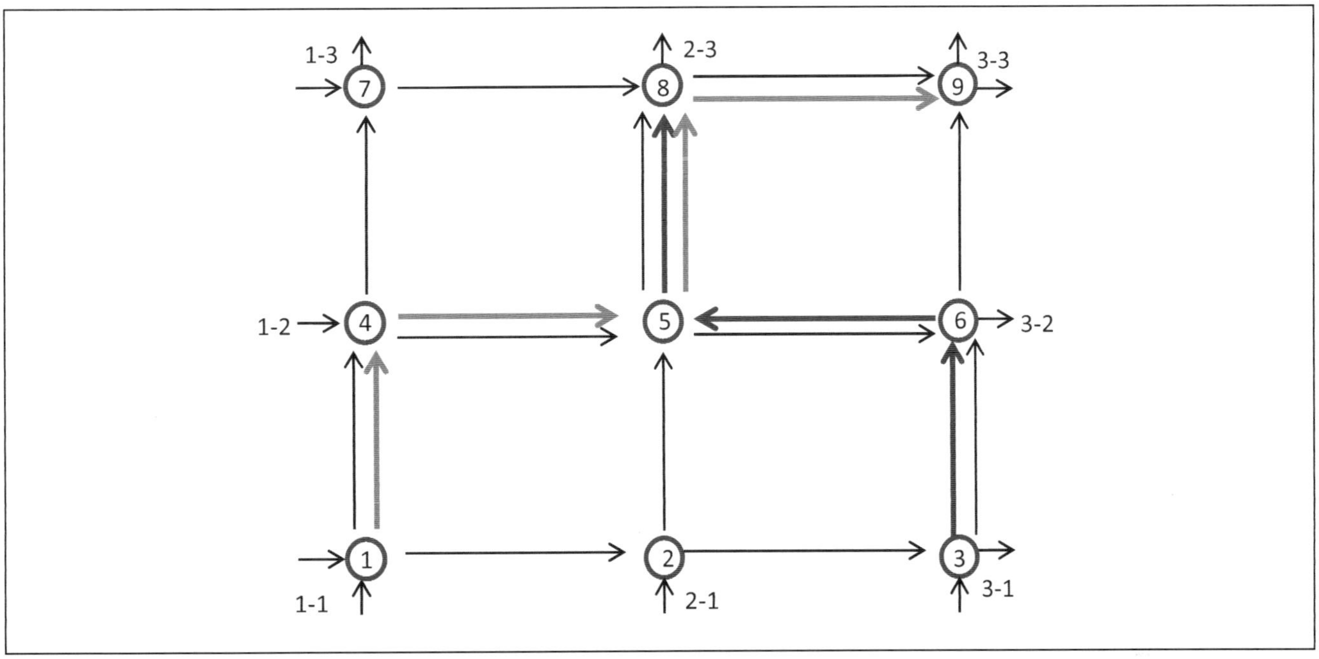

Figure 2. Illustration of paths.

$S(\bar{S})$ = outbound (inbound) saturation flow rate

$S_{(o,d)}(\bar{S}_{(o,d)})$ = outbound (inbound) saturation flow of specific OD pair

n = total number of intersections

$N_s(N_e)$ = Intersection that progression band started (ended)

i = index variable represents intersection number and corresponding downstream (upstream) section for outbound (inbound) traffic

j = index variable represents the note for specific OD pair

p = number of OD pairs for which dedicated bands are sought

$y_{o,d}(\bar{y}_{o,d})$ represent progression bands created for major ODs that join an arterial from side streets. The subscript "o, d" indicates the origin and destination of an OD pair. A lag left turn is considered for side streets with major OD streams. Therefore, the progression bands corresponding to these ODs start in the middle of a red interval for arterial direction. As shown in Equations 1 and 2, the objective function of OD-NETBAND is defined to maximize the sum of progression bandwidths weighted by both the volume to saturation flow ratio and the total number of road segments traveled. $b, \bar{b}, z, w_i, \bar{w}_i, \bar{w}'_i, t_i, \bar{t}_i, x_{o,d,i}, \bar{x}_{o,d,i}, y_{o,d}, \bar{y}_{o,d}, \delta_i, \bar{\delta}_i$ are decision variables that maximize the objective function (Equation 1):

$$MaxB = \underbrace{\sum_{i=1}^{n-1} (a_i{*}b) + (\bar{a}_i{*}\bar{b})}_{I} + \underbrace{\sum_{j=1}^{p} \left[\underbrace{\sum_{i=1}^{N_e-N_s} a_{(o,d)}(i){*}y_{N_s,N_e}}_{a^*} + \underbrace{\sum_{i=1}^{N_s-N_e} \bar{a}_{(o,d)}(i){*}\bar{y}_{N_s,N_e}}_{b} \right]}_{II} \quad (1)$$

(Part a^* is for OD pairs that travel along outbound direction, part b^* is for OD pairs that travel along inbound direction.)

where

$$a_i = \frac{V_i}{S_i} \quad \bar{a}_i = \frac{\bar{V}_i}{\bar{S}_i} \quad a_{od}(i) = \frac{V_{o,d}}{S_{o,d}} \quad \bar{a}_{od}(i) = \frac{\bar{V}_{o,d}}{\bar{S}_{o,d}} \quad (2)$$

subject to

$$1/C_2 \le z \le 1/C_1 \quad (3)$$

$$(1-k){*}\bar{b} \ge (1-k){*}k{*}b \quad (4)$$

$$(1-k_{od}(j)){*}y_{N_s,N_e} \ge (1-k_{od}(j)){*}k_{od}(j){*}b \text{ where } j = \begin{cases} 1 & (for\ OD_1) \\ \dots & \dots \\ n & (for OD_n) \end{cases} \quad (5)$$

$$(1-\bar{k}_{od}(j))^{*}\bar{y}_{N_s,N_e} \geq (1-\bar{k}_{od}(j))^{*}\bar{k}_{od}(j)^{*}\bar{b} \text{ where } j = \begin{cases} 1 & (for\ OD_1) \\ \dots & \dots \\ n & (for OD_n) \end{cases} \tag{6}$$

$$\begin{cases} \boldsymbol{w_i + b_i \leq 1 - r_i} \\ \boldsymbol{\bar{w}_i + \bar{b}_i \leq 1 - \bar{r}_i} \\ \boldsymbol{\bar{w}`_i + \bar{b}_i \leq 1 - \bar{r}_i} \end{cases} \quad \boldsymbol{i} = 1, \dots, \boldsymbol{n}; \tag{7}$$

$$\begin{cases} (\boldsymbol{d_i/f_i})z \leq \boldsymbol{t_i} \leq (\boldsymbol{d_i/e_i})z \\ (\boldsymbol{\bar{d}_i/\bar{f}_i})z \leq \boldsymbol{\bar{t}_i} \leq (\boldsymbol{\bar{d}_i/\bar{e}_i})z \end{cases} \quad \boldsymbol{i} = 1, \dots, \mathbf{n} - 1; \tag{8}$$

$$\begin{cases} (\boldsymbol{d_i/h_i})z \leq (\boldsymbol{d_i/d_{i+1}})\boldsymbol{t_{i+1}} - \boldsymbol{t_i} \leq (\boldsymbol{d_i/g_i})z \\ (\boldsymbol{\bar{d}_i/\bar{h}_i})z \leq (\boldsymbol{\bar{d}_i/\bar{d}_{i+1}})\boldsymbol{\bar{t}_{i+1}} - \boldsymbol{\bar{t}_i} \leq (\boldsymbol{\bar{d}_i/\bar{g}_i})z \end{cases} \boldsymbol{i} = 1, \dots, \mathbf{n} - 2; \tag{9}$$

$$\begin{aligned} &(\boldsymbol{w_i + \bar{w}_i}) - (\boldsymbol{w_{i+1} + \bar{w}_{i+1}}) + (\boldsymbol{t_i + \bar{t}_i}) \\ &+ \boldsymbol{\delta_i L_i - \bar{\delta}_i \bar{L}_i - \delta_{i+1} L_{i+1} + \bar{\delta}_{i+1} \bar{L}_{i+1} - m_i} \\ &= (\boldsymbol{r_{i+1} - r_i}) \quad \boldsymbol{i} = 1, \dots, \boldsymbol{n} - 1; \end{aligned} \tag{10}$$

$$\bar{w}`_i + \bar{b} + \bar{w}_i = 1 - \bar{r}_i \quad i = 1, \dots, n; \tag{11}$$

$$\bar{r}_i - \bar{r}_{i+1} + \bar{w}`_i - \bar{w}`_{i+1} = w_{i+1} - w_i \quad i = 1, \dots, n-1; \tag{12}$$

$$\boldsymbol{x_{N_s,N_e,i} \geq r_i} \quad \boldsymbol{i = N_s} + 1, \dots, \boldsymbol{N_e}; \tag{13}$$

$$\boldsymbol{\bar{x}_{N_s,N_e,i} \geq \overline{r_i}} \quad \boldsymbol{i = N_s} - 1, \dots, \boldsymbol{N_e}; \tag{14}$$

$$\begin{cases} \boldsymbol{y_{N_s,N_e} + x_{N_s,N_e,i} \leq r_i} \\ \boldsymbol{\bar{x}_{N_s,N_e,i}} + \bar{}_{N_s,N_e} \leq \boldsymbol{\bar{r}_i} \end{cases} \quad \boldsymbol{i = N_s}; \tag{15}$$

$$\begin{aligned} &\boldsymbol{x_{N_s,N_e,i} - x_{N_s,N_e,i+1} = w_i - w_{i+1} - r_{i+1} + r_i} \\ &\boldsymbol{i = N_s}, \dots, \boldsymbol{N_e} - 1; \end{aligned} \tag{16}$$

$$\begin{aligned} &\boldsymbol{\bar{x}_{N_s,N_e,i+1} - \bar{x}_{N_s,N_e,i} = \bar{w}`_{i+1} - \bar{w}`_i + \bar{r}_{i+1} - \bar{r}_i} \\ &\boldsymbol{i = N_e}, \dots, \boldsymbol{N_s} - 1; \end{aligned} \tag{17}$$

$$\sum \boldsymbol{y_{N_s,N_e} + b \leq} 1 - \boldsymbol{r_i} \quad \boldsymbol{i = max(N_s)} + 1, \dots, \boldsymbol{min(N_e)}; \tag{18}$$

$$\sum \bar{y}_{N_s,N_e} + \bar{b} \leq 1 - \bar{r}_i \quad i = min(N_s) - 1, \dots, max(N_e); \tag{19}$$

$$\begin{aligned} &b, \bar{b}, z, w_i, \bar{w}_i, \bar{w}`_i, t_i, \bar{t}_i, x_{N_s,N_e,i}, \bar{x}_{N_s,N_e,i}, y_{N_s,N_e}, \\ &\bar{y}_{N_s,N_e}, \delta_i, \bar{\delta}_i \geq 0 \quad i = 1, \dots, n. \end{aligned} \tag{20}$$

Objective function (1) aims to maximize a weighted sum of progression bandwidths. The weights are calculated as the sum of each OD flow divided by its corresponding saturation flow rate at each link. It maximizes the progression bands both along arterials (see Part I) and for specific major OD pairs (see Part II). $a_i(\bar{a}_i), a_{od}(\bar{a}_{od})$ in Equation 2 are the volume to saturation flow ratios of each traffic stream that are defined for individual segments; constraint (3) defines the upper and lower limits for cycle time; constraints (4), (5) and (6) are to provide a wider band to one of the two arterial directions and to the OD pair with a higher traffic volume; constraints (7) and (11) ensure that all through bands along arterial direction are within the available green time; constraint (8) defines upper and lower travel speeds; constraint (9) controls the speed changes and ensures that these changes are not too drastic; constraints (10) and (12) are the loop integer constraints that provide synchronization of traffic signals along the arterial direction; constraints (13), (14), (18) and (19) assure that any one of the OD bands does not infringe upon the red intervals along arterial directions; constraint (15) makes sure that vehicles join the arterial from a side street only when the arterial direction is subject to a red signal; constraints (16) and (17) provide synchronization between intersections for major ODs; and constraint (20) ensures all decision variables are non-negative.

The OD-BAND model is formulated for a sample arterial shown in Figure 3. As an example, progression

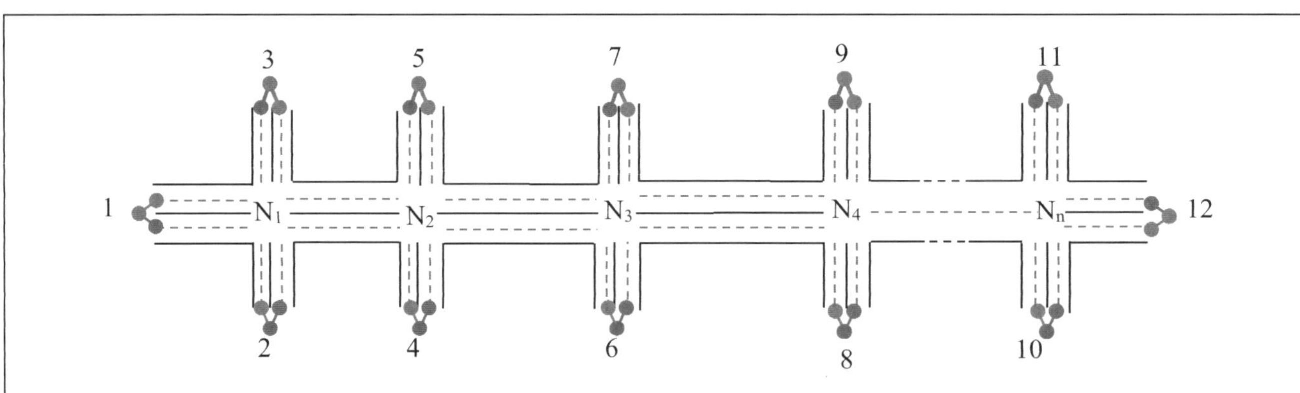

Figure 3. Sample arterial scheme for OD-BAND model formulation.

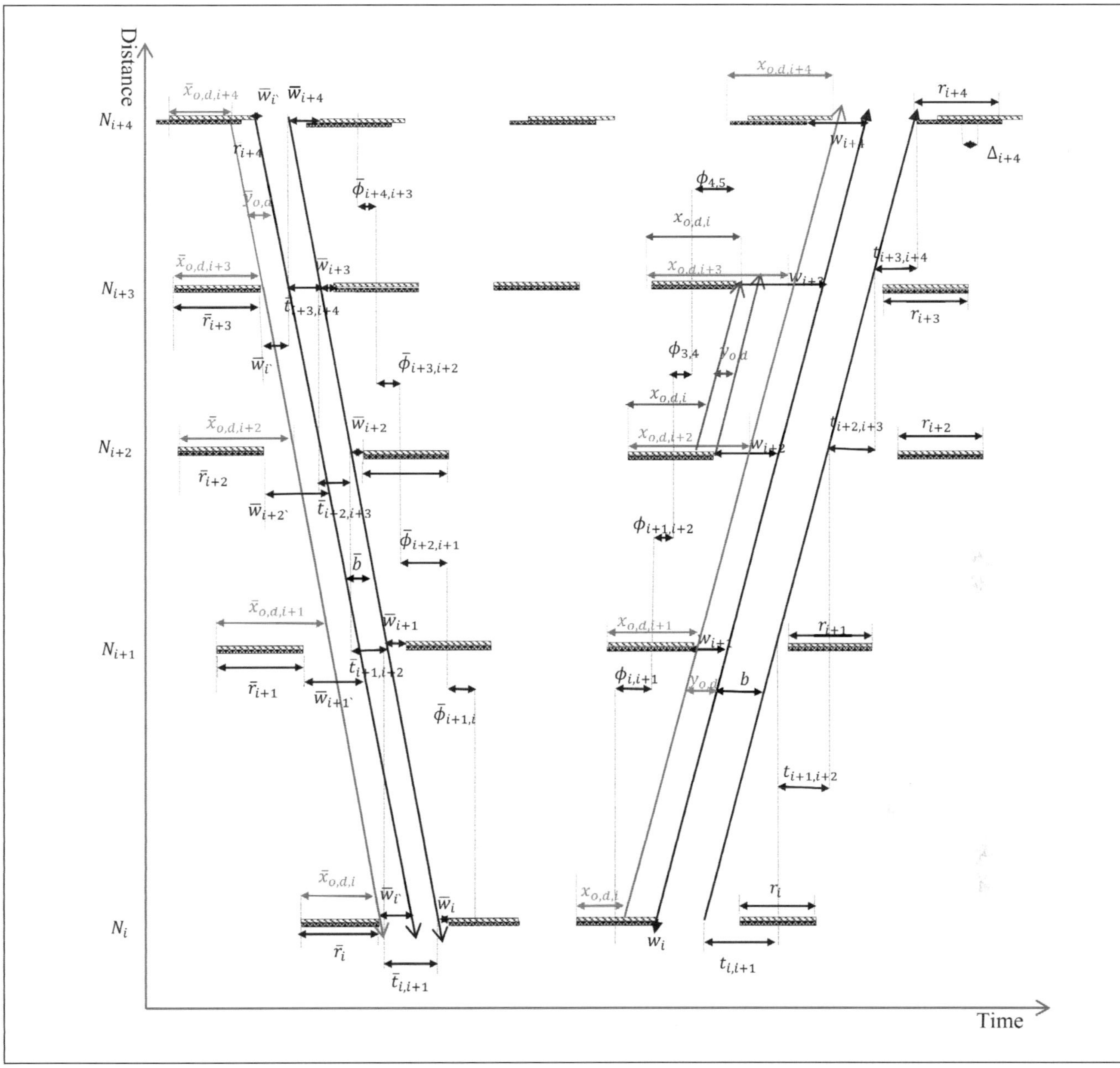

Figure 4. Time–space diagram for OD-BAND model.

bands are generated for additional major OD pairs 3→12, 6→9 (for outbound direction) and 10→1 (for inbound direction). Figure 4 presents a time–space diagram to illustrate formulation of the OD-BAND model.

OD-NETBAND: Network Model

As in the OD-BAND model, separate progression bands for major OD pairs are dedicated in the network model. Different from OD-BAND, network loop constraints are included to ensure all signals in the network are synchronized and operate with a common cycle time. The loop constraints are formulated separately based on the time–space diagrams for traffic movements from east to west and north to south. Previously defined variables are modified and converted into a matrix version for ease of formulation of the network model. As an example, the red times of outbound and inbound through traffic are represented by $r_i(\bar{r}_i)$ in OD-BAND. In OD-NETBAND, these red times are denoted by $r_m[n](\bar{r}_m[n])$, where m indicates column number and n indicates the row number. For instance, $r_2[3]$ is the outbound (N–S) red time for intersection 8 in Figure 3, since it is located

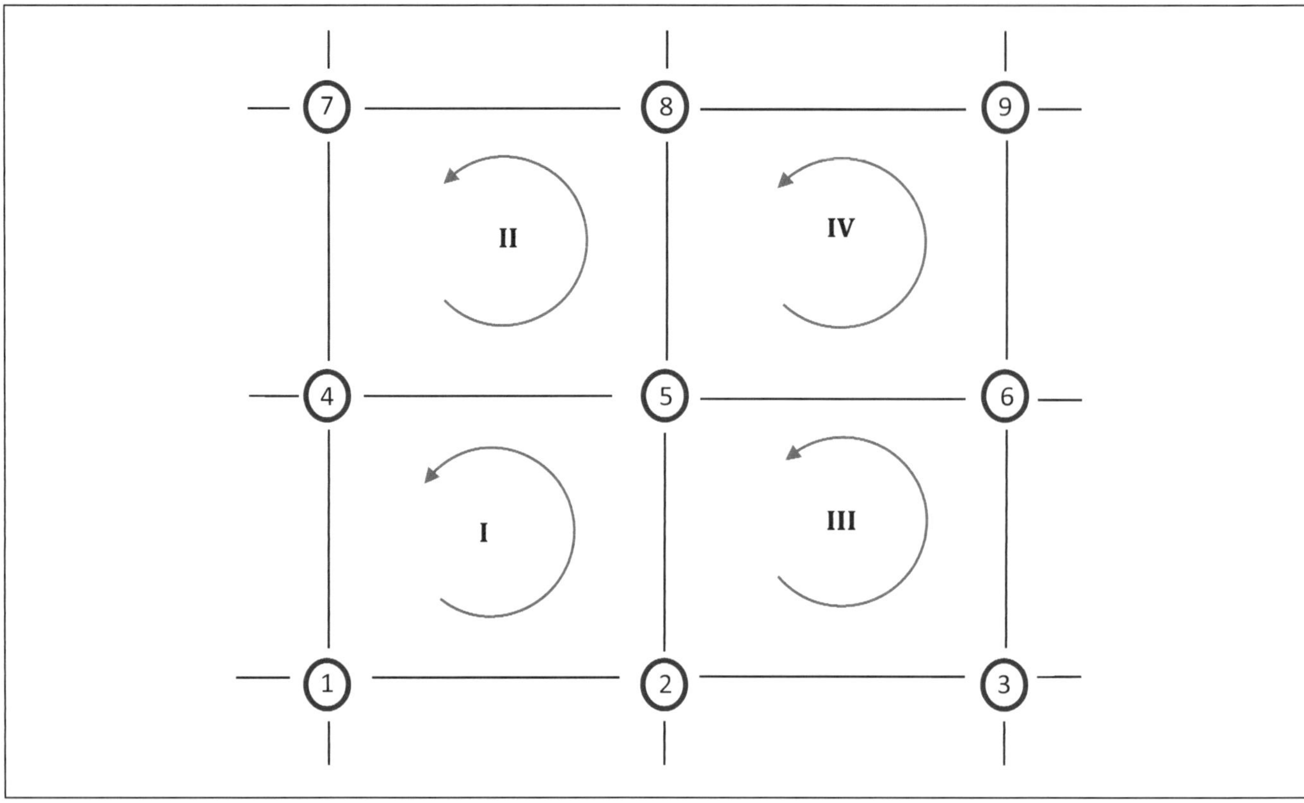

Figure 5. Possible loop constraints for three-by-three grid network.

at the intersection point of the second column and the third row. Similarly, other variables defined in OD-BAND are adopted by adding a notation that indicates column and row numbers with an exception for travel time. In OD-NETBAND, travel times between intersections are represented by $t_{ns}[m,n]$ and $t_{ew}[n,m]$, where "ns" and "ew" indicate north to south and east to west directions, respectively. Additionally, "m" and "n" indicate the column and row numbers, respectively.

Four loop constraints are defined for the three-by-three case study network. Each loop constraint is formulated to connect four intersecting arterials. As shown in Figure 5, Loop constraint I connects intersections 1, 2, 5, and 4; loop constraint II connects intersections 4, 5, 8, and 7; loop constraint III connects 2, 3, 6, and 5, and the last loop constraint (IV) connects intersections 5, 6, 9, and 8. Since any two loop constraints have at least one intersection in common and collectively they cover all intersections, there is no need to define additional loop constraints.

In Figure 6, the movements of traffic flows for both north to south and east to west directions are presented. These time–space diagrams are used to derive the loop constraints for OD-NETBAND. The two sets of time–space diagrams share some common points and they are marked as green stars. These common points provide the continuity needed in developing the loop constraints. Using the beginning of the red time of northbound through traffic as the start point, Loop Constraint I is formulated as follows:

$$
\begin{aligned}
&R_1[1] + \bar{L}_1[1] + w_1[1] + t_{ns}[1,1] - w_1[2] - \bar{L}_1[2] - R_1[2] \\
&= \bar{L}_{1,1}[1] + w_{1,1}[1] + t_{ew}[1,1] \\
&- w_{2,2}[1] - \bar{L}_{2,2}[1] + R_2[1] + \bar{L}_2[1] + w_2[1] \\
&+ t_{ns}[2,1] - w_2[2] - \bar{L}_2[2] - R_2[2] \\
&+ \bar{L}_{2,2}[2] + w_{2,2}[i+1] - t_{ew}[2,1] - w_{1,1}[2] - \bar{L}_{1,1}[2]
\end{aligned}
\quad (21)
$$

$$
r_m[n] = R_m[n] + \bar{L}_m[n] \quad m = 1,2,3; n = 1,2,3 \quad (22)
$$

In Figure 6, $r_m[n]$ is calculated using Equation 22. The left side of Equation 21 connects intersections 1 and 4, whereas the right side of the equation connects intersections 1, 2, 5, and 4. Both sides of Equation 21 end at the beginning of through traffic red interval of intersection 4. By making the two sides equal, the four intersections (i.e., 1, 2, 4, and 5) are synchronized. Similarly, other loop constraints (II, III, and IV) are formulated.

Properly setting offsets can significantly improve the network traffic control performance and provide efficient platoon movements through several intersections.

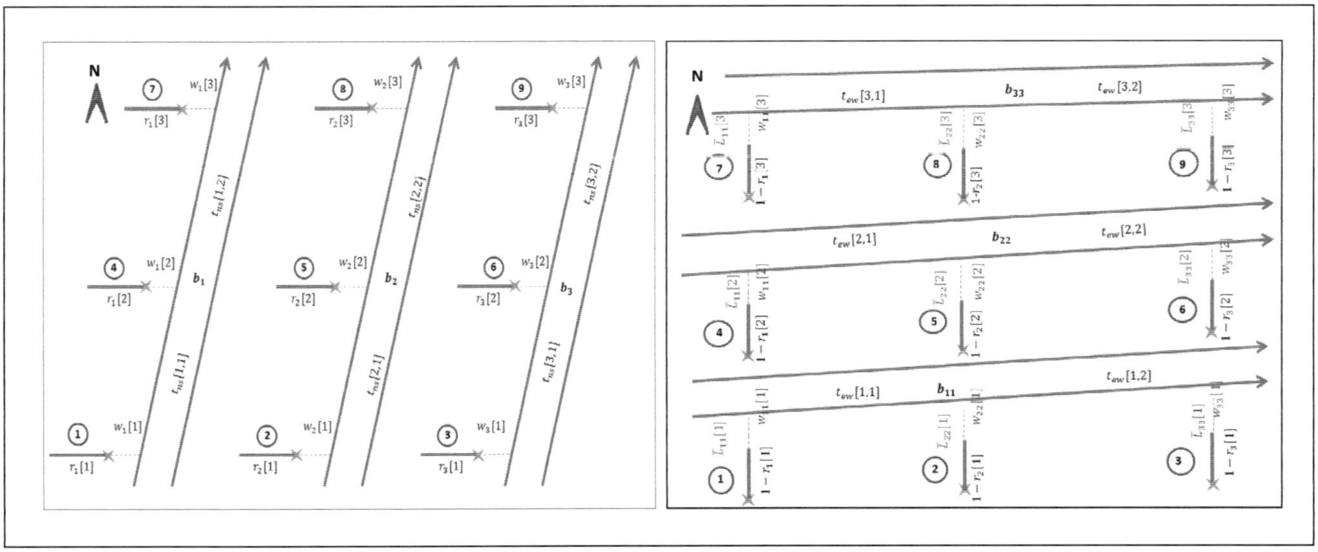

Figure 6. Time–space diagrams of traffic flows along north to south and east to west directions.

Table 2. Summary of Optimal Cycle Lengths and Offsets

Case	Model	Cycle length	Offsets for all nodes								
			1	2	3	4	5	6	7	8	9
I	MAXBAND-86	76	0.0	9.7	30.5	9.7	16.8	55.2	15.2	66.0	57.4
	Synchro	65	0.0	5.0	22.0	10.0	22.0	34.0	38.0	36.0	50.0
	OD-NETBAND	60	0.0	2.1	18.7	15.0	1.5	35.8	18.0	46.0	36.9
II	MAXBAND-86	61	0.0	1.3	30.3	2.3	18.6	32.6	18.6	31.6	39.6
	Synchro	65	0.0	10.0	27.0	21.0	21.0	38.0	38.0	33.0	3.0
	OD-NETBAND	60	0.0	1.3	33.3	13.0	16.3	43.3	24.3	38.3	50.0

The determination of offsets for a closed network is different from those for a single arterial. In a network, the existence of loops requires that adding all offsets along any loop should give the same offset at the beginning intersection. The offsets in OD-NETBAND are derived based on the time–space diagrams shown in Figure 6. The beginning of northbound through red interval of the first intersection is selected as the reference point. Since there are no left turns from any approaches at this intersection, this reference point is fixed for both inbound and outbound traffic movements. As an example, the offset constraint of intersection 5 is derived as follows:

$$\begin{aligned} &R_1[1] + \bar{L}_1[1] + w_1[1] + t_{ns}[1,1] - w_1[2] - \bar{L}_1[2] - R_1[2] \\ &\quad + \bar{L}_{1,1}[2] + w_{1,1}[2] + t_{e,w}[2,1] \\ &\quad - w_{2,2}[2] - \bar{L}_{2,2}[2] = \bar{L}_{1,1}[1] + w_{1,1}[1] + t_{e,w}[1,1] \\ &\quad - w_{2,2}[1] - \bar{L}_{2,2}[1] + R_2[1] \\ &\quad + \bar{L}_2[1] + w_2[1] + t_{ns}[2,1] - w_2[2] - \bar{L}_2[2] - R_2[2] \end{aligned} \tag{23}$$

The left side of Equation 23 calculates the offset through intersections 1→4→5, while the right side of the equation calculates the offset through intersections 1→2→5. By ensuring that the two sides are equal to each other, we can be confident that the offset of intersection 5 is correctly specified, which is a key requirement in network-wide traffic signal coordination. Similarly, the offset constraints for the remaining intersections are formulated.

Study Design

OD-NETBAND is formulated as a mixed integer linear program. The IBM CPLEX Optimization Studio (*19*) is used to find the optimal control plan (e.g., cycle length, phase sequence, offsets) for the grid network in Figure 1. Additional control plans for the same network are calculated using MAXBAND-86 (*8*) and Synchro (*20*). These control plans are summarized in Tables 2 and 3 and are evaluated using AIMSUN (*21*). Each of these plans is

Table 3. Phase Sequences for Case I and Case II

Case I			
Nodes	MAXBAND-86	Synchro	OD-NETBAND
1,2,3,6,7, and 9	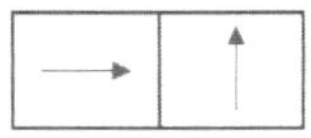		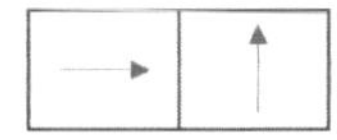
4 and 8	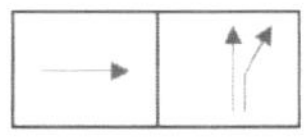	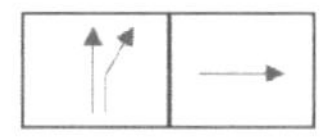	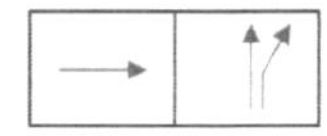
5	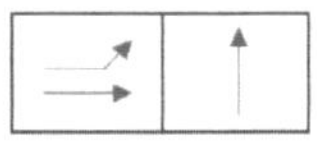		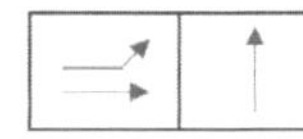
Case II			
Nodes	MAXBAND-86	Synchro	OD-NETBAND
1,2,3,6,7, and 9	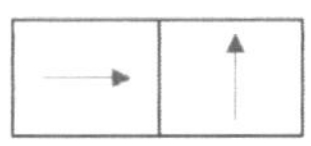	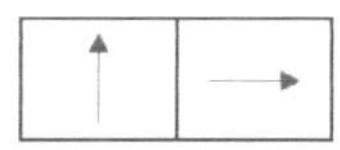	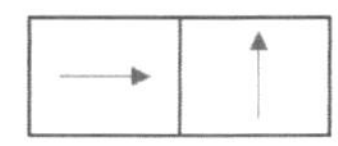
4 and 8	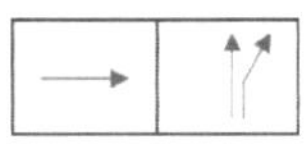	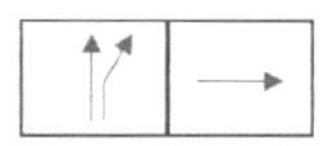	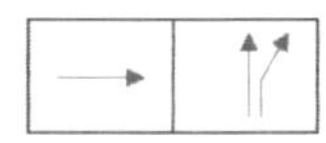
5	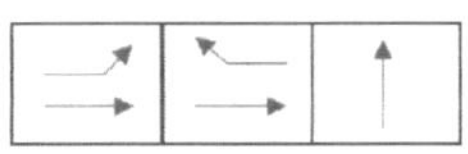	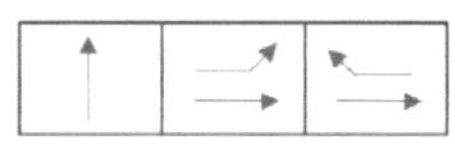	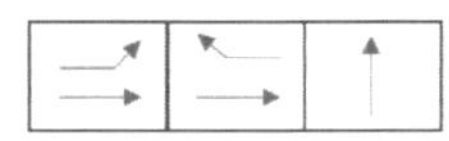

simulated 20 times based on the same set of random seeds and each simulation run represents 60 minutes of traffic operations.

In addition to the data provided in Figure 1 and Table 1, four seconds of loss time are considered in the AIMSUN simulation, including 3 seconds of yellow and 1 second of all red intervals. Speed limits on all road segments are set to 34 mph (about 55 kilometers per hour) and default turning speeds in AIMSUN are used. The saturation flow rate is set as 1,900 vehicles/hour green time/lane. Two through lanes and additional exclusive lane(s) are coded for each approach.

Evaluation Results

The performance of OD-NETBAND is compared with the performance of Synchro and MAXBAND-86 for the same grid network and traffic demand. Two case studies with different traffic flow patterns and lane configurations are considered. In Case I, in addition to through traffic along arterials, an additional OD stream (OD1) is included with one turning lane. In Case II, two additional OD streams (OD1 and OD2) are considered with two turning lanes (one for each OD). For both cases, fixed paths are assumed for the two major ODs and the paths are selected based on the shortest free flow travel times.

The simulation results are analyzed in terms of overall network performance, through traffic performance and major OD performance. Delay time, speed, stop time, and travel time are selected as measures of effectiveness (MOEs) for model comparison. Delay time is defined as the average delay time in seconds per vehicle mile, which represents the time difference between the expected and the actual travel time; speed is calculated as the average journey speed of each vehicle that has left the system within simulation duration, in miles per hour; stop time is described as the average time that is spent by vehicles at stand still position per vehicle mile, and travel time is represented in hours as the sum of all travel times experienced by vehicles that have exited the network within simulation duration.

Case I Result

Case I considers one major OD (i.e., OD1) with 800 vehicles per hour from node 1-1 to 3-3 (Figure 7), in addition to the through traffic along six arterials (500 vehicles/hour). Table 4 shows the performance of control plans generated by MAXBAND-86, Synchro, and OD-NETBAND. It can be seen that the latter generates the lowest overall network and major path (OD1) delays.

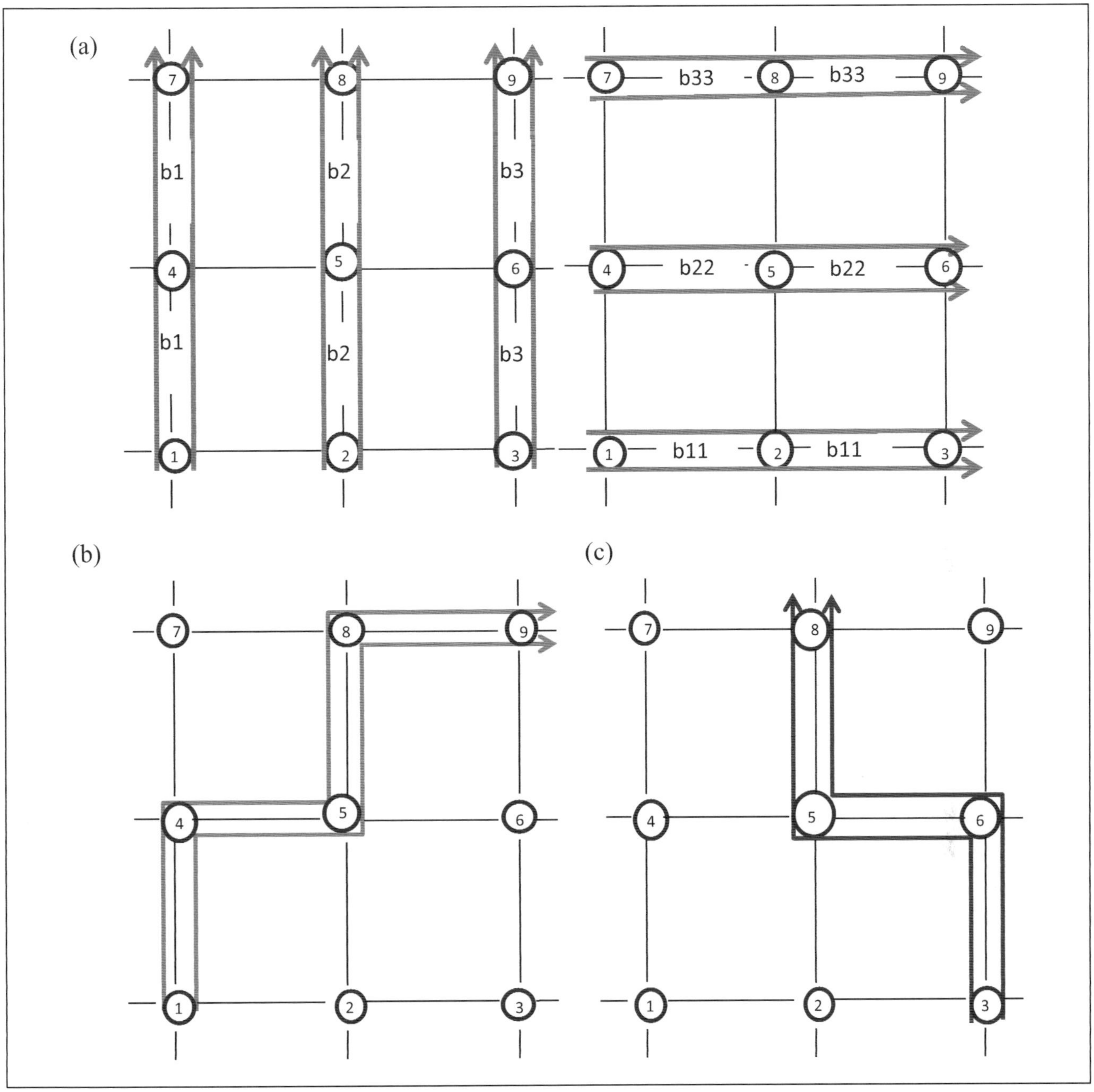

Figure 7. Illustration of continuous uniform progression bands created by OD-NETBAND: (*a*) bands created for through traffic; (*b*) bands for OD1; (*c*) bands for OD2.

Additionally, in terms of almost all remaining MOEs, OD-NETBAND performs the best. Its control plan generates the lowest stop time, travel time, and delay with the highest travel speed.

The three network signal coordination models are also compared in terms of progression bandwidths. Since MAXBAND-86 and Synchro cannot create separate progression bands for specific major ODs, their progression bands for through traffic streams are illustrated in Figure 7*a*. For OD-NETBAND, in addition to through progression bands, a dedicated progression band OD1 is also presented in Figure 7*b*.

In Table 5, progression bandwidths of the three models are compared in terms of bandwidth to cycle ratio. It can be seen that OD-NETBAND generates dedicated progression bands for each traffic stream, while Synchro and MAXBAND-86 generate progression bands just for through traffic along the arterials and fail to create a

Table 4. Performance Results for Cases I and II

	MAXBAND-86		Synchro		OD-NETBAND		
	Mean	SD	Mean	SD	Mean	SD	Units
Case I Results							
Delay							
Overall network	69.92	0.38	53.67	0.17	51.30	0.19	sec/mile
Through traffic	70.15	0.46	48.45	0.22	49.80	0.22	
OD 1	69.06	0.57	73.68	0.40	56.99	0.42	
Speed							
Overall network	21.85	0.06	23.70	0.03	23.81	0.03	mph
Through traffic	22.19	0.07	24.66	0.04	24.30	0.04	
OD 1	20.56	0.07	20.01	0.04	21.95	0.05	
Stop time							
Overall network	53.53	0.33	39.51	0.14	36.02	0.15	sec/mile
Through traffic	56.01	0.40	38.30	0.19	37.89	0.17	
OD 1	44.10	0.46	44.16	0.28	28.89	0.31	
Travel time							
Overall network	176.34	0.44	160.08	0.19	157.71	0.21	hour
Through traffic	175.99	0.50	154.28	0.25	155.63	0.23	
OD 1	177.67	0.58	182.30	0.35	165.60	0.34	
Case II Results							
Delay							
Overall network	68.55	2.77	60.18	0.29	57.00	0.45	sec/mile
Through traffic	55.00	1.29	52.52	0.34	49.40	0.28	
OD 1 + OD 2	94.16	7.07	74.79	0.77	71.48	1.09	
Speed							
Overall network	21.88	0.22	22.65	0.04	23.04	0.06	mph
Through traffic	23.73	0.13	24.03	0.06	24.37	0.05	
OD 1 + OD 2	18.37	0.52	20.01	0.08	20.50	0.14	
Stop time							
Overall network	50.43	2.25	42.73	0.22	39.16	0.40	sec/mile
Through traffic	41.88	1.03	38.97	0.30	36.23	0.25	
OD 1 + OD 2	66.59	5.89	49.89	0.58	44.76	0.99	
Travel time							
Overall network	175.19	2.77	166.78	0.28	163.62	0.49	hour
Through traffic	160.90	1.30	158.40	0.38	155.28	0.33	
OD 1 + OD 2	202.20	7.08	182.77	0.73	179.51	1.16	

Table 5. Progression Bandwidths for Cases I and II (Cycles)

	b1	b11	b2	b22	b3	b33	OD1	OD2	Total band
Case I									
OD-NETBAND	0.25	0.31	0.27	0.27	0.23	0.21	0.44	-	1.98
MAXBAND-86	0.50	0.30	0.22	0.22	0.30	0.00	-	-	1.54
Synchro	0.49	0.26	0.26	0.22	0.31	0.22	-	-	1.76
Case II									
OD-NETBAND	0.47	0.31	0.21	0.40	0.26	0.31	0.13	0.34	2.43
MAXBAND-86	0.50	0.31	0.20	0.34	0.31	0.31	-	-	1.97
Synchro	0.48	0.31	0.22	0.25	0.34	0.22	-	-	1.82

progression band for OD1. Such a failure may contribute to the higher overall network delays for Synchro and MAXBAND-86 in Table 4.

Case II Result

Case II considers two major ODs (OD1 and OD2) with 800 vehicles per hour from node 1-1 to 3-3 and from node 3-1 to 2-3, in addition to the through traffic along arterials (500 vehicles/hour). The performance of the proposed model is compared with the performance of MAXBAND-86 and Synchro in Table 4. Consistent with the Case I results, OD-NETBAND generates the lowest overall network and major path (OD1 + OD2) delays. The new model performs the best in terms of all selected MOEs. The result confirms the advantages of the new model's unique ability to generate separate progression bands for major OD flows. It also suggests the importance of considering vehicular OD data instead of individual intersection turning movement count data as Synchro and MAXBAND-86 do.

To compare the three models further, the progression bandwidths of each traffic stream are presented in Table 5. The result again indicates that OD-NETBAND is able to generate balanced progression bands for each traffic stream compared with MAXBAND-86 and Synchro.

Conclusions and Discussion

This research develops a network-wide progression bandwidth optimization model OD-NETBAND, which utilizes vehicular origin–destination (OD) information as the model input. Previous network-wide progression bandwidth optimization methods are only able to optimize progression bands along single arterials by considering individual intersection traffic counts and ignore the important information contained in major traffic OD flows. Ignoring OD information may lead to unnecessary wait time and increase the overall network delay, especially when there are major OD flows starting from one arterial and traversing to other arterials. This study aims to fill this knowledge gap by developing a network-wise OD based progression bandwidth optimization method that allocates dedicated progression bands for each major OD pair.

The proposed new model is formulated as a mixed integer linear program and solved by IBM CPLEX Optimization Studio. The performance of the proposed model is then compared with optimum control plans generated by MAXBAND-86 and Synchro using AIMSUN microscopic simulation. Two case studies are conducted with different traffic demand patterns and lane configurations. The results of both case studies are analyzed and compared based on delay, stop time, speed, travel time, and progression bandwidth. In terms of all selected MOEs, OD-NETBAND is found to be very effective when the traffic volumes of some cross-street OD streams are significant. This demonstrates the importance of allocating dedicated progression bands to major OD streams within a network.

Note that MAXBAND-86 has the worst overall performance in both cases but Synchro has the smallest bandwidths in Case II. A possible explanation is that there is not a straightforward and linear relationship between progression bandwidth and delay. However, we have seen, in general, that maximizing bandwidth improves MOEs such as delay and travel speed. Also, MAXBAND maximizes progression bandwidths without considering the corresponding traffic demands. Some of the progression bands may not be effectively utilized due to low demand. On the other hand, Synchro takes delay minimization into consideration while maximizing progression bandwidths. This is why in some cases it is difficult for bandwidth maximization models to outperform Synchro in terms of delay, though significantly improving stop time and the smoothness of the OD traffic flows.

This research considers a three-by-three network and represents an initial attempt to develop a formulation for more general networks in the future. General networks would require additional considerations depending on the major OD flows to be modeled and optimized. The loop constraints, specifically, can be derived automatically for a general network using graph-theoretic concepts (see [*7*, *22*]). The proposed OD-NETBAND is well suited to address the forthcoming introduction of connected and autonomous vehicles, which will provide the opportunity of obtaining accurate and detailed OD information for signal coordination and for integrated route guidance and progression band optimization.

Author Contributions

The authors confirm contribution to the paper as follows: study conception and design: Nathan H. Gartner, Tugba Arsava, Yuanchang Xie; data collection: Tugba Arsava; analysis and interpretation of results: Tugba Arsava, Nathan H. Gartner, Yuanchang Xie; draft manuscript preparation: Tugba Arsava, Nathan H. Gartner, Yuanchang Xie. All authors reviewed the results and approved the final version of the manuscript.

References

1. Yang, X. K. Comparison among Computer Packages in Providing Timing Plans for Iowa Arterial in Lawrence. *Journal of Transportation Engineering*, Vol. 127, No. 4, 2001, pp. 311–318.
2. Little, J. D. The Synchronization of Traffic Signals by Mixed-Integer Linear Programming. *Operations Research*, Vol. 14, No. 4, 1966, pp. 568–594.
3. Little, J. D. C., M. D. Kelson, and N. H. Gartner. MAXBAND: A Program for Setting Signals on Arterials and Triangular Networks. *Transportation Research Record: Journal of the Transportation Research Board*, 1981. 795: 40–46.
4. Gartner, N. H., S. F. Assmann, F. Lasaga, and D. L. Hou. MULTIBAND—A Variable-Bandwidth Arterial Progression Scheme. *Transportation Research Record: Journal of the Transportation Research Board*, 1990. 1287: 212–222.
5. Gartner, N. H., S. F. Assmann, F. L. Lasaga, and D. L. Hou. A Multi-Band Approach to Arterial Traffic Signal Optimization. *Transportation Research Part B: Methodological*, Vol. 25, No. 1, 1991, pp. 55–74.
6. Zhang, C., Y. Xie, N. H. Gartner, C. Stamatiadis, and T. Arsava. AM-Band: An Asymmetrical Multi-Band Model

for Arterial Signal Coordination. *Transportation Research Part C: Emerging Technologies*, Vol. 58, 2015, pp. 515–531.
7. Gartner, N. Constraining Relations among Offsets in Synchronized Signal Networks. *Transportation Science*, Vol. 6, No. 1, 1972, pp. 88–93.
8. Chang, E. C. P., S. L. Cohen, C. Liu, N. A. Chaudhary, and C. Messer. MAXBAND-86: Program for Optimizing Left-Turn Phase Sequence in Multi-Arterial Closed Networks. *Transportation Research Record: Journal of the Transportation Research Board*, 1988. 1181: 61–67.
9. Chaudhary, N. A., and C. J. Messer. PASSER-IV: A Program for Optimizing Signal Timing in Grid Networks. *Transportation Research Record: Journal of the Transportation Research Board*, 1993. 1421: 82–93.
10. Stamatiadis, C., and N. H. Gartner. MULTIBAND-96: A Program for Variable Bandwidth Progression Optimization in Multi-Arterial Traffic Networks. *Transportation Research Record: Journal of the Transportation Research Board*, 1996. 1554: 9–17.
11. Gartner, N. H., and C. Stamatiadis. Arterial-Based Control of Traffic Flow in Urban Grid Networks. *Mathematical and Computer Modeling*, Vol. 35, No. 5–6, 2002, pp. 657–671.
12. Gartner, N. H., and C. Stamatiadis. Progression Optimization in Large Scale Urban Traffic Networks: A Heuristic Decomposition Approach, Transportation and Traffic Theory. In A. Ceder (ed.), Proc.,14th International Symposium on Transportation and Traffic Theory, Pergamon, New York, NY, 1999, pp. 645–661.
13. Gartner, N. H., and C. Stamatiadis. Progression Optimization Featuring Arterial- and Route-Based Priority Signal Networks. *Journal of Intelligent Transportation Systems*, Vol. 8, No. 2, 2004, pp. 77–86.
14. Tian, Z. Z., M. Varun, and H. C. Liu. Effectiveness of Lead-Lag Phasing on Progression Bandwidth. *Transportation Research Record: Journal of the Transportation Research Board*, 2008. 2080: 22–27.
15. Park, B., C. J. Messer, and T. Urbanik. Traffic Signal Optimization Program for Oversaturated Conditions: Genetic Algorithm Approach. *Transportation Research Record: Journal of the Transportation Research Board*, 1999. 1683: 133–142.
16. Gomes, G. Bandwidth Maximization Using Vehicle Arrival Functions. *IEEE Transactions on Intelligent Transportation Systems*, Vol. 16, No. 4, 2015, pp. 1977–1988.
17. Arsava, T., Y. Xie, N. H. Gartner, and J. Mwakalonge. Arterial Traffic Signal Coordination Utilizing Vehicular Traffic Origin-Destination Information. Proc., 17th International IEEE Conference on Intelligent Transportation Systems (ITSC), Qingdao, China, IEEE, New York, NY, 2014, pp. 2132–2137.
18. Arsava, T., Y. Xie, and N. H. Gartner. Arterial Progression Optimization Using OD-BAND: Case Study and Extensions. *Transportation Research Record: Journal of the Transportation Research Board*, 2016. 2558: 1–10.
19. *IBM ILOG CPLEX Optimization Studio*. https://www.ibm.com/us-en/marketplace/ibm-ilog-cplex/. Accessed July 30, 2017.
20. Trafficware Synchro 6. http://trafficware.infopop.cc/synchro.htm. Accessed July 30, 2017.
21. *TSS-Transport Simulation Systems*. AIMSUN 8. https://www.aimsun.com/aimsun/editions/. Accessed July 30, 2017.
22. Chang, E. C.-P., S. L. Cohen, C. Liu, N. A. Chaudhary, and C. Messer. MAXBAND-86: Program for Optimizing Left-Turn Phase Sequence in Multiarterial Closed Networks. *Transportation Research Record: Journal of the Transportation Research Board*, 1988. 1181: 61–67.

The Standing Committee on Traffic Signal Systems (AHB25) peer-reviewed this paper (18-05652).

Article

TRR
JOURNAL OF THE TRANSPORTATION RESEARCH BOARD

Transportation Research Record
2018, Vol. 2672(18) 71–80

DOI: 10.1177/0361198118772691
journals.sagepub.com/home/trr
SAGE

Introducing a Cost-Effective Approach for Improving the Arterial Traffic Performance Operating Under the Semi-Actuated Coordinated Signal Control

Sina Dabiri[1], Kianoush Kompany[2], and Montasir Abbas[1]

Abstract

The semi-actuated coordinated operation mode is a type of signal control where minor approaches are placed with detectors to develop actuated phasing while major movements are coordinated without using detection systems. The objective of this study is to propose a cost-effective approach for reducing delay in the semi-actuated coordinated signal operation without incurring any extra costs in terms of installing new detectors or developing adaptive controller systems. We propose a simple approach for further enhancing a pre-optimized timing plan. In this method, the green splits of non-coordinated phases are multiplied by a factor greater than one. In the meantime, the amount of green time added to the non-coordinated phases is subtracted from the coordinated phases to keep the cycle length constant. Thus, if the traffic demand on the side streets exceeds the expected traffic flow, the added time in the non-coordinated phase enables the non-coordinated phases to accommodate the additional traffic demand. A regression analysis is implemented so as to identify the optimal value of the mentioned factor, called actuated factor (ActF). The response variable is the average delay reduction (seconds/vehicle) of the simulation runs under the proposed signal timing plan compared with the simulation runs under the pre-optimized timing plan, obtained through the macroscopic signal optimization tools. External traffic movements, left-turn percentage, and ActF are the explanatory variables in the model. Results reveal that the ActF is the only significant variable with the optimal value of 1.15 that is applicable for a wide range of traffic volumes.

Implementing optimal signal timing plans at intersections is one of the effective solutions for managing traffic operations. By convention, an appropriate method of traffic signal control is selected depending on traffic characteristics and roadway geometry conditions. Traffic signal control schemes usually involve a pre-timed or an actuated mode, or a combination of the two (*1*). As opposed to the pre-timed mode, the actuated control dynamically responds to the fluctuations in the traffic demand information collected by detectors. The actuated operation is divided into two categories: semi-actuated and fully actuated. In the semi-actuated signal control, the detectors are installed only on the minor traffic movements. Green time is assigned by default to the major approaches unless a call is placed on one of the minor approaches. Coordination is another strategy for programming the traffic signals, which provides a progressive traffic flow between multiple closely spaced intersections along an arterial to reduce the traffic performance indices such as the number of stops, delay, and travel time (*1*). Establishing a temporal relationship between the green times of similar phases in the adjacent arterial intersections so that the vehicles move in platoons with a minimum number of stops is the main purpose of a coordinated system. Integrating semi-actuated and coordinated operation modes generates a semi-actuated coordinated control architecture which

[1]Charles Edward Via, Jr. Department of Civil and Environmental Engineering, Virginia Polytechnic Institute and State University, Blacksburg, VA
[2]Atkins North America, Raleigh, NC

Corresponding Author:
Address correspondence to Sina Dabiri: sina@vt.edu

outperforms the semi-actuated and fully actuated signal timing systems. Such a method engenders significant improvements of vehicular travel times and stops (*2, 3*).

In arterials in which the speed limits on the coordinated streets are less than 40 mph, coordinated phases on the major street can turn green without using any detector, while non-coordinated green phases, on the side streets, are requested through actuations of detection systems. The green splits of coordinated phases are at least corresponding to the split time in the optimal timing plan. However, the signal states for non-coordinated phases remain green unless their designated detectors gap out or max out. The unused green time as a result of either gap out or phase skip of the non-coordinated phases is transferred to the coordinated phases, which returns the green to the coordinated phases earlier than the normal setup. Sophisticated understanding of the factors and parameters related to the semi-actuated coordinated signal operation offers us an opportunity to improve such a traffic signal control with a simple yet efficient method.

In this paper, a simple and straightforward procedure is introduced for enhancing the optimal signal timing plan for arterials that practitioners have already decided to be operated with the semi-actuated operation mode. The optimal signal timing plan has been already obtained with macroscopic traffic analysis tools. In this study, the actuated parameters and detection elements are configured according to available traffic guidelines and manuals, whereas the major optimal signal timing parameters are generated using VISTRO, as a powerful macroscopic signal optimization software package. While offsets, phase sequence, and background cycle length are kept constant, the green splits for non-coordinated phases are multiplied by a factor greater than one. Note that no detectors are utilized for the coordinated movements as the operation mode is semi-actuated. The extra green time of non-coordinated phases and the multiplication factor are called added green time (AGT) and actuated factor (ActF), respectively. Then, to preserve the background cycle length, AGT is subtracted from the green time of the coordinated phases. The specific objective of this study is to identify the applicability of using such a factor (ActF) and evaluate its optimal value in different traffic conditions. The proposed methodology creates a signal timing plan so that the total delay in an arterial is reduced compared with the optimal timing plan produced by VISTRO. Furthermore, we assess the significance of the major and side streets traffic volumes in determining the optimal ActF.

This methodology is simple and cost-effective for two principal reasons: (1) it does not incur additional costs for installing new detectors, changing detection configurations, or developing new software to dynamically optimize signal timing plans; and (2) the recommended method requires only basic math calculation, which can be easily adopted by practitioners. The approach does not even need to deploy the data-driven and advanced learning techniques that have recently been much utilized in transportation fields (*4, 5*).

Actuated and Coordinated Control Literature Review

To date, a considerable amount of research has been conducted on a wide range of actuated and coordinated control aspects since the study of Jovanis and Gregor, who introduced this type of traffic signal control (*6*). A majority of methodologies for achieving an optimal actuated and coordinated operation can be categorized in two groups. The methods in the first group have concentrated on the arterial control strategies, and in the second group have focused on optimizing the signal timing parameters. Each of these viewpoints is discussed in the following paragraphs.

A summary of recommended guidelines and strategies for implementing the actuated-coordinated signal systems has been provided by Chang (*7*). In addition to the detector configuration and intersection timing parameters, the study provided a step-by-step procedure for selecting the optimal strategy including pre-timed conversion, early return, and v/c ratio control strategy. The users are able to select the appropriate strategy based on the level of saturation and the importance of the previous timing plan settings. Developing alternatives to improve an arterial performance using the actuated and coordinated concepts was the main objective of the research conducted by Messer and Nageswara (*8*). First, the optimal pre-timed coordinated timing plan was transformed into the actuated-coordinated timing plan via vehicle actuations. The effectiveness of this change was evaluated by traffic simulation. The results indicated that reducing the green time of the optimal pre-timed coordinated phases results in no performance improvement of the actuated system, whereas increasing the green split of the pre-timed coordinated phase deteriorated the performance. The rationale behind this is that the extra green time of the coordinated phases is not transferred to non-coordinated phases when demand in the minor street is high, as detection systems are not installed on the major street to implement the gap-out operation. To propose an optimal strategy for one-way or two-way progression in the congested or uncongested direction, Chaudhary and Balke assessed five coordination strategies by applying a simulation study when the degree of saturation for each intersection varies from 0.86 to 1.12 (*9*). Findings revealed that the one-way progression in the congested direction was an appropriate method when the network was experiencing oversaturation. Abbas et al. outlined a framework for the design of arterial control strategies for oversaturated conditions (*10*). The set of network critical links and their traffic volumes were essential factors in determining the appropriate control strategy. In a study by He et al., multiple priority requests from various modes of transportation (e.g., emergency vehicles, buses, and pedestrians) were accommodated in a coordinated and actuated signal control system (*11*). Considering an isolated semi-actuated intersection, Lin et al. investigated the effect of side-street traffic volumes on the major-street green time ratio. They sought to find a relationship between traffic volumes and number of stops as the base for making the optimal decision in signal coordination (*12*). In comparison with adaptive control strategies, Chia et al. demonstrated that conventional actuated-coordinated time-of-day plans work better during the peak hours (*13*).

Assessing the actuated and coordinated control parameters was the focus of the second category. Early return to green and the formation of the stop bar waiting queues occur frequently in the actuated-coordinated signal systems. Consequently, designing a constant offset on the basis of free-flow speed and the arterial link length causes a considerable amount of delay, which reveals the need for an adaptive offset algorithm. To overcome this shortcoming, a real-time offset transitioning algorithm was proposed to dynamically adjust the offset with the purpose of providing smooth progression of the vehicles moving on the corridor (*14*). The issue of early return to green can also be addressed by considering the stochastic distribution of starts/ends of green for the coordinated phases according to a large signal state data obtained from a field data collection operation (*15*). Taking the advantage of these distributions, an offline offset refiner was proposed to fine-tune the offset after designing the signal timing plans. Defining force-off points is a common method for running the coordination in the actuated-coordinated signal operation. Comparing the performance of the floating and fixed force-off modes for various volume-to-capacity ratios revealed that the fixed force-off mode outperformed the other modes in all levels of the traffic demand (*16*). In the fixed force-off mode, in contrast to the floating force-off mode, the available unused green time owing to gapping out or phase skipping of the non-coordinated phases can be added to the next phase, given that the traffic volume on the next phase is excessive. No matter whether the next phase is coordinated or not, "it may use the extra time up to the force-off point" (*1*). However, in the floating force-off mode, the only upcoming movement that can utilize the unused green time, of the extra green time that gapped out earlier, is the coordinated phase. Therefore, floating force-off limits the flexibility of the amount of the green times assigned to the non-coordinated phases (*1*). In addition to the benefits of traditional actuated-coordinated signal systems, a neuro-fuzzy signal controller can also improve the progression by establishing a real-time secondary coordination between the upstream coordinated through-movement, and the downstream non-coordinated left-turn movement (*17*). The proposed controller allows for alleviating queue spillback and upstream link blockage at closely spaced intersections, which are the two traffic issues that stem from the early return to green and the demand variation of the side-street links.

The inability of commercial macroscopic traffic signal optimization tools to take into account traffic uncertainties suggests the need to develop microscopic simulation-based traffic signal optimization methods (*18*). Moreover, these existing optimization software packages primarily focus on major timing plan parameters, (e.g., green splits) rather than actuated parameters, (e.g., vehicle extension). Considering the retrospective approximation concept, Markov monotonic search algorithm as the optimizer, and VISSIM as the simulation engine, a framework for optimizing the maximum green time as one of the actuated parameters was developed (*19*). Park et al. presented a stochastic-optimization method for simultaneously optimizing the basic signal timing parameters and actuated settings including green time, vehicle extension, and recall mode (*20, 21*). The Genetic Algorithm and Shuffled Frog-Leaping Algorithm were used as the optimizers, which were interfaced with CORSIM as the microscopic simulation platform.

Although many studies have been conducted to investigate various aspects of the actuated-coordinated systems, no simple and practical methodology has been proposed to improve the available optimal signal timing plan in a practically feasible computation time. To address such a lack of research, this paper aims to propose a straightforward methodology on the basis of the signal timing plan so as to ameliorate the arterial performance.

Methodology

In the semi-actuated coordinated systems, detectors are only installed along the non-coordinated movements on the side streets and the left-turn movements on the major street. Non-coordinated phases serve for at least their designated minimum green time only if the controller receives a call from their associated detectors; otherwise, the phase is skipped. The green time for the actuated phases is extended by a defined unit extension value if the gap between two consecutive vehicle actuations is less than the vehicle extension parameter; if not, the phase is gapped out. The maximum green time, which is equal to the optimal green split created by an optimization tool, is reached if there is continuous demand on the actuated phase. The unused green time of the actuated phases owing to gapping out or phase skipping is added to the green time of coordinated phases. Thus, the green split of the coordinated phases may be extended by the unused green time of the actuated phases. To synchronize the coordinated phases, the controller is forced to operate within the constraint of a constant cycle length.

Using the concept of the semi-actuated coordinated signal control, which means increasing the green split in the non-coordinated phases while keeping the cycle length constant, might be an efficient solution for improving the arterial performance. In the case of demand fluctuations on the side streets, increasing the green split time of the side streets also gives room to handle the extra traffic demand of the side streets more effectively. In the case that there is no extra demand, the additional green time would be transferred to the coordinated phases. It should be noted that the increase in the green splits of the non-coordinated phases needs to be accompanied by the decrease in the green splits of the coordinated phases so as to maintain the designated cycle length constant. Short-term traffic fluctuations within an hour need to be considered using peak hour factor (PHF), which is the ratio of the average hourly volume to four times the peak 15- minute volume. In the absence of the field measurements, a value of 0.92 is recommended for the urban area by the Highway Capacity Manual, which is employed in this study (*22*).

To implement the proposed method, first of all, the optimal timing plan parameters including offsets, cycle length, and green splits are determined using VISTRO, a commercial macroscopic simulation-optimization tool with a high computational speed. In this study, pedestrian phases are not considered during the development of the timing plans. In the next step, the green splits of the actuated non-coordinated phases, calculated by VISTRO for each intersection, are multiplied by a factor greater than one, which is called ActF. The optimal value of ActF is determined by a regression analysis, which is explained in the following sections. Then, the total added seconds in each ring diagram are deducted from the coordinated phase in that ring in order to keep the cycle length constant. Similar to the cycle length, the offsets of the new timing plan are the same as the VISTRO timing plan. For the sake of simplicity, the green splits of the timing plan created by VISTRO and the new timing plan are called "Base TP" and "Factored TP," respectively. Employing the NEMA phase numbering scheme and the standard ring-barrier diagram, the Factored TP is determined as follows:

$$g_{i,\,\text{Factored TP}} = ActF^{*}g_{i,\,\text{Base TP}} \quad \forall i \in \text{non-coordinated phases} \tag{1}$$

$$g_{j1,\,\text{Factored TP}} = g_{j1,\,\text{Base TP}} - \left(\sum_i g_{i,\,\text{Factored TP}} - \sum_i g_{i,\,\text{Base TP}}\right) \quad i \in \text{non-coordinated phases in ring 1} \tag{2}$$

$$g_{j2,\,\text{Factored TP}} = g_{j2,\,\text{Base TP}} - \left(\sum_i g_{i,\,\text{Factored TP}} - \sum_i g_{i,\,\text{Base TP}}\right) \quad i \in \text{non-coordinated phases in ring 2,} \tag{3}$$

Where g_i is the green time of the non-coordinated phase i in the corresponding timing plan and ActF is the actuated factor that the non-coordinated phases need to be multiplied by. g_{j1} and g_{j2} denote the green splits of coordinated phases in ring 1 and 2, respectively. Equation 1 shows how to convert the green splits of the non-coordinated phases in the Base TP to the Factored TP, and Equations 2 and 3 represent how to obtain the new values for the coordinated phases in rings 1 and 2, respectively. It is worth mentioning that if the amount of $g_{j1,\text{Factored TP}}$ and $g_{j2,\text{Factored TP}}$ become lower than their associated minimum green, an adjustment must be applied to keep all green splits equal to or greater than their minimum values. The procedure of obtaining Factored TP is illustrated through an example in Figure 1 by using Equations 1–3 when the ActF is 1.10. The first and the second diagrams show the Base TP and Factored TP, respectively, with a cycle length of 60 seconds. In this example, phases 2 (in ring 1) and 6 (in ring 2) are the coordinated phases.

Statistical Analysis

The specific objective of this study is to investigate whether applying the ActF to the signal timing plan offered by VISTRO can improve the arterial performance measure. Furthermore, the study seeks to determine an optimal value for the ActF under different levels of traffic demand. If the hypothesis of the existence of such a factor holds true, then practitioners can implement the optimal value of ActF to the timing plans offered by VISTRO (or any other signal timing optimization tool) and expect a better arterial performance. To achieve this goal, a statistical model needs to be developed to capture the variation of a defined performance index according to distinct values of ActF and various traffic volumes in the arterial. Such a goal can be accomplished through regression analysis. The response variable in the regression model is denoted by delay reduction percentage (DRP) and is formulated in Equation 4.

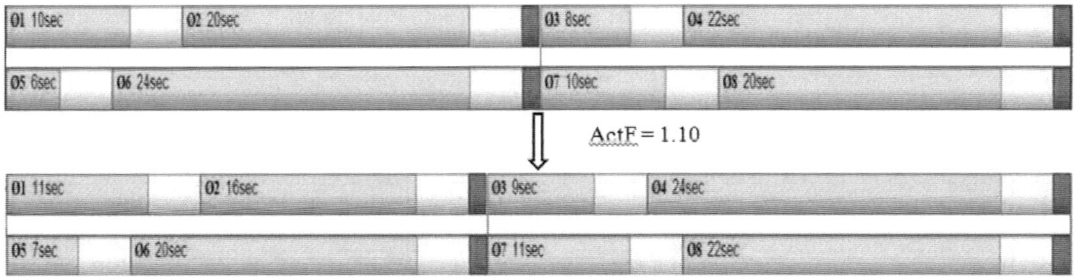

Figure 1. Obtaining the factored timing plan from the base timing plan generated by VISTRO using the actuated factor equal to 1.10.

$$DRP = \frac{Average\ Delay\ based\ on\ Base\ TP - Average\ Delay\ based\ on\ Factored\ TP}{Average\ Delay\ based\ on\ Base\ TP} \times 100 \quad (4)$$

The above equation represents the percentage in the delay reduction when the arterial is operated under the Factored TP obtained by the proposed method, in comparison with the Base TP calculated by VISTRO. To account for the traffic fluctuations, the proposed regression model comprises the external traffic volumes as the explanatory variables, in addition to the left-turn rates and ActF.

Study Site

An arterial with three intersections located in Prices Fork Road, Blacksburg, Virginia is selected as the test site, as illustrated in Figure 2. All intersections are assumed to be operating by 8 phases with protected left-turn signal configuration on all approaches. Eastbound and westbound directions are the coordinated routes. According to the NEMA phasing configuration, phases 2 and 6 are the coordinated phases along the major street. Traffic volumes of external movements on the side and the major street are a group of explanatory variables in the statistical model that are presented in Figure 2. To compute the traffic volume on each movement using the external volumes, the rate of both the left-turn and right-turn volumes must be identified. In this study, a fixed value equal to 10% of the total volume of an approach is considered as the right-turn proportion, and the left-turn proportion is involved as another explanatory variable of the regression model. In summary, the ActF, left-turn rates (LT), and external traffic volumes are all selected to be the explanatory variables and DRP in Equation 4 is the response variable.

Actuated Settings

As the speed limit of the side streets and left-turn movements is low (less than 35 mph), the detection zones are assumed to be located only at the stop lines. The

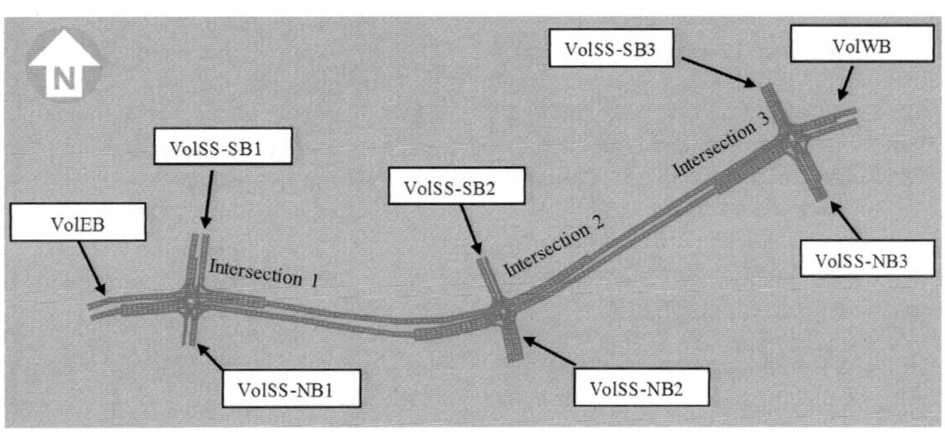

Figure 2. Real-world arterial with three intersections, and the external traffic volumes (explanatory variables) in the arterial.

Table 1. Description of the Explanatory Variables

Variable	Description	Type	Input values	Unit
VolSS_SBi	External traffic volume, southbound toward intersection i (i = 1, 2, 3).	continuous	200 to 600	vehicles/hour
VolSS_NBi	External traffic volume, northbound toward intersection i (i = 1, 2, 3).	continuous	200 to 600	vehicles/hour
VolEB	External traffic volume entering the simulation network from the west side of the arterial.	continuous	400 to 1200	vehicles/hour
VolWB	External traffic volume entering the simulation network from the east side of the arterial.	continuous	400 to 1200	vehicles/hour
LT	The left-turn rate.	discrete	10, 20, 30	percent (unitless)
ActF	Actuated Factor: Used for defining the amount of the green time to be transferred to non-coordinated phases from the coordinate phases of the signal timing plan optimized by VISTRO.	discrete	1.00 to 1.30	–

actuated timing plan settings were selected according to the values recommended by Traffic Signal Timing Manual (*1*). Assuming the speed limits on the side streets and the major street equal 25 and 35 mph, respectively, the minimum green times for the coordinated phases and non-coordinated phases are set as 2 and 7 seconds. This arrangement meets the drivers' expectancy (*1*).

Yellow change and red clearance intervals are set to 3 and 1 seconds. Following the guidelines of the Traffic Signal Timing Manual and careful observation of simulation runs lead to choosing 2 seconds as the most suitable value for the vehicle extension parameter.

Data Description

To examine the impact of ActF on the arterial performance measure (e.g., DRP) for various traffic conditions, a regression analysis is implemented. It is worth mentioning that the purpose of developing a regression analysis is not to predict the DRP, yet to purely test the effect of ActF on DPR. However, first data need to be collected for performing every type of regression model. The first step in the process of data collection for this problem is to perform design of experiments (DOE) using JMP Pro 11.0.0 statistical software package. DOE is a cost-effective statistical method which is contingent on a response variable, explanatory variables, and the type of model (*23*). In this study, the response surface model (RSM) is selected for conducting the experiment. In addition to the explanatory variables (i.e., ActF, *LT*, and external traffic volumes in Table 1), RSM contains two-factor interactions (multiplication of every two variables) and quadratic terms of all explanatory variables.

According to the type and number of factors as well as the chosen model, DOE suggests a specific number of runs. For each experimental run in DOE, the response variable (i.e., DPR) must be computed according to specific values of the explanatory variables. The external volumes for the side streets are continuous variables ranging from 200 to 600 vehicles/hour, and the westbound and eastbound traffic volumes in the major street range continuously from 400 to 1200 vehicles/hour. Left-turn percentage (LT) is a discrete variable including three levels: 10%, 20%, and 30%. ActF is also a discrete variable with 7 levels from 1 to 1.30 in 0.05 increments. To minimize the possibility of the green splits of coordinated phases in the Factored TP being lower than their minimum values, the values of ActF are constrained to 1.30 as the upper bound. The number of scenarios recommended by JMP Pro for such a DOE is 72. Each scenario, (i.e., each combination of the variables), is a run with 10 different random seeds to account for the network stochasticity. Therefore, considering the base scenarios operated under the VISTRO signal timing plans, a total of 1440 runs are simulated. The description of the basic explanatory variables used in this study is presented in Table 1. The simulation runs are performed in VISSIM, a microscopic traffic simulation model, to obtain the average delay (seconds/vehicle) as the arterial performance index. After modeling the study site in VISTRO, the following steps are taken in order to compute the response variable for each experimental simulation run:

1. Having the external volumes and LT rate, calculate the traffic volumes for all phases at each intersection.
2. By importing the calculated volumes to VISTRO, generate the base optimal timing plan parameters including cycle length, offsets, and green splits.
3. Determine the Factored TP in accordance with Equations 1–3.
4. Employing the Ring Barrier Controller, first, import the calculated volumes and the created test site from VISTRO to VISSIM; then, run the simulation for the Base TP and Factored TP in 1-hour analysis period. Note that the simulation network is warmed up for 15 minutes before this 1-hour analysis duration.
5. Record the average delay (seconds/vehicle) corresponding to the Base TP and Factored TP.

Table 2. Calibration Results of the Parent Scenario; NA = Not Applicable

Link description	Count	Output	GEH	GEH < 5	GEH<4
Intersection 1 - Eastbound	826	832	0.10	Yes	NA
Intersection 1 - Northbound	215	220	0.17	Yes	NA
Intersection 1 - Southbound	636	635	0.02	Yes	NA
Intersection 1 - Westbound	333	331	0.05	Yes	NA
Intersection 3 - Eastbound	448	454	0.14	Yes	NA
Intersection 2 - Northbound	217	212	0.17	Yes	NA
Intersection 2 - Southbound	652	660	0.16	Yes	NA
Intersection 3 - Northbound	217	221	0.14	Yes	NA
Intersection 3 - Westbound	434	426	0.19	Yes	NA
Intersection 3 - Southbound	217	212	0.17	Yes	NA
Between intersections 1 & 2 - Eastbound	678	706	0.53	Yes	NA
Between intersections 1 & 2 - Westbound	339	334	0.14	Yes	NA
Between intersections 2 & 3 - Eastbound	609	625	0.32	Yes	NA
Between intersections 2 & 3 - Westbound	348	343	0.13	Yes	NA
Sum	6169	6211	0.27	NA	Yes

6. By using the average delays in step 4, calculate DRP using the Equation 4.

Model Calibration

After modeling the arterial in VISSIM and before simulating all the DOE scenarios, volume calibration was undertaken for one scenario, called the parent scenario. All other scenarios are built based on the calibrated parent scenario. Travel time counters are coded on all links of the study site to calculate hourly volumes and compare the volumes with the input traffic counts. Volume calibration is achieved by using the GEH formula, invented by Geoffrey E. Havers in the 1970s:

$$GEH = \sqrt{\frac{(E - V)^2}{(E + V)/2}} \tag{5}$$

where E and V are model estimated volume and traffic counts obtained from DOE, respectively.

The GEH criterion requires the GEH value to be less than 5 and 4 for 85% of the network links and for the entire network, respectively (*24*). Table 2 summarizes the results of the volume calibration of the parent scenario, which indicates that the parent scenario has satisfied all requirements.

Results and Discussion

Once the experiment procedure was completed by collecting the DPR for all the 72 simulation runs, a regression analysis was developed using JMP Pro 11.0.0. As DOE is based on RSM, the main effects, the two-way interactions, and the quadratic terms needed to be added in the regression model.

Table 3 reports the *F*-test results of the regression analysis. For the sake of brevity, only the main effects of explanatory variables and $ActF^2$ are presented in Table 3, as all the other variables yield insignificant values. The F-test examines the hypothesis that all the

Table 3. Results of the *F*-test for the Main Effects and the Significant Variable

Source	Number of parameters	Degree of freedom	Sum of squares	*F* ratio	Prob. > *F*
VolEB(400,1200)	1	1	23.242	0.815	0.401
VolWB(400,1200)	1	1	1.129	0.040	0.849
VolSS_NB1(200,600)	1	1	0.725	0.025	0.879
VolSS_SB1(200,600)	1	1	0.078	0.003	0.960
VolSS_NB2(200,600)	1	1	25.888	0.908	0.377
VolSS_SB2(200,600)	1	1	7.348	0.258	0.630
VolSS_NB3(200,600)	1	1	26.546	0.931	0.372
VolSS_SB3(200,600)	1	1	2.091	0.073	0.796
LT(10,30)	1	1	22.707	0.796	0.407
ActF (1,1.3)	1	1	33.011	1.158	0.323
$ActF^2$	1	1	471.447	**16.535**	**0.007**

Note: Bold numbers indicate significant variables in the 95% confidence interval.

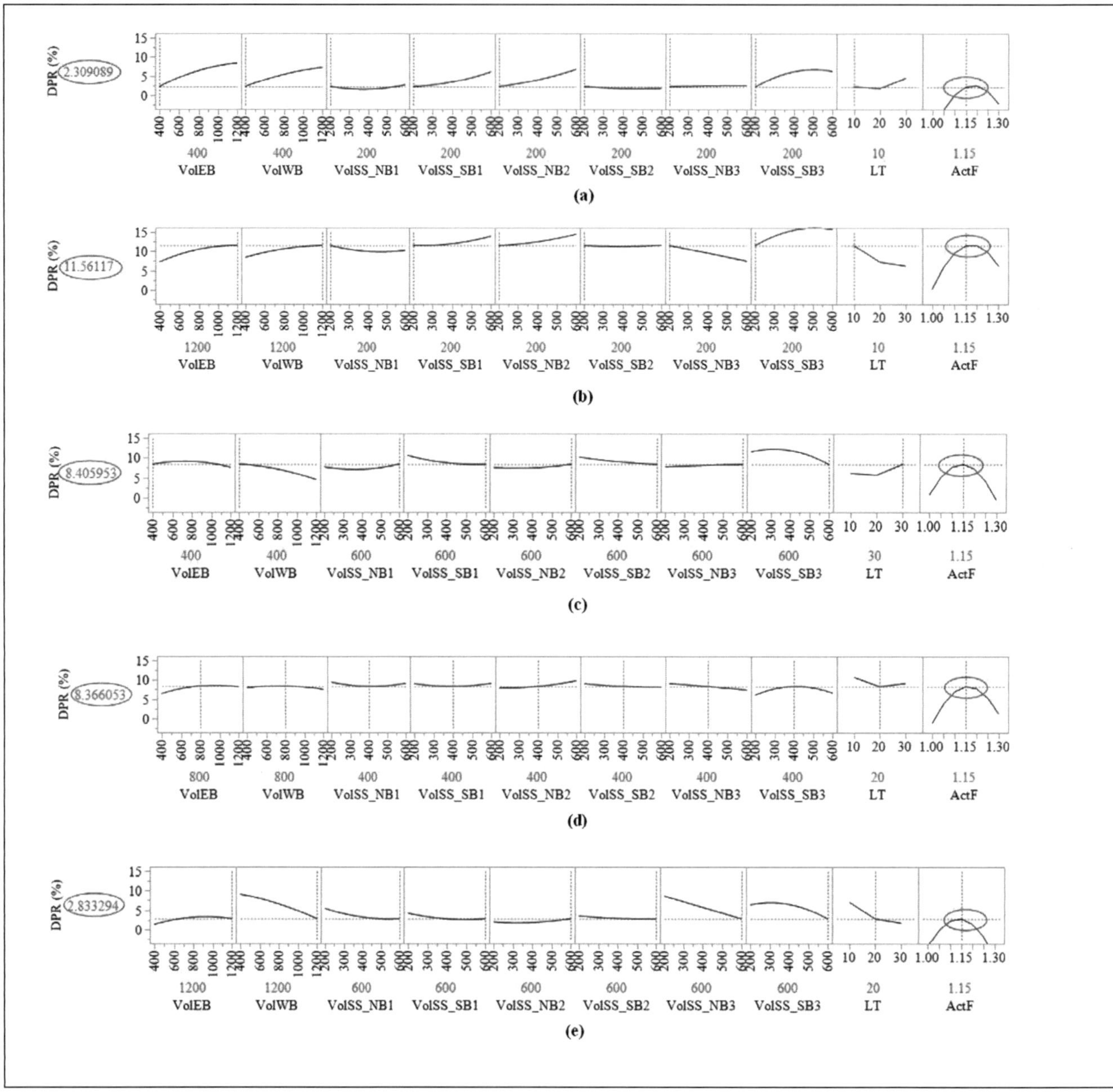

Figure 3. Prediction profilers for all the explanatory variables: (*a*) minimum volumes in all approaches; (*b*) maximum volumes on the major street and minimum volumes on the side streets; (*c*) minimum volumes on the major street and maximum volumes on the side street; (*d*) moderate volumes in all approaches; (e) maximum volumes in all approaches.

parameters associated with one particular effect in the model are zero; otherwise, the parameter would have the significant effect in the model. Results of the *F*-test demonstrate that the only significant factor affecting the percentage of delay reduction is the quadratic term of ActF. A very small *p*-value indicates that the term $(\text{ActF} - 1.15)^2$ is highly significant even for the significance level of 0.01, as opposed to all combinations of the traffic volumes and the left-turn rate variables that are insignificant.

To better realize and visualize the model and the interesting findings regarding the existence of an optimal value for the actuated factor, Figure 3 illustrates the prediction profilers for all explanatory variables. The prediction profile of the explanatory variable X depicts how the response variable varies when the X variable changes, as shown with black lines in Figure 3. The vertical and horizontal red-dotted lines depict the current values for the explanatory variable X and the response variable.

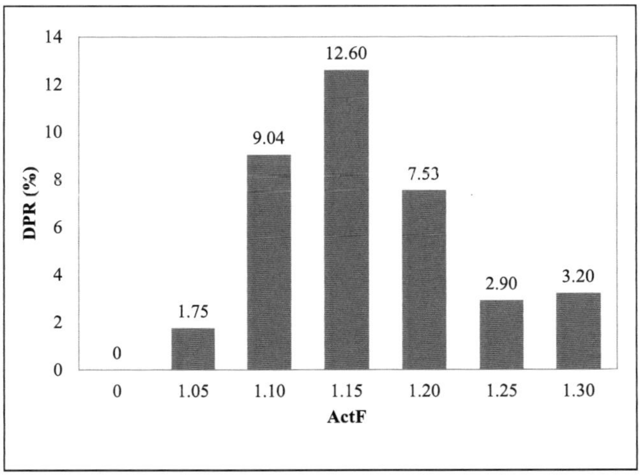

Figure 4. Average DPR(%) versus Actuated Factor (ActF).

Figure 3 includes five different combinations of traffic volumes on the major and side streets as examples of various traffic conditions. Figure 3, *a* and *e* show the minimum and maximum traffic volumes of all external movements, respectively. Figure 3*b* presents the maximum traffic volumes for the major street and the minimum traffic volumes for the side streets, and the traffic combination of Figure 3*c* is the other way around of Figure 3*b*. Finally, Figure 3*d* is a representative of medium-level traffic volumes in all approaches. The region of the optimal ActF has been highlighted in the very right block of Figure 3, *a–e*. Notwithstanding that the traffic fluctuations considerably affect the basic timing plans (i.e., the Base TP), they have a negligible impact on the optimal value of ActF. As can be seen in Figure 3, *a–e*, the optimal factor is either 1.15 or very close to 1.15 for our test site. Comparing DPR with ActF = 1.15 and the exact optimal ActF indicates the delay reduction for the ActF of 1.15 and the exact optimal ActF is nearly the same. Therefore, it can be concluded that the optimal ActF value of 1.15 is appropriate for a wide range of traffic conditions. Furthermore, the value of 1.15 is not a large value to decrease the green split of the coordinated phases beyond their minimum green time.

Another interesting finding is that the proposed method is more efficient for the arterials with medium traffic volumes, (i.e., neither operating under the congested nor free-flow conditions). Accordingly, the highest values of delay reduction (red values highlighted with green ovals) correspond to traffic patterns in Figure 3, *b–d*. Nonetheless, with regards to Figure 3, *a* and *e*, the proposed method is less efficient when both the side and the major streets are subjected to either low or high levels of traffic volume.

It is also of interest to examine how the mean of DPR is distributed among different values of ActF. To this end, DPR is plotted against ActF in Figure 4. Similar to the pattern noticed in Figure 3, the average of DPR is maximized on ActF = 1.15, suggesting that this is the value that works best for different traffic conditions for our test site.

Conclusion

In this paper, we proposed an efficient and cost-effective procedure to improve the traffic arterial performance operating under the semi-actuated coordinated signal control method. The proposed method only changes optimal green splits while keeping the cycle length and offsets unchanged. The non-coordinated phases are multiplied by ActF, which is a factor greater than one. In the meantime, the extra added amount of time in the non-coordinated phases is deducted from the coordinated phases to keep the cycle length constant. The statistical analysis indicated that the ActF of 1.15 is a reasonable value for all traffic conditions in our study site. Our method does not require either installation of detectors for the coordinated movements or dynamic optimization systems. Furthermore, by having the optimal value of ActF in hand, implementing our method is simple enough to be adopted by practitioners.

However, the scope of this study was limited in terms of an arterial with specific geometry conditions, (e.g., the number of lanes and the number of intersections), and pre-defined traffic characteristics, (e.g., traffic mode, speed distribution, and PHF). Moreover, in terms of traffic flow conditions, our reported results are applicable as long as traffic volumes in major and minor approaches do not exceed the maximum traffic volumes in Table 1. Future studies need to be carried out to investigate the robustness of the proposed method under a variety of traffic conditions and multi-modal traffic situations. Since considering pedestrians adds more complexity to the semi-actuated coordinated signal control, accounting for pedestrian calls while using our proposed method would be an interesting research area to explore in the future. Moreover, developing a universal regression model that can report the optimal value of the ActF in all traffic and network conditions would be of utmost interest.

Author Contributions

The authors confirm contribution to the paper as follows: Research Supervision: Dr. Montasir Abbas; study conception and design: Sina Dabirir, Dr. Montasir Abbas; data collection: Sina Dabiri; analysis and interpretation of results: Sina Dabiri, Kianoush Kompany; draft manuscript preparation: Sina

Dabiri, Kianoush Kompany. All authors reviewed the results and approved the final version of the manuscript.

References

1. Koonce, P. *Traffic Signal Timing Manual*, 2nd ed. Federal Highway Administration, 2008.
2. Chang, E.C.-P, and J. Koothrappally. Field Verification of Coordinated Actuated Control. *Transportation Research Record: Journal of the Transportation Research Board*, 1994. 1456: 83.
3. Park, B. B., Y. Chen, H. Cho, and I. Yun. Quantifying the Benefits of Adaptive Split Feature on the Operation of Coordinated Actuated Traffic Signal System. *KSCE Journal of Civil Engineering*, Vol. 19, No. 1, 2015, pp. 311–317.
4. Dabiri, S., and K. Heaslip. Inferring Transportation Modes from GPS Trajectories Using a Convolutional Neural Network. *Transportation Research Part C: Emerging Technologies*, Vol. 86, 2018, pp. 360–371.
5. Dabiri, S, and M. Abbas. Evaluation of the Gradient Boosting of Regression Trees Method on Estimating the Car following Behavior. *Transportation Research Record: Journal of the Transportation Research Board*, accepted for publication, 2018.
6. Jovanis, P. P, and J. A. Gregor. Coordination of Actuated Arterial Traffic Signal Systems. *Journal of Transportation Engineering*, Vol. 112, No. 4, 1986, pp. 416–432.
7. Chang, E. Guidelines for Actuated Controllers in Coordinated Systems. *Transportation Research Record: Journal of the Transportation Research Board*, 1996. 1554: 61–73.
8. Messer, C. J, and R. V. Nageswara. *Improved Traffic Signal Coordination Strategies for Actuated Control*. Southwest Region University Transportation Center, Center for Transportation Research, University of Texas, 1996.
9. Chaudhary, N. A, and K. N. Balke. *Real-time Coordinated-actuated Traffic Control During Congested Conditions*, Report FHWA/TX-99/1288-S. Texas Transportation Institute, 1997.
10. Abbas, M., Z. Adam, and D. Gettman. Development and Evaluation of Optimal Arterial Control Strategies for Oversaturated Conditions. *Transportation Research Record: Journal of the Transportation Research Board*, 2011. 2259: 242–252.
11. He, Q., K. L. Head, and J. Ding. Multi-modal Traffic Signal Control with Priority, Signal Actuation and Coordination. *Transportation Research Part C: Emerging Technologies*, Vol. 46, 2014, pp. 65–82.
12. Lin, D., N. Wu, T. Zong, and D. Mao. (eds). *Modeling the Impact of Side-street Traffic Volume on Major-street Green Time at Isolated Semi-actuated Intersections for Signal Coordination Decisions*. No. 16-1308, 2016.
13. Chia, I., X. Wu, S. S. Dhaliwal, J. Thai, and X. Jia. Evaluation of Actuated, Coordinated, and Adaptive Signal Control Systems: A Case Study. *Journal of Transportation Engineering, Part A: Systems*, Vol. 143, No. 9, 2017. 05017007.
14. Abbas, M., D. Bullock, and L. Head. Real-time Offset Transitioning Algorithm for Coordinating Traffic Signals. *Transportation Research Record: Journal of the Transportation Research Board*, 2001. 1748: 26–39.
15. Yin, Y., M. Li, and A. Skabardonis. Offline Offset Refiner for Coordinated Actuated Signal Control Systems 1. *Journal of Transportation Engineering*, Vol. 133, No. 7, 2007, pp. 423–432.
16. Lee, J. Phase Force-off Modes in Coordinated-Actuated Signal Operations. *Procedia Computer Science*, Vol. 83, 2016, pp. 600–606.
17. Sun, X. L., T. Urbanik, and L. D. Han. (eds). Neurofuzzy Control to Actuated-coordinated System at Closely-spaced Intersections. *Applied Mechanics and Materials*, Vol. 321, 2013 pp. 1249–1258.
18. Dabiri, S, and M. Abbas. (eds). *Arterial Traffic Signal Optimization using Particle Swarm Optimization in an Integrated VISSIM-MATLAB Simulation Environment*. 19th International Conference on Intelligent Transportation Systems (ITSC), Rio de Janeiro, 2016. IEEE.
19. Li, P., M. Abbas, R. Pasupathy, and L. Head. Simulation-based Optimization of Maximum Green Setting Under Retrospective Approximation Framework. *Transportation Research Record: Journal of the Transportation Research Board*, 2010. 2192: 1–10.
20. Yun, I, and Park, B. Stochastic Optimization for Coordinated Actuated Traffic Signal Systems. *Journal of Transportation Engineering*, Vol. 138, No. 7, 2012 pp. 819–829.
21. Park, B, and J. Lee. Optimization of Coordinated-actuated Traffic Signal System: Stochastic Optimization Method Based on Shuffled Frog-leaping Algorithm. *Transportation Research Record: Journal of the Transportation Research Board*, 2009. 2128: 76–85.
22. *Highway Capacity Manual*. TRB, National Research Council, Washington, D.C., 2000.
23. Inc SI. *JMP Pro 11 Design of Experiments Guide*. SAS Institute Inc., Cary, NC, 2013.
24. Dowling, R., A. Skabardonis, and V. Alexiadis. *Traffic Analysis Toolbox Volume III: Guidelines for Applying Traffic Microsimulation Modeling Software*, Federal Highway Adminstration, 2004.

The Standing Committee on Traffic Signal Systems (AHB25) peer-reviewed this paper (18-02770).

TRR
JOURNAL OF THE TRANSPORTATION RESEARCH BOARD

Article

Transportation Research Record
2018, Vol. 2672(18) 81–92

DOI: 10.1177/0361198118786842
journals.sagepub.com/home/trr
SAGE

Signal Timing Optimization with Connected Vehicle Technology: Platooning to Improve Computational Efficiency

Xiao (Joyce) Liang[1], S. Ilgin Guler[1], and Vikash V. Gayah[1]

Abstract

This paper develops a real-time traffic signal optimization algorithm in the presence of connected and autonomous vehicles (CAVs). The proposed algorithm leverages information from connected vehicles (CVs) arriving at an intersection to identify naturally occurring platoons that consist of both CVs and non-CVs. Signal timings are then selected to optimize the sequence at which these platoons are allowed to discharge through the intersection to minimize total vehicle delay. Longitudinal trajectory guidance that explicitly accounts for vehicle acceleration and deceleration behavior is provided to the lead autonomous vehicle (AV) in any platoon to minimize the total number of stopping maneuvers performed by all vehicles. Simulation tests reveal that the proposed platoon-based algorithm provides superior computational savings (over 95%) compared with a previously developed algorithm that focuses on optimizing departure sequences of individual vehicles, with negligible changes in operational performance. The computational savings allow the platoon-based algorithm to accommodate intersections with four multi-lane approaches and left turns, whereas large computational costs limited the previous vehicle-based algorithm to only two single-lane approaches without conflicting left turns. Additional simulation tests of the platoon-based algorithm on these more realistic intersection configurations show that intersection performance increases as the penetration rate of CAVs in the vehicle fleet increases. However, the marginal benefits decrease rapidly after the fleet is composed of 40% CAVs.

Traffic signals are large contributors to travel delays, vehicle emissions, and wasted fuel in urban environments. There have been many efforts to optimize traffic signal timings to reduce these negative effects. A summary of contemporary signal control strategies can be found in Papageorgiou et al. (*1*). Recent advances in vehicular, communications, and mobile computing technologies offer the potential for even more flexible traffic signal control paradigms. New communication technologies allow vehicles to collect and share more detailed information about existing traffic patterns. The introduction of autonomous vehicles (AVs) will also allow vehicles and signals to cooperate as a means to improve signalized intersection operations.

One recent line of research has proposed using real-time information from connected and automated vehicles (CAVs) to minimize total vehicle delay at signalized intersections by optimizing individual vehicle departure sequences (*2*) while also minimizing total vehicle stopping maneuvers by modifying the longitudinal trajectories of AVs (*3*). However, these algorithms were developed only for intersections with two single-lane approaches that could be controlled with simple two-phase signals and did not have conflicting left turns. Extension to more realistic intersections (e.g., those with four approaches, multiple lanes on each approach, or conflicting left-turning maneuvers) with more vehicles and phasing options was not possible due to the computational demands imposed.

The present paper proposes an algorithm that uses CV information to identify platoons of vehicles arriving or queuing at the intersection rather than individual vehicles. The signal control algorithm then identifies the optimal sequence that these platoons are allowed to discharge through the intersection to minimize total vehicle delay. Longitudinal trajectory guidance that explicitly accounts for vehicle acceleration and deceleration behavior is then provided only to the lead AV in any platoon to minimize the total number of stopping maneuvers performed by all vehicles. The platoon-based algorithm significantly improves upon the computational demands of existing strategies that rely on

[1]Department of Civil and Environmental Engineering, The Pennsylvania State University, University Park, PA

Corresponding Author:
Address correspondence to Vikash V. Gayah: gayah@engr.psu.edu

determining individual vehicle departure sequences, allowing it to accommodate more realistic intersections comprised of four multi-lane approaches and left turns.

The remainder of the paper is organized as follows. First, a summary of the literature is provided. Next, the algorithm used in this paper is presented. Then, results of simulation tests on simplistic intersection configurations are discussed. Next, the algorithm is expanded to accommodate more realistic intersection configurations. Finally, some concluding remarks are made.

Literature Review

Many studies have provided in-depth summaries of existing traffic signal control strategies. Florin and Olariu (*4*) categorized these strategies based on the type of information used as an input to signal control. Li et al. (*5*) instead considered the control philosophies employed by various traffic signal systems. Finally, Rios-Torres and Malikopoulos (*6*) provided a review of the research related to the coordination of CAVs.

With regard to signal control in a connected environment, several studies propose methods to use information on individual vehicle arrivals, speeds, and locations to optimize signal timing and phasing plans (*7–16*). Each of these has considered varying objectives, including vehicle/person/platoon delay (*8–12*), total queue length (*13*), cumulative travel time (*14–16*), and a combination of various criteria (*7, 10*). Various methods of CV-based adaptive traffic control systems were proposed to optimize the cycle length (*7*), order of phases (*9, 10, 13*), green splits (*8, 11, 15*), or offsets (*7, 8, 16*) based on the previous objectives. Most of these studies considered different traffic demands and market penetration rates of connected vehicles and concluded that CV-based control schemes out-perform traditional actuated signal control. Table 1 provides a summary of this literature in terms of penetration rates of CVs considered.

Other studies have utilized CV information to optimize discharge sequences of individual vehicles. Common objective functions considered in this method are the minimization of evacuation time (*17, 18*), travel time (*19*), the sum of queue lengths (*20*), and total vehicle delay (*2, 3, 21, 22*). These efforts were typically implemented for a single, isolated intersection and generally assume that all vehicles arriving at the intersection were connected. Few of these efforts specifically considered the impact of different penetration rates of CV and only two studies proposed methods to estimate arrival information for unequipped vehicles (i.e., non-CVs) (*2, 3*).

In recent years, studies of autonomously-controlled vehicles have become more prevalent due to technological advances that allow a vehicle's longitudinal trajectory to be modified by changing its speed and acceleration in real time. A summary of these studies, including the range of penetration rates that can be accommodated and the longitudinal trajectory guidance method, is also provided in Table 1. Reservation-based strategies consider requests from drivers reserving a block of space-time in the intersection and decides whether to grant or reject this request in real time. Dresner and Stone (*24*) considered this method with 100% AV penetration while a follow up study (*25*) considered both AVs and non-AVs. The tradeoff between the precision at which AVs can maintain their trajectories and the efficiency of the intersection controller was also studied (*23*). Hausknecht et al. (*26*) extended this approach from an isolated intersection to a network of interconnected intersections. Au et al. (*32*) furthered this by considering semi-autonomous vehicles with limited self-driving automation. Other, non-reservation-based strategies also exist (*3, 27–31*). In addition to not considering multimodal traffic (e.g., bicycles or pedestrians), a major limitation of most of the literature is the assumption of 100% AV penetration rate. Only three papers consider AV penetration rates lower than 100%. Two of these employ the reservation-based method, which is not compatible with existing signal controllers and is subject to significant safety concerns, while the last does not incorporate vehicle acceleration/deceleration behaviors into the longitudinal trajectory guidance process. These previous AV-based signal control strategies are also limited by the computational demands imposed during the optimization and trajectory design process which limits their practical application.

To alleviate these issues, this paper proposes a platoon-based algorithm that can significantly reduce computational complexity while maintaining operational performance. Platoon-based algorithms have previously been used in the literature for signal control (*33*); however, these previous algorithms did not consider CAV data as input. The proposed platoon-based algorithm reduces computational demands to ensure that the signal control optimization can be performed in real time and allows the algorithm to be extended to more realistic scenarios that are not otherwise possible. This study also develops a longitudinal trajectory control algorithm for AVs that are the lead vehicles of platoons while accounting for more realistic vehicle dynamics, including acceleration and deceleration behaviors. Finally, this study considers and discusses the impact of various market penetration ratios of CVs and AVs.

Signal Control Algorithm

This section describes the platoon-based signal control algorithm proposed in this paper. The methodology builds upon the vehicle-based algorithms proposed in Guler at al. (*2*) and Yang et al. (*3*). However, the proposed platoon-based algorithm provides the following unique contributions:

1. CV information is used to identify platoons of vehicles arriving or stopped at an intersection. These vehicle platoons are then used to determine the optimal signal timing plan that minimizes the total delay experienced by these platoons as they discharge through the intersection. The use of platoons over

Table 1. Summary of Literature

Paper	Penetration rate of CVs	Penetration rate of AVs	Longitudinal trajectory control method for AVs
Gradinescu et al. (*7*)	100%	N/A	N/A
Beak et al. (*8*)	25–100%	N/A	N/A
He et al. (*9*)	20–100%	N/A	N/A
Priemer and Friedrich (*13*)	10–100%	N/A	N/A
Lee et al. (*14*)	10–100%.	N/A	N/A
Goodall et al. (*10*)	10–100%	N/A	N/A
Hu et al. (*11*)	0% for autos, only buses are connected	N/A	N/A
Choi et al. (*15*)	0–100%	N/A	N/A
Day et al. (*16*)	<1%	N/A	N/A
Wu et al. (*17*)	100%	N/A	N/A
Yan et al. (*18*)	100%	N/A	N/A
Cai et al. (*19*)	100%	N/A	N/A
Ahmane et al. (*20*)	100%	N/A	N/A
Pandit et al. (*21*)	100%	N/A	N/A
Guler et al. (2)	0 ~ 100%	N/A	N/A
Stebbins et al. (22)	100%	N/A	N/A
Au and Stone (*23*)	0 ~ 100% (Semi-automated)	0 ~ 100%	Reservation-based system
Dresner and Stone (*24*)	N/A	100%	Reservation-based system
Dresner and Stone (*25*)	N/A	0, 90, 95, 99, and 100%	Reservation-based system
Hausknecht et al. (26)	N/A	100%	Reservation-based system
Kamal et al. (27)	N/A	100%	Vehicle–intersection coordination scheme (VICS)
Lee and Park (*28*)	N/A	100%	Cooperative vehicle–intersection control (CIVC)
Li and Wang (*29*)	N/A	100%	Enumerate all departure sequences, select the collision-free least-travel-time one
Li et al. (*30*)	N/A	100%	Multiple cases considered
Yang et al. (3)	0 ~ 100%	0 ~ 100%	Sudden change of speed
Zohdy and Rakha (*31*)	N/A	100%	Cooperative adaptive cruise control (CACC) & Game Theory

individual vehicles significantly reduces the computational demands of the signal control algorithm when compared with using individual vehicles, which allows the algorithm to be applied to intersections with higher vehicle flows, more complex geometries, and more flexible phasing schemes.

2. Computational demands are further reduced by providing longitudinal trajectory guidance for only the lead AV in each platoon (as opposed to every AV within a platoon).
3. The longitudinal trajectory guidance explicitly accounts for vehicle acceleration and deceleration and thus more realistically models vehicle dynamics near the intersection.
4. Methods are developed to extend the algorithm from highly simplistic cases (intersections with two single-lane approaches) to more realistic scenarios consisting of four multi-lane approaches with conflicting left turns. These extensions leverage the reduced computational effort provided by points 1 and 2.

The first three improvements are discussed below in the description of the updated algorithm, while the last is described in a later section. In this section, we consider the same setup used in the original vehicle-based algorithms (*2*, *3*), which consists of a signalized intersection with two single-lane approaches (see Figure 2*a*). This scenario will be extended in a later section for application to a more realistic intersection with four multi-lane approaches and conflicting left turns.

The general framework and logic of the proposed platoon-based control algorithm is graphically illustrated in Figure 1, which also demonstrates how the control algorithm relates to the simulation framework used to test it (described in the next section). The control algorithm takes the form of a bi-level optimization problem with a rolling horizon. The upper level aims to select the optimal platoon departure sequence at the intersection to minimize anticipated vehicle delays, whereas the lower level aims to minimize the total number of stopping maneuvers by providing longitudinal trajectory guidance to a subset of AVs as they pass through the

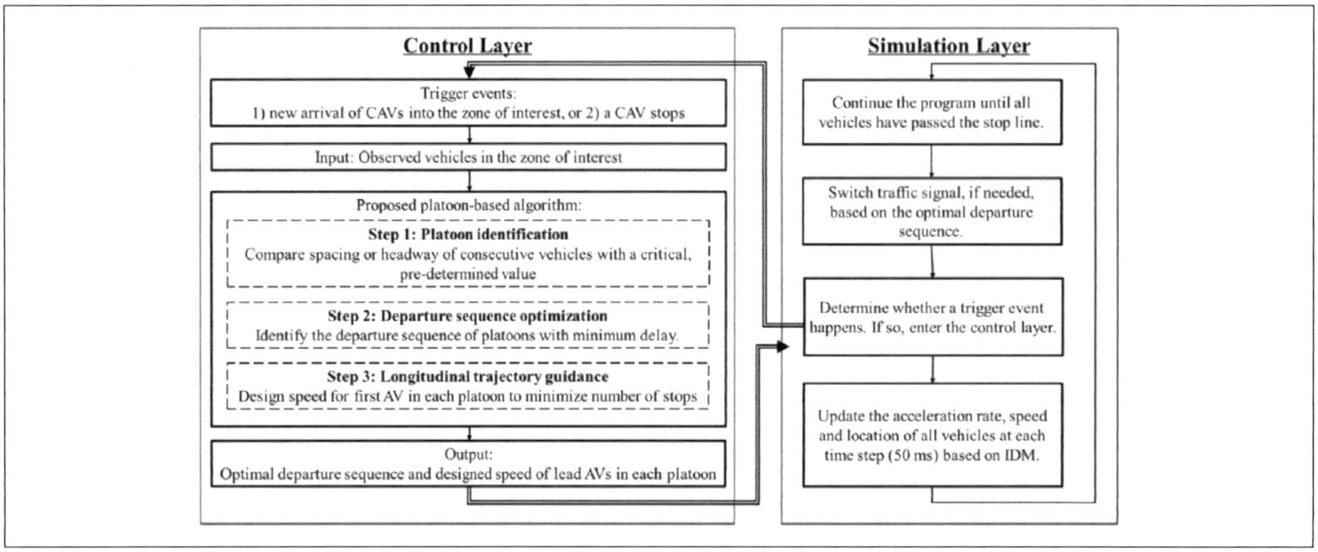

Figure 1. Flowchart of the algorithm.

intersection. Three types of vehicles are considered: (a) Traditional vehicles, which do not communicate and cannot be directly controlled; (b) CVs, which provide their location and speed to the signal controller but cannot be directly controlled; and, (c) AVs, which can both provide information to and receive trajectory guidance from the controller.

The input to the algorithm is the set of estimated virtual arrival times at the intersection stop line for all observed CVs or AVs (CAVs) within a zone of interest around the intersection (see left-hand side of Figure 2*a*). The set of vehicles and their virtual arrival times are updated every time one of two events occurs: (a) a new CAV arrives at the zone of interest; or (b) a CAV stops within the zone of interest. Information from stopped CAVs is used to identify the presence of any traditional vehicles that are stopped downstream of a stopped CAV vehicle. These traditional vehicles are identified when the distance between a stopped CAV and the intersection stop bar or nearest downstream CAV in the same lane is greater than a vehicle length, which suggests the presence of an unconnected vehicle(s). Traditional vehicles identified in this manner are then included in the set of vehicles considered in the algorithm (see middle part of Figure 2*a*).

The remainder of this section describes the steps performed to identify vehicle platoons from the individual CAV data, optimize the platoon departure sequences, and provide longitudinal trajectory guidance. These steps are repeated every time a CAV stops within or arrives at the zone of interest.

Step 1: Platoon Identification

In this step, information obtained from individual CAVs and identified traditional vehicles is used to identify vehicle platoons arriving or stopped at the intersection. Platoons are identified based on the headways and spacings of CAVs or identified traditional vehicles on each approach. Each vehicle's headway and spacing are compared with a critical, predefined threshold value. Vehicles with either headways or spacings lower than at least one of these critical threshold values are assumed to be traveling together in a platoon. The spacing threshold value is used to identify platoons of stopped vehicles already waiting at the intersection, while the headway threshold value is used to identify platoons of moving vehicles that are approaching the intersection. Both CAVs and identified traditional vehicles are considered in the platooning process; however, since only stopped traditional vehicles are identified using the CAV information, the traditional vehicles are only incorporated using the spacing threshold. Note that some identified platoons may consist of only a single vehicle. This would occur when this vehicle is traveling by itself or if the vehicle is the only CAV traveling in a group of moving traditional vehicles.

Step 2: Optimal Platoon Departure Sequence Selection

The goal of this step is to determine the sequence in which platoons are allowed to depart through the intersection such that vehicle delay is minimized. Unlike previous studies, which considered vehicle departure sequences, the potential departure sequences considered here are based only on the vehicle platoons identified in Step 1. This assumes that cars within a platoon will discharge through the intersection together. The only exception to this is when the maximum green time is triggered, in which case the platoon may be broken. Using platoons implicitly reduces the number of vehicle departure sequences considered since vehicles within one platoon must be allowed to move through the intersection before vehicles in another are allowed to pass,

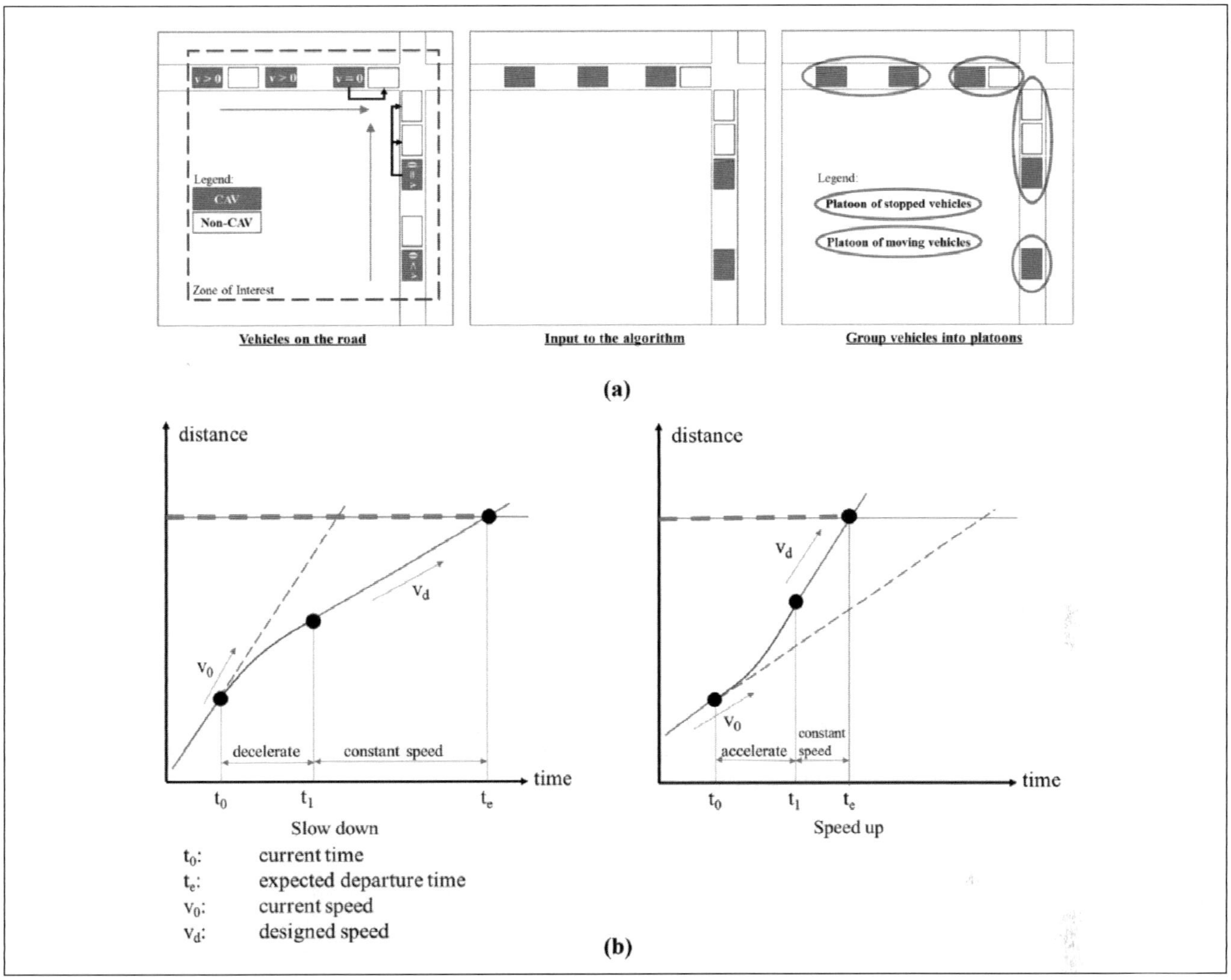

Figure 2. Illustration of: (*a*) intersection configuration and inputs to the algorithm and (*b*) trajectory design calculations.

which is more restrictive than previous algorithms. However, previous results suggest that the vehicle-based algorithm tends to naturally group vehicles together into platoons since having vehicles move through the intersection as a platoon maximizes the efficiency of the intersection and minimizes the acceleration time of stopped vehicles. Thus, considering vehicle platoons in the optimization process is not expected to have a significant effect on operational efficiency. On the other hand, the computational time required to determine the optimal departure sequence/signal timing plan increases exponentially with the number of vehicles at the intersection in the vehicle-based approach. By constraining the possible vehicle departure sequences using the identified platoons, the platoon-based approach is expected to provide significant computational advantages since fewer departure sequences need to be examined. The tradeoff between computational efficiency and operational efficiency using the proposed platooning method is examined in the following section.

The same two methods used to identify the optimal departure sequence in the previous vehicle-based algorithm (*3*) may be applied in the platoon-based algorithm. The first is an enumeration method, which simply identifies all possible combinations of platoon departure sequences. The second is a branch-and-bound method that uses an intelligent tree search strategy to identify the optimal platoon departure sequence. More details on the specifics of the branch-and-bound methodology are provided by Yang et al. (*3*). Both methods estimate delays of departure sequences, as proposed in Guler et al. (*2*), to identify the optimal departure sequence that will result in minimum delay. Both also incorporate minimum and maximum green time constraints to ensure that identified solutions are feasible to implement at the intersection. Furthermore, the signal timing reverts to a fixed-time plan with all approaches receiving the maximum green time when no CVs are present.

Step 3: Longitudinal Trajectory Guidance

The goal of this step is to provide longitudinal trajectory guidance for a subset of AVs to minimize the total number of stopping maneuvers performed at the intersection without significantly increasing vehicle delays. The trajectory design is performed for all departure combinations identified using either the enumeration or branch-and-bound algorithm in Step 2. Trajectory guidance is considered only for an AV identified as the lead of a vehicle platoon. The trajectories of other vehicles in the platoon are indirectly influenced since they must follow the lead AV. By only providing trajectory guidance to the lead AV instead of all AVs, computational effort is significantly reduced without sacrificing operational performance.

To design the trajectory of any given AV, we first calculate the time it is expected to enter the intersection (t_e) depending on the given departure sequence considered. This expected time is equal to the maximum of the start of the green time or the sum of the departure time of the last vehicle to discharge through the intersection (on any approach) and some minimum discharge headway. With this information, the optimal speed at which the vehicle needs to travel to reach the intersection by t_e is determined. Yang et al. (*3*) suggested providing similar trajectory guidance but assumed AVs could accelerate or decelerate instantaneously, which is not possible. In this work, the trajectory design method is made more practical by considering any acceleration that a vehicle must perform before reaching an optimal speed. New equations (omitted here for brevity) are developed to identify the optimal speed at which AVs should travel to depart through the intersection at t_e while accounting for acceleration or deceleration. Illustrative examples are provided in Figure 2*b*. The left-hand side of Figure 2*b* ("Slow down") illustrates the case in which an AV should decelerate to avoid arriving at the intersection before its departure time and having to stop, while the right-hand side ("Speed up") illustrates the case in which an AV should accelerate to arrive at the intersection earlier and not waste time green. The latter case occurs when a previously-controlled AV must adjust its speed in response to a new departure sequence. Note also that since AVs with designed trajectories enter the intersection at a non-zero initial speed, the delay calculations are updated to reflect this change.

Intersections with Two Single-Lane Approaches

In this section, we apply the proposed signal control algorithm to an intersection with two single-lane approaches. This is done to compare the proposed platoon-based algorithm with the vehicle-based algorithm to assess changes in computational and operational efficiency. The remainder of this section describes the simulation framework and the sensitivity analysis.

Simulation Framework

The proposed signal control algorithm was tested using a simulation framework implemented in Java. The entire framework is illustrated in Figure 1 and consists of two interacting layers: (a) the control layer, which implements the signal control algorithm described in the previous section; and, (b) the simulation layer, which emulates realistic traffic behavior.

The simulation layer generates a random set of vehicle arrivals and simulates the longitudinal movement of each vehicle using the intelligent driver model (IDM) car-following model (*34*). Within the simulation layer, the location, speed, and acceleration of each vehicle is updated at every time step (25 ms), until the vehicle completely discharges through the intersection. The simulation layer uses the optimal departure sequence from the control layer and converts it to a signal timing plan that incorporates constraints on minimum and maximum green times for each approach. An imaginary stopped vehicle is placed a distance of one minimum spacing downstream of the intersection stop line during all red signal periods to model how drivers will decelerate (and eventually stop) when approaching a red signal. All vehicles—non-CAVs, CVs, and AVs—are modeled using the IDM. The target free-flow speeds of non-CAVs and CVs are fixed at a maximum free-flow speed value while the target free-flow speed of AVs is modified dynamically based on the control layer. Specifically, any AV for which a trajectory is designed will receive the speed guidance as an input and will use this new speed as its target free-flow speed in the IDM calculations. This allows AVs to receive speed guidance while ensuring that vehicle dynamics are consistent across all vehicle types.

For the two-approach intersection tested in this section, 100 random iterations are considered with 500 unique vehicles simulated in each iteration. The total input flow (combined flow of the two approaches) is set to 1,200 veh/h and vehicles are evenly distributed across the two competing approaches, which represents an undersaturated intersection. All vehicles are assumed to be both connected and automated for this battery of tests and the enumeration optimization method is used, which provides the highest computational complexity within the control algorithm. The former assumption will be relaxed in the next section.

Sensitivity to Critical Headway and Spacing Threshold Values

This section examines the sensitivity of the proposed platoon-based signal control algorithm to the critical headway

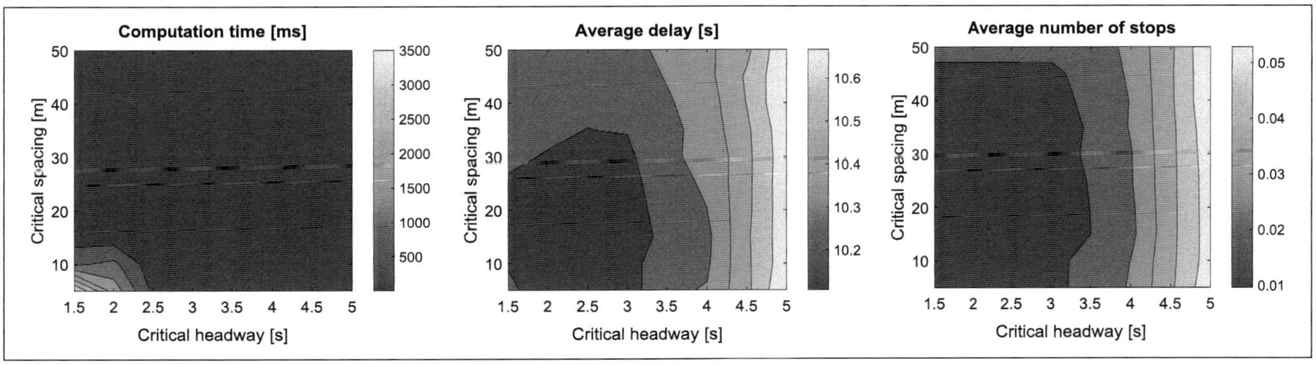

Figure 3. Contour plots of computation time, average delay, and average number of stops per vehicle as a function of critical spacing and critical headway.

and spacing threshold values used to group individual vehicles into platoons. Figure 3 provides contour plots that demonstrate how the computational time required by the signal control algorithm and operational performance (measured by the average vehicle delay and number of stopping maneuvers performed) changes with these critical threshold values. In all plots, the horizontal axis represents the critical headway threshold while the vertical axis represents the critical spacing threshold. The bottom left of each figure represents the threshold values in which each platoon consists of only an individual vehicle. For these threshold values (spacing of 5 m or headway of 1.5 s), the platoon-based control operates exactly the same as the vehicle-based control.

Notice that the computational effort required decreases as both the critical headway and critical spacing threshold increases. This occurs because larger threshold values cause vehicles to be grouped into fewer—but larger—platoons. The presence of fewer platoons results in fewer departure sequences considered in the optimization and thus significant computational savings. In this set of tests, the platoon-based optimization can reduce computation time by over 95%, when the largest thresholds are applied, compared with the vehicle-based control. However, doing so limits the flexibility of the algorithm, potentially resulting in suboptimal departure sequences compared with using individual vehicles as input to the optimization. Interestingly, the average delay seems to be rather insensitive to the critical headway or critical spacing values with a maximum increase of just about 5% when the largest thresholds are applied compared with the vehicle-based control. This confirms that the vehicle-based algorithm naturally groups vehicles together into platoons to pass through the intersection. Therefore, considering only vehicle platoons in the optimization does not significantly impact average vehicle delay but provides a noteworthy decrease in computational time required for the optimization. The number of stops seems to be more sensitive to the critical headway and spacing values and can increase by up to 300%. However, the absolute magnitude of the changes appears to be rather small.

Figure 4 provides the computation time, average delay, and average number of stops as a function of critical spacing for a fixed critical headway of 1.5 sec (left-hand side) and as a function of critical headway for a fixed critical spacing of 5 m (right-hand side). These plots represent cross-sections of the contour plots along the horizontal and vertical axes shown in Figure 3. Error bars representing one standard error are included within Figure 4 to provide further insight into the variation of these metrics across individual simulation runs. The results indicate that decreases in computational time and increases in the number of stops are statistically significant as the critical spacing increases, while differences in delays are not statistically significant. Most changes tend to be statistically significant as the critical headway increases.

Based on these results, a critical spacing of 15 m and critical headway of 2.5 sec are selected to identify platoons for all simulation results in the remainder of this paper to balance between computational and operational efficiency. However, this choice is not necessarily a recommendation; the parameters selected can be made depending on the specific computationally and efficiency needs.

Intersections with Four Multi-Lane Approaches and Conflicting Left Turns

The previous section demonstrates that using platoons significantly reduces the computational effort necessary for the proposed signal control algorithm with minimal negative impacts to operational performance. This reduction in computational effort allows the algorithm to accommodate more complex intersections that serve many more vehicles. In this section, we describe how the platoon-based algorithm can be extended to intersections that have four multilane approaches and conflicting left turns served in protected phases. We first describe the extension to four-approach intersections with multiple lanes but without left turns, and then show how conflicting left turns can be incorporated.

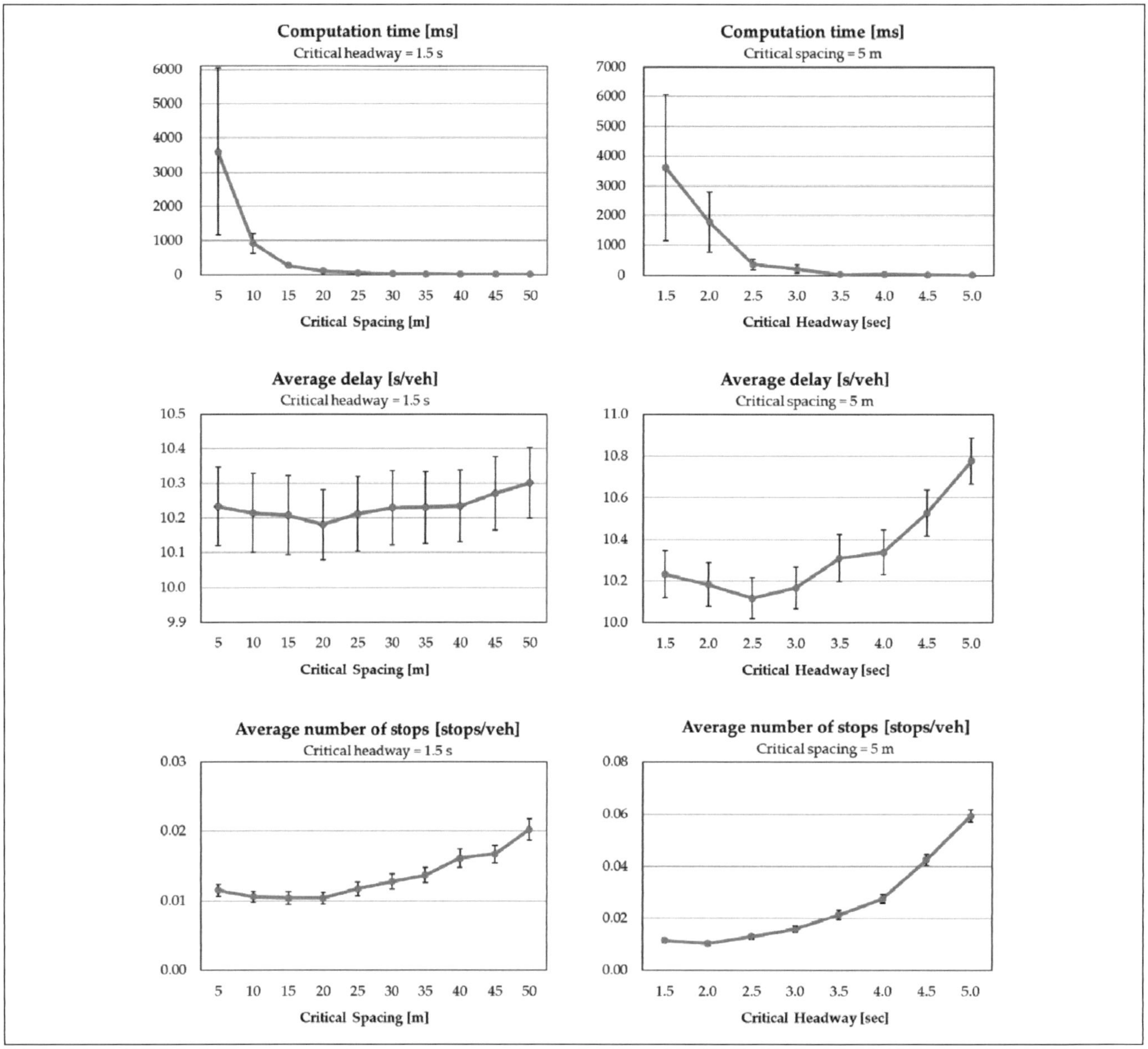

Figure 4. Sensitivity of computation time, average delay, and average number of stops to critical spacing (left) and critical headway (right).

Extensions to Accommodate More Complex Intersection Configurations

Four-approach intersections with multiple lanes per approach are accommodated by converting the more complex intersection into an equivalent, hypothetical intersection with single-lane approaches and then applying the previous algorithm. A graphical example of this conversion is presented in Figure 5*a*. First, vehicles on adjacent lanes of the same approach are modeled as if they were in the same positions, but traveling on a single lane. By only focusing on each vehicle's longitudinal position along each approach, lane changes near the intersection should be accommodated easily as long as they do not impact vehicles' longitudinal trajectories or the order in which vehicles arrive at the intersection.

The four approaches are then "folded" into two approaches, which represents the aggregation of cars that would pass through the intersection during the same phase without conflict. In this folding process, vehicles on opposite approaches are aggregated onto one approach (e.g., vehicles on the north-bound and south-bound approaches are placed onto a single, combined approach). Each vehicle is placed onto the combined approach relative to its location from the stop bar. The locations of cars in these hypothetical lanes are

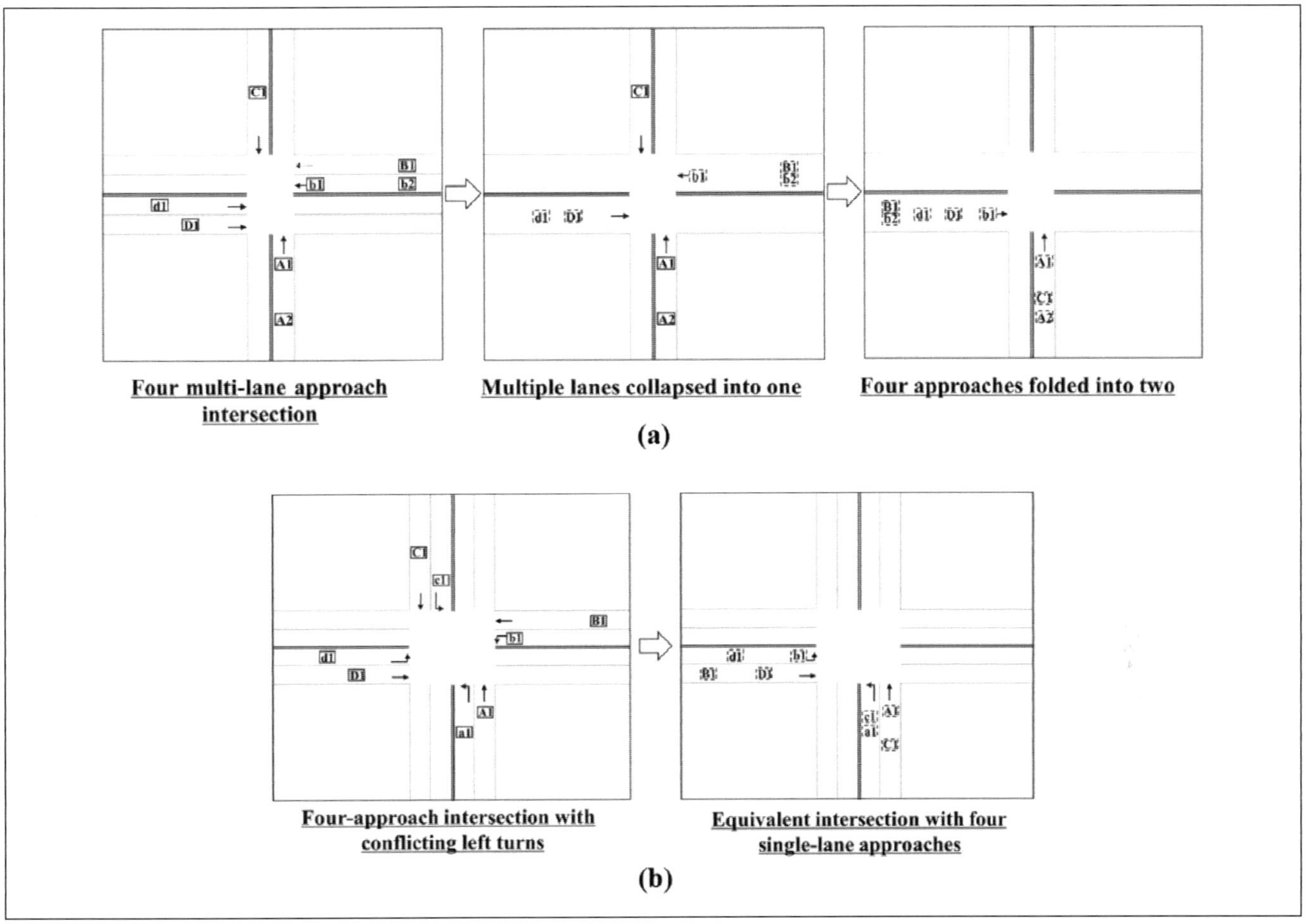

Figure 5. Illustration of hypothetical intersections created to extend algorithm to more realistic intersections: (*a*) extension to four multi-lane approaches and (*b*) extension to accommodate left turns.

the same as their real locations, hence vehicles can overlap locations on these hypothetical lanes. Hypothetical lanes are used to ensure that cars that could pass the intersection together will be considered simultaneously in the algorithm.

This hypothetical single-lane, two-approach intersection is used to identify platoons (Step 1) and determine the optimal departure sequence (Step 2). However, the trajectory guidance (Step 3) and delay calculation is performed using the original, multi-lane, four-approach intersection to ensure that the vehicle dynamics are accurately captured.

Conflicting left turns are only accommodated within this framework when left turns are physically separated from through-moving vehicles (i.e., have their own designated lane near the intersection) and are served in a protected fashion. In this case, four possible signal phases exist: (a) north–south through,(b) east–west through, (c) north–south left turn, and (d) east–west left turn. A hypothetical intersection with four single-lane approaches is used to describe this situation; see Figure 5*b*. Each of these hypothetical approaches represents a single movement/phase (either left-turning or through-moving for each of the two conflicting directions). To create this hypothetical intersection, both opposite through movements and opposite left-turning movements are "folded" in a similar manner as described above. The optimization procedure is then expanded to consider four potential phases instead of two.

Simulation Tests

This section examines how well the platoon-based strategy performs on these more complex intersection configurations.

Operational Efficiency under Constant Demand. We now apply the platoon-based optimization to an intersection with conflicting left turns to examine its performance for various CAV penetration rates. In this set of tests, the total input flow to the intersection (combined flow of all four approaches) is set to 2,400 veh/h (representing an undersaturated intersection) and vehicles are evenly distributed across all approaches. A randomly selected 20% of the vehicles on each approach are assumed to turn left. Six CV and AV penetration rates (0.0, 0.2, 0.4, 0.6, 0.8 and 1.0) were considered and the

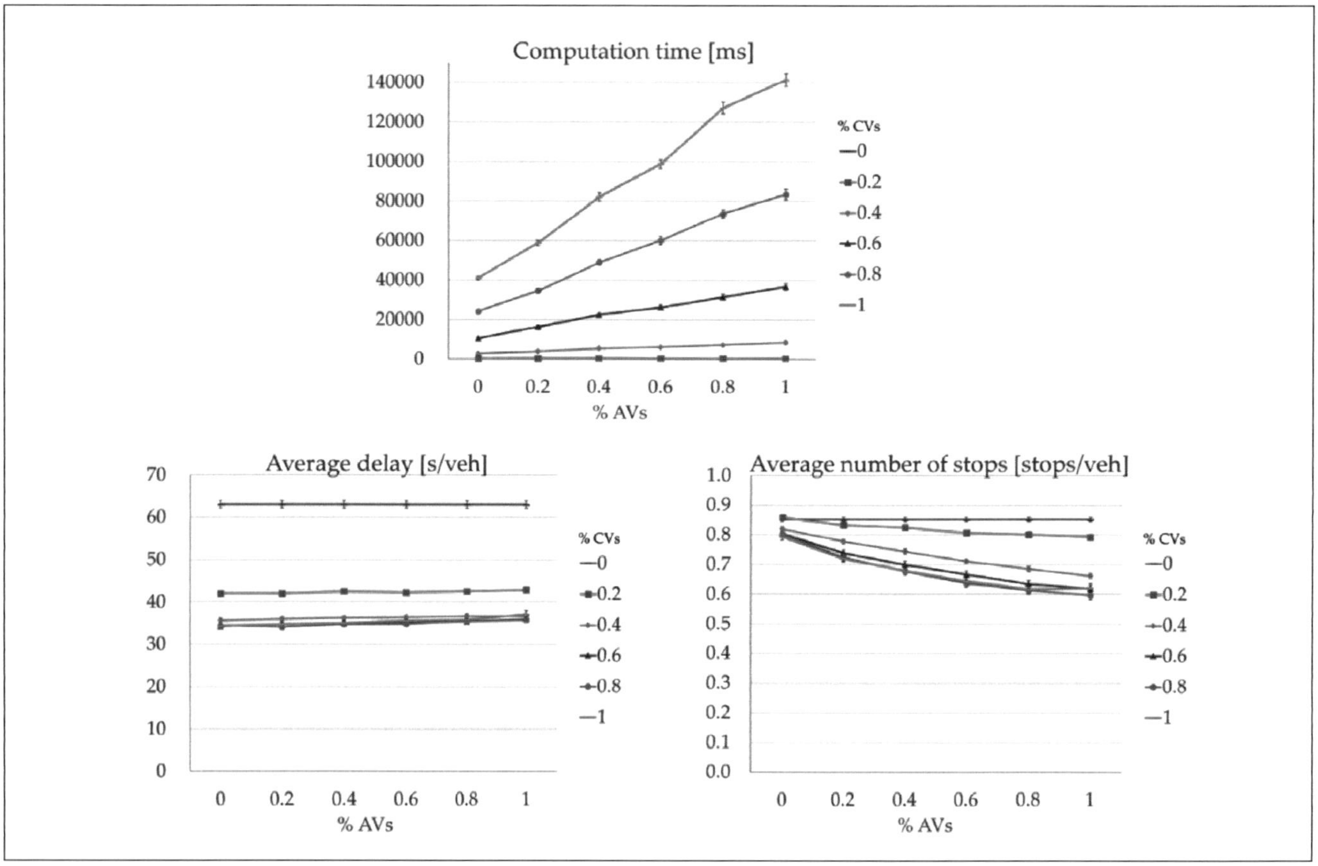

Figure 6. Sensitivity of algorithm to penetration ratios (undersaturated case).

algorithm repeated for each of the 36 combinations of CV/AV penetration rates. Note that the penetration of AVs refers to the proportion of CVs that are also autonomous. Figure 6 illustrates the computation time, average individual delay, and average number of stops for all combinations of CAV penetration rates when the proposed platoon-based signal control algorithm is applied.

Figure 6 reveals that computation time increases with the penetration ratio of CAVs. This is expected since a larger number of CAVs triggers more events and causes the control algorithm to be applied more frequently. A larger number of CAVs also increases the number of vehicles considered in the optimization, which further increases computational requirements. Both average vehicle delay and the number of stopping maneuvers generally decrease with increase in the penetration ratio of CVs. With higher penetration rates of CVs, more information is provided to the control algorithm, which results in departure sequences with lower vehicle delays. However, the marginal benefits diminish once the CV penetration rate exceeds 0.4. For higher CV penetration rates, platoons are already well-identified and additional information does not much improve algorithm performance. Similarly, as the penetration rate of AVs increases, trajectory guidance for them can provide significant reductions in number of stops. It is interesting to note that the delay increases slightly when AV trajectory guidance is provided since the AVs must enter the intersection at a lower speed; overall, however, this effect seems to be minimal.

Operational Efficiency under Fluctuating Demand with Oversaturated Period. The performance of the proposed algorithm was also tested in a fluctuating demand case in which the intersection experienced periods of oversaturation. The results, omitted here for brevity, follow the same general trends as shown in Figure 6. Examination of the algorithm performance reveals that the operational benefits are almost entirely obtained during the periods when the intersection is undersaturated. During the periods of oversaturation, the optimization algorithm tends toward a fixed-time strategy in which oversaturated approaches receive the maximum green time. These results suggest that the proposed algorithm is most effective when the intersection is undersaturated.

Computational Efficiency. Simulation tests were also performed to compare the computational effort required by the proposed platoon-based control scheme with the vehicle-based approach. For these tests, only four-approach intersections with single lanes and no conflicting left turns were considered. The inclusion of left turns was not computationally feasible for the vehicle-based method used for comparison, which provides further evidence of the computational benefits of the proposed platoon-based method. Multiple (50) random iterations were performed, each with 1,000 simulated vehicles at the intersection. Both the CV and AV penetration rates were assumed to be 50% and an undersaturated flow of 2,400 veh/h was assumed.

The computational time was obtained for three different optimization methods: (M1) vehicle-based control with trajectory guidance provided to all AVs; (M2) platoon-based control with trajectory guidance provided to all AVs; and (M3) platoon-based control with trajectory guidance for only the lead AV in each platoon. The simulation tests reveal that the platoon-based optimization method M2 reduced the computational effort required by three orders of magnitude (from approximately 37,000 ms to 63 ms) compared with the vehicle-based method M1. Furthermore, providing trajectory guidance only to the lead AV in each platoon (method M3) further reduced the computational effort by an additional 20%. The average vehicle delay was insensitive to the optimization method used (platoon- or vehicle-based). Thus, the platoon-based strategy reduced computational effort without significant increases in delay.

Conclusion

This paper develops a real-time traffic signal optimization algorithm in the presence of CAVs. The proposed algorithm identifies platoons of vehicles (including both CAVs and traditional vehicles) using information available from the CAVs only. The signal timing pattern is then determined to optimize the sequences in which these vehicle platoons are able to discharge through the intersection. The algorithm uses similar logic to a previously proposed algorithm that optimizes the discharge sequence of individual vehicles. However, the platoon-based algorithm provides significant computational benefits by first grouping vehicles into platoons because fewer departure sequences need to be examined to identify the optimal sequence. The algorithm also provides longitudinal trajectory guidance to a subset of AVs (specifically, those leading any platoons) to reduce the total number of stopping maneuvers performed at the intersection. This provides further computational benefits compared with existing algorithms that provide longitudinal trajectory guidance to all AVs. These computational savings allow the platoon-based algorithm to accommodate more realistic intersections that serve more vehicles than the vehicle-based algorithm. Specifically, the vehicle-based algorithm was proposed only for intersections with two single-lane approaches while the platoon-based algorithm can be readily extended to accommodate intersections with four multi-lane approaches and conflicting left turns.

The computational and operational efficiency of the proposed algorithm are tested using a simulation platform that emulates driver behavior using the IDM. The simulation results reveal that the platoon-based algorithm provides significant reductions in computational time over existing vehicle-based methods with minimal changes to average vehicle delay or number of stopping maneuvers. Furthermore, the algorithm provides larger and more significant operational benefits as the penetration rate of CAVs increases. However, the marginal benefits are much smaller after the fleet is composed of 40% CAVs, since platoons are well-identified at this penetration rate. Furthermore, the operational benefits are primarily realized during periods of undersaturation.

The algorithm is limited in that left turns are only accommodated when they are served in protected phases and from dedicated left-turn lanes. Future extensions will examine more flexible phasing schemes and left turns served from shared lanes. Additional tests will also be performed to determine how well the algorithm performs on an arterial when multiple intersections are closely spaced. Finally, extensions will be considered to improve performance during periods of oversaturation.

Author Contributions

The authors confirm contribution to the paper as follows: All authors participated in the study conception and design, analysis and interpretation of results, and draft manuscript preparation. All authors reviewed the results and approved the final version of the manuscript.

References

1. Papageorgiou, M., C. Diakaki, V. Dinopoulou, A. Kotsialos, and Y. Wang. Review of Road Traffic Control Strategies. *Proceedings of the IEEE*, Vol. 91, No. 12, 2003, pp. 2043–2067.
2. Guler, S. I., M. Menendez, and L. Meier. Using Connected Vehicle Technology to Improve the Efficiency of Intersections. *Transportation Research Part C: Emerging Technologies*, Vol. 46, 2014, pp. 121–131.
3. Yang, K., S. I. Guler, and M. Menendez. Isolated Intersection Control for Various Levels of Vehicle Technology: Conventional, Connected, and Automated Vehicles. *Transportation Research Part C: Emerging Technologies*, Vol. 72, 2016, pp. 109–129.
4. Florin, R., and S. Olariu. A Survey of Vehicular Communications for Traffic Signal Optimization. *Vehicular Communications*, Vol. 2, 2015, pp. 70–79.
5. Li, L., D. Wen, and D. Yao. A Survey of Traffic Control with Vehicular Communications. *IEEE Transactions on Intelligent Transportation Systems*, Vol. 15, No. 1, 2014, pp. 425–432.
6. Rios-Torres, J., and A. Malikopoulos. A Survey on the Coordination of Connected and Automated Vehicles at Intersections and Merging at Highway On-Ramps. *IEEE Transactions on Intelligent Transportation Systems*, Vol. 18, No. 5, 2017, pp. 1066–1077.
7. Gradinescu, V., C. Gorgorin, R. Diaconescu, V. Cristea, and L. Iftode. Adaptive Traffic Lights Using Car-to-Car Communication. *Proc., 65th IEEE Vehicular Technology*, Dublin, Ireland, IEEE, New York, 2007, pp. 21–25.

8. Beak, B., K. L. Head, and Y. Feng. Adaptive Coordination Based on Connected Vehicle Technology. *Transportation Research Record: Journal of the Transportation Research Board*, 2017. 2619: 1–12.
9. He, Q., K. L. Head, and J. Ding. PAMSCOD: Platoon-Based Arterial Multi-Modal Signal Control with Online Data. *Transportation Research Part C: Emerging Technologies*, Vol. 20, 2012, pp. 164–184.
10. Goodall, N., B. Smith, and B. B. Park. Traffic Signal Control with Connected Vehicles. *Transportation Research Record: Journal of the Transportation Research Board*, 2013. 2381: 65–72.
11. Hu, J., B. B. Park, and Y. J. Lee. Coordinated Transit Signal Priority Supporting Transit Progression under Connected Vehicle Technology. *Transportation Research Part C: Emerging Technologies*, Vol. 55, 2015, pp. 393–408.
12. Yang, K., M. Menendez, and S. I. Guler. Multimodal Traffic Control using Connected Vehicle Technology. *Transportmetrica Part B*, 2018. DOI: 10.1080/21680566.2018.1434019.
13.Priemer, C., and B. Friedrich. A Decentralized Adaptive Traffic Signal Control using V2I Communication Data. *Proc., IEEE International Conference on Intelligent Transportation Systems*, St. Louis, Mo., IEEE, New York, 2009, pp. 1–6.
14. Lee, J., B. Park, and I. Yun. Cumulative Travel-Time Responsive (CTR) Real-Time Intersection Control Algorithm under the Connected Vehicle Environment. *Journal of Transportation Engineering*, Vol. 139, No. 10, 2013, pp. 1020–1029.
15. Choi, S., B. B. Park, J. Lee, H. Lee, and S. H. Son. Field Implementation Feasibility Study of Cumulative Travel-Time Responsive (CTR) Traffic Signal Control Algorithm. *Journal of Advanced Transportation*, Vol. 50, No. 8, 2016, pp. 2226–2238.
16. Day, C. M., H. Li, L. Richardson, J. Howard, T. Platte, J. R. Sturdevant, and D. M. Bullock. Detector-Free Optimization of Traffic Signal Offsets with Connected Vehicle Data. *Transportation Research Record: Journal of the Transportation Research Board*, 2017. 2620: 54–68.
17. Wu, J., A. Abbas-Turki, A. Correia, and A. El Moudni. Discrete Intersection Signal Control. *Proc., IEEE International Conference on Service Operations and Logistics, and Informatics*, Philadelphia, Pa., IEEE, New York, 2007, pp. 1–6.
18. Yan, F., M. Dridi, and A. El Moudni. Autonomous Vehicle Sequencing Algorithm at Isolated Intersections. *Proc., IEEE International Conference on Intelligent Transportation Systems*, St. Louis, Mo., IEEE, New York, 2009, pp. 1–6.
19.Cai, C., Y. Wang, and G. Geers. Adaptive Traffic Signal Control Using Wireless Communications. Presented at 91st Annual Meeting of the Transportation Research Board, Washington, D.C., 2012.
20.Ahmane, M., A. Abbas-Turki, F. Perronnet, J. Wu, A. E. Moudni, J. Buisson, and R. Zeo. Modeling and Controlling an Isolated Urban Intersection based on Cooperative Vehicles. *Transportation Research Part C: Emerging Technologies*, Vol. 28, 2013, pp. 44–62.
21. Pandit, K., D. Ghosal, H. M. Zhang, and C. N. Chuah. Adaptive Traffic Signal Control with Vehicular Ad Hoc Networks. *IEEE Transactions on Vehicular Technology*, Vol. 62, 2013, pp. 1459–1471.
22. Stebbins, S., J. Kim, M. Hickman, and H. L. Vu. Combining Model Predictive Intersection Control with Green Light Optimal Speed Advisory in a Connected Vehicle Environment. Presented at 96th Annual Meeting of the Transportation Research Board, Washington, D.C., 2017.
23.Au, T. C., and P. Stone. Motion Planning Algorithms for Autonomous Intersection Management. *Proc., AAAI 2010 Workshop on Bridging the Gap between Task and Motion Planning*, 2010.
24. Dresner, K., and P. Stone. Multiagent Traffic Management: A Reservation-Based Intersection Control Mechanism. *Proc., 3rd International Joint Conference on Autonomous Agents and Multiagent Systems*, IEEE, New York, Vol. 2, 2004, pp. 530–537.
25.Dresner, K., and P. Stone. Human-Usable and Emergency Vehicle-Aware Control Policies for Autonomous Intersection Management. *Proc., 4th International Workshop on Agents in Traffic and Transportation (ATT)*, Hakodate, Japan, 2006.
26Hausknecht, M., T. C. Au, and P. Stone. Autonomous Intersection Management: Multi-Intersection Optimization. *Proc., IEEE/RSJ International Conference on Intelligent Robots and Systems (IROS)*, San Francisco, Calif., IEEE, New York, 2011, pp. 4581–4586.
27. Kamal, M., J. I. Imura, T. Hayakawa, A. Ohata, and K. Aihara. A Vehicle-Intersection Coordination Scheme for Smooth Flows of Traffic Without using Traffic Lights. *IEEE Transactions on Intelligent Transportation Systems*, Vol. 16, No. 3, 2015, pp. 1136–1147.
28.Lee, J., and B. Park. Development and Evaluation of a Cooperative Vehicle Intersection Control Algorithm under the Connected Vehicles Environment. *IEEE Transactions on Intelligent Transportation Systems*, Vol. 13, No. 1, 2012, pp. 81–90.
29. Li, L., and F. Y. Wang. Cooperative Driving at Blind Crossings using Intervehicle Communication. *IEEE Transactions on Vehicular Technology*, Vol. 55, No. 6, 2006, pp. 1712–1724.
30Li, Z., L. Elefteriadou, and S. Ranka. Signal Control Optimization for Automated Vehicles at Isolated Signalized Intersections. *Transportation Research Part C: Emerging Technologies*, Vol. 49, 2014, pp. 1–18.
31.Zohdy, I. H., and H. Rakha. Game Theory Algorithm for Intersection-Based Cooperative Adaptive Cruise Control (CACC) Systems. *Proc., IEEE International Conference on Intelligent Transportation Systems*, Anchorage, Alaska, IEEE, New York, 2012, pp. 1097–1102.
32. Au, T. C., S. Zhang, and P. Stone. Autonomous Intersection Management for Semi-Autonomous Vehicles. In *The Routledge Handbook of Transportation*, 2015, pp. 88–104, Routledge.
33.Xie, S., and B. Lu. Schedule-Driven Intersection Control. *Transportation Research Part C: Emerging Technologies*, Vol. 24, 2012, pp. 168–189.
34. Triber, M., A. Hennecke, and D. Helbing. Congested Traffic States in Empirical Observations and Microscopic Simulations. *Physical Review E*, Vol. 62, No. 2, 2000, pp. 1805–1824.

The Standing Committee on Traffic Signal Systems (AHB25) peer-reviewed this paper (18-01415).

Article

TRR
JOURNAL OF THE TRANSPORTATION RESEARCH BOARD

Transportation Research Record
2018, Vol. 2672(18) 93–106

DOI: 10.1177/0361198118784168
journals.sagepub.com/home/trr
SAGE

Evaluation of Multiple Hardware and Software in the Loop Signal Controllers in Simulation Environment

Sharmin-E-Shams Chowdhury[1], Aleksandar Stevanovic[1], and Nikola Mitrovic[1]

Abstract

This study evaluates two groups of methods to model traffic signal operations in microscopic simulation: hardware-in-the-loop simulation (HILS) and software-in-the-loop simulation (SILS). These methods have become standards for accurate modeling of traffic signal operations, but in spite of the large number of available options there are no studies that have conducted relevant comparative evaluations. This study bridges this gap by investigating signal timing and operational differences of these two methods in basic actuated operations of a single signalized intersection. The emphasis is given to broad examination of various platforms as opposed to more complex experiments done with individual platforms. A representative number of 65-minute simulation runs was executed for each experimental scenario. The results showed that differences between various HILS and SILS platforms are large enough that one cannot confidently switch between the platforms without affecting the final outcomes. The study confirmed previous findings about the impact of the initialization process on the simulation results, but the initialization itself does not seem to be the major source of discrepancy. Further investigation is needed to reveal role of consistency of internal NEMA-based controller logics among various controllers. These findings put a considerable dilemma/restriction on how various HILS and SILS platforms, either alone or in conjunction with other higher forms of traffic control strategies, can be used in joint fashion.

Traffic signal operations are often simulated in the context of three distinctive methods: the simulation model's controller emulator, hardware-in-the-loop simulation (HILS), and software-in the-loop simulation (SILS). Each of these methods consists of two components: a simulator of traffic flows and a generator of traffic signal states. Although the simulation of traffic flows is performed in a common way, the traffic states are generated in different manners for each platform. For instance, the simulation model's controller emulator is embedded in the simulation tool (e.g., CORSIM, VISSIM) and based on the well-known ring-barrier controller (RBC) platform that has successfully reproduced the National Electrical Manufacturers Association (NEMA) concept (*1*). On the other hand, HILS involves the presence of an actual traffic controller that is connected to the simulator of traffic flows with the help of a physical controller interface device (CID) (*2, 3*). Finally, SILS is considered as an advanced version of HILS, where the actual/physical controller is replaced with its software version (*4*).

Several manufacturers have recently developed/enabled their own HILS and SILS systems to further contribute to realistic and accurate modeling of traffic signal operations. Whereas there were only a few of these platforms available on the market just few years ago, today there are at least a dozen of these HILS and SILS systems, developed by various manufacturers. The evaluation of various traffic operations and strategies has been performed utilizing these systems individually in many studies (*5–20*). However, the cross-evaluation of the existing HILS and SILS applications is still missing. Such cross-comparison might reveal the potential operational differences (if any) among various HILS and SILS platforms. In addition, such comparison might answer the question of whether an identical set of traffic flow and signal timing inputs in various HILS and SILS platforms would lead to identical/similar outputs in terms of signal- and vehicle-based performance measures.

A goal of this study is to address the above questions by investigating operational differences of five HILS and five SILS systems by examining how each method operates in a test scenario. The evaluated HILS systems include: (i) Intelight (2070-LDX) (*21*), (ii) Econolite (Cobalt) (*22*), (iii) McCain (ATC eX) (*23*), (iv) Siemens (m50) (*24*), and

[1]Department of Civil, Environmental, and Geomatics Engineering, Florida Atlantic University, Boca Raton, FL

Corresponding Author:
Address correspondence to Nikola Mitrovic: nmitrovi@fau.edu

(v) PEEK (ATC-1000) (*25*). The SILS platforms considered in this study are: (i) RBC in VISSIM (*26*), (ii) ASC/3 Econolite (*27*), (iii) McCain 2033 (*28*), (iv) Siemens Virtual Next Phase (VNP) (*29, 30*), and (v) Fourth Dimension (Virtual D4) (*31*). The test scenario encompasses a single intersection, with moderate traffic and a modest percentage of left- and right-turning movements. Two additional scenarios with increased traffic demand (10% and 20%) are analyzed to investigate if the same trends can be observed with each signal control platform. Traffic signal parameters are calculated by following the common Webster procedure and recommendations from Signal Timing and Highway Capacity Manuals (*32*). Simulation runs, with predefined random seed and 100-ms simulation resolution time, are executed for each of ten signal control modeling alternatives. Various sets of vehicular and signal performance measures are extracted from VISSIM models and properly evaluated.

The rest of the paper is organized as follows. First, a brief overview of the relevant research is presented. Then, the authors present the data flow and major components of the HILS and SILS frameworks, and discuss how the signal timing parameters are embedded in each of the ten simulation alternatives. In the subsequent sections, evaluations of the ten signal control alternatives are presented. The final section closes with concluding remarks and ideas for future research.

Literature Review

The recent advancement in traffic signal simulation enables researchers to evaluate various signal control strategies and test controller hardware and software in different HILS and SILS environments, which eventually results in improved accuracy of modeled traffic signal operations.

The HILS concept and its components are introduced in several studies (*2, 5, 6*). Engelbrecht et al. presented the HILS development while highlighting its benefits to research organizations, universities, and so forth. (*5*). Engelbrecht, in another study, showed how HILS can be utilized to assess the effect of traffic signal control on traffic performance, and expressed that the HILS concept could only be a way to evaluate different vendor-specific features among controllers (*6*). Li et al. provided a sketch of real-time HILS and discussed the hardware design of CID. This study explains HILS interface software design for three different simulation models: CORISM, VISSIM, and SimTraffics (*2*).

The HILS concept has been used by many authors in order to test a variety of traffic applications, strategies, and systems, and to review their effectiveness. For instance, Bullock et al. presented the approach of using HILS procedures for applications such as transit priority or adaptive control (*7*). Another study focused on the simulation of a single intersection by using HILS to evaluate the effects of transit signal priority (TSP) signal control strategies on transit performance (*8*). Yun et al. used VISSIM-based HILS to examine the feasibility and robustness of the adaptive maximum feature available in the EPAC300 actuated controller (*9*). Hunter et al. performed HILS evaluation of ACTRA (i.e., semi actuated, coordinated, time-of-day) and SCATS (i.e., adaptive) signal control on a multi-intersection arterial segment (*10*). An adaptive traffic control system which incorporates traffic flow prediction models was tested and appraised through a case study in which HILS was employed (*11*). In another study, HILS was utilized for assessing hardware components running the ATSC MARLIN software in a simulation environment (*12*).

The SILS concept overcame several drawbacks of HILS, such as its need to run in real time, its incapability of synchronization between the controller's and computer's clock, and so forth. (*4*). In the context of SILS, Zlatkovic et al. investigated the use and benefit of the VISSIM traffic simulation software in evaluating TSP strategies using RBC and Advanced System Controller series 3 (ASC/3) SILS (*13*). Another study showed an application of the ASC/3 SILS simulation to implement and analyze TSP using built-in TSP features, and the controller logic processor (*14*). He et al. proposed a platoon-based multimodal signal control model combined with the actuated control logic in VISSIM using SILS to coordinate a system of actuated signals (*15*). Day and Bullock used SILS to investigate the impacts of two force-off modes (fixed and floating) and two methods of determining coordinated phase split allocation (semi actuated and fully actuated coordination) in four potential coordination strategies (*16*). In another study researchers used VISSIM–ASC/3 SILS to optimize the signal timings obtained from the field by using an integrated approach of VISSIM, SSAM, and VISSIM-based genetic algorithm optimization of signal timings (VISGAOST) (*17*). Day et al. described an emerging methodology for the evaluation of adaptive signal control, named "system-in-the-loop simulation." The methodology incorporated existing SILS by linking virtualized traffic controllers with real-world adaptive-control systems (*18*). Moreover, a methodology to obtain in-depth information on controller capabilities based on SILS was introduced in the study by Mladenovic et al., which showed that SILS is able to assess unidentified controller software capabilities (*19*).

Comparison of HILS and SILS is addressed in very few studies. Stevanovic et al. evaluated the coordinated-actuated signal systems executed through emulation-in-the-loop simulation (EILS), HILS, and SILS environments, and concluded that EILS may generate significantly different results from SILS and HILS, which both performed in a very similar way (*20*).

One can conclude that HILS and SILS concepts are widely utilized to evaluate various signal traffic operations in the simulation environment. However, the existing literature does not provide a comprehensive comparison of various HILS and SILS options. This paper bridges this gap in the existing literature by investigating the performance of various HILS/SILS setups for an intersection exposed to identical traffic flows and signal parameters.

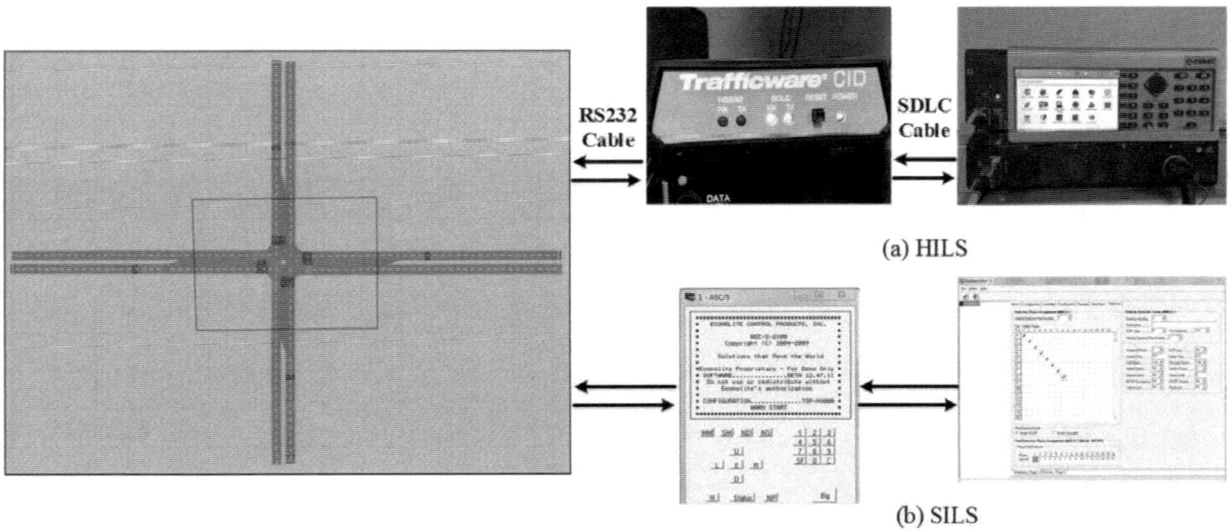

SG Number	1	2	3	4	5	6	7	8
SG Name	EBL	WBT	SBL	NBT	WBL	EBT	NBL	SBT
Min Green	5	7	4	4	5	7	4	4
Veh Extension	3	3	2.5	2.5	3	3	2.5	2.5
Max1	20	55	15	25	20	55	15	25
Yellow	3	3	3	3	3	3	3	3
Red Clearance	1	1	1	1	1	1	1	1
TMC	120	600+**80***	45	225+**30***	120	600+**80***	120	225+**30***
Right Turn movement for each approaches are represented by bold number and *								

Figure 1. Experimental network and major components of HILS and SILS frameworks: (*a*) HILS; (*b*) SILS.

Research Methodology

Study Network

The hypothetical test network used in this study is essentially a single signalized intersection, shown in Figure 1. The authors had to restrict experiments to a single intersection because modeling of multiple intersections would require multiple controllers of the same type and/or interfacing various controllers to work in connected/coordinated fashion, which are research topics outside of the scope of this paper. Figure 1 shows a VISSIM model of the test intersection (left) along with the signal timing and traffic volumes (base case scenario) for each movement (bottom). As it can be observed from Figure 1, the test intersection operates in a common eight-phase regime (all left turns are protected) and moderate traffic demand. Right turns on red are allowed on each approach of this intersection. Figure 1 also shows the relevant signal parameters that are used in HILS and SILS platforms and conceptual flow diagrams of HILS and SILS frameworks. The test network is developed in both VISSIM 5.4 and VISSIM 7 versions, to support various operational requirements of HILS and SILS platforms.

Experimental Setup

A single timing plan (no coordinated operations), network layout (lanes, detectors, signal heads), and input traffic volumes were kept identical for all experiments with various HILS and SILS platforms. In addition, to minimize potential discrepancies resulting from the start process of the various controllers, starting phases (2 & 6) were kept identical. The same applies to minimum recalls (2 & 6), dual entries, and other parameters that may have an impact on the actuated controller's operations.

The following ten (five HILS and five SILS) signal control platforms are compared in VISSIM:

- HILS
  - Econolite (Cobalt)
  - Intelight (2070-LDX)

  - McCain (ATC eX)
  - PEEK (ATC-1000)
  - Siemens (m50)

- SILS

  - ASC/3 Econolite
  - Fourth Dimension (Virtual D4)
  - McCain 2033
  - Ring-Barrier Controller (RBC)
  - Siemens Virtual Next Phase (VNP)

It should also be noted that the authors tried to evaluate another HILS (Trafficware) and another SILS platform (MaxTime). However, because of technical issues with a controller (HILS) and interface with simulation program (SILS), in spite of all efforts, the authors could not enable these two platforms to be ready for the experiments performed in this study. It should also be noted that the VISSIM's RBC platform is not a true SILS as it is more an emulation of the true controller software. However, this platform was included as the base case scenario which has been based on a "true" SILS platform (D4), and which has been proven to give very realistic traffic signal outputs in several previous studies (*33*). It is also a standard platform for modeling coordinated-actuated operations in the United States. The same applies to VISSIM, which was obviously a choice for simulation software as no other simulation tool can support all of the HILS and SILS interfaces investigated in this study.

Hardware-in-the-Loop Simulation Setup

The hardware-in-the-loop concept relies on the fact that the status of the detectors generated by simulation models is sent through the CID to the actual traffic controller (*2*, *3*). The traffic controller determines the status of signal control, based on the detector input data, and sends the data back to the simulation model through the CID. All of the experiments are performed by connecting each HILS controller with Trafficware CID, which is a synchronous data link control (SDLC)-type of CID used in signal controller test labs around the country. The CID hardware is coupled with Trafficware TS2 testbox software, which ensures that the CID simulates BIUs according to the NEMA TS2 standards. Before each experiment, operations of the CID were actually tested through the testbox software to ensure proper communications between computer and the controller.

As the PTV VISSIM and Trafficware CID are developed by different manufacturers, and require different formats of input and output data, a special program was developed to establish the interface between these two devices (*33*). This tool is referred to as "Virtual CID" and has been described elsewhere (*33*).

Figure 1*a* depicts the two-way communication between the major components of the HILS framework. The detector actuation from the VISSIM microsimulation model is converted using the developed software tool and further sent to CID through RS232 cable (see Figure 1*a*). The converted detector data are then sent from Trafficware CID to the actual controller (e.g., Econolite) through a SDLC cable (see Figure 1*a*). This setup is applied for all tested controllers. Once the controller determines the status of the signals, this information is sent from controller to the PTV VISSIM microsimulation model through the Trafficware CID device and Virtual CID (see Figure 1*a*).

Figure 2 documents, through screenshots of controller displays, that all of the HILS signal timings were set in a consistent way. In VISSIM, the HILS controllers are executed through an external control option and relevant DLL and WTT files are provided as required (see top left corner of Figure 2).

Software-in-the-Loop Simulation Setup

Microscopic simulation model (VISSIM), virtual traffic controller, and interface module are three main components of the SILS framework. These components, in the context of ASC3/ Econolite SILS, are shown in Figure 1*b*. Microscopic simulation model (e.g., VISSIM) emulates the vehicular and driver behavior. It constantly updates the position of the vehicles in the network and status of the detectors, based on vehicle actuation/presence. The status of the detectors is sent from the simulation module to the virtual traffic controller (see Figure 1*b*). The virtual traffic controller takes as input the detector's status and predefined signal timing parameters and returns the status of the signal to the simulation module. This data exchange is performed at every simulation step. The interface module allows the user to set the signal timing parameters in the virtual traffic controller. In addition, the interface module provides data exchange between the simulation module and the virtual traffic controller. All three components have to be located on the same PC.

Figure 3 shows how the test signal plan is embedded within each of the tested SILS platforms, and the setup of each SILS. VISSIM has predefined controller options for the following SILS controllers: RBC, Econolite ASC/3, and McCain 2033. However, Siemens and Virtual D4 SILS are emulated by selecting VAP as a controller type in VISSIM controller's dialog box (see Figure 3). Figure 3 also shows signal timing values in the respective databases of the individual controllers.

Results

Because of the very time-consuming simulation process (each HILS simulation can run only in real time) the number of simulation runs was limited to ten for each controller group in base case scenario and five for increased volume (10% and 20%) scenarios. All simulations were done with the same (single) random seed to reduce stochasticity, which

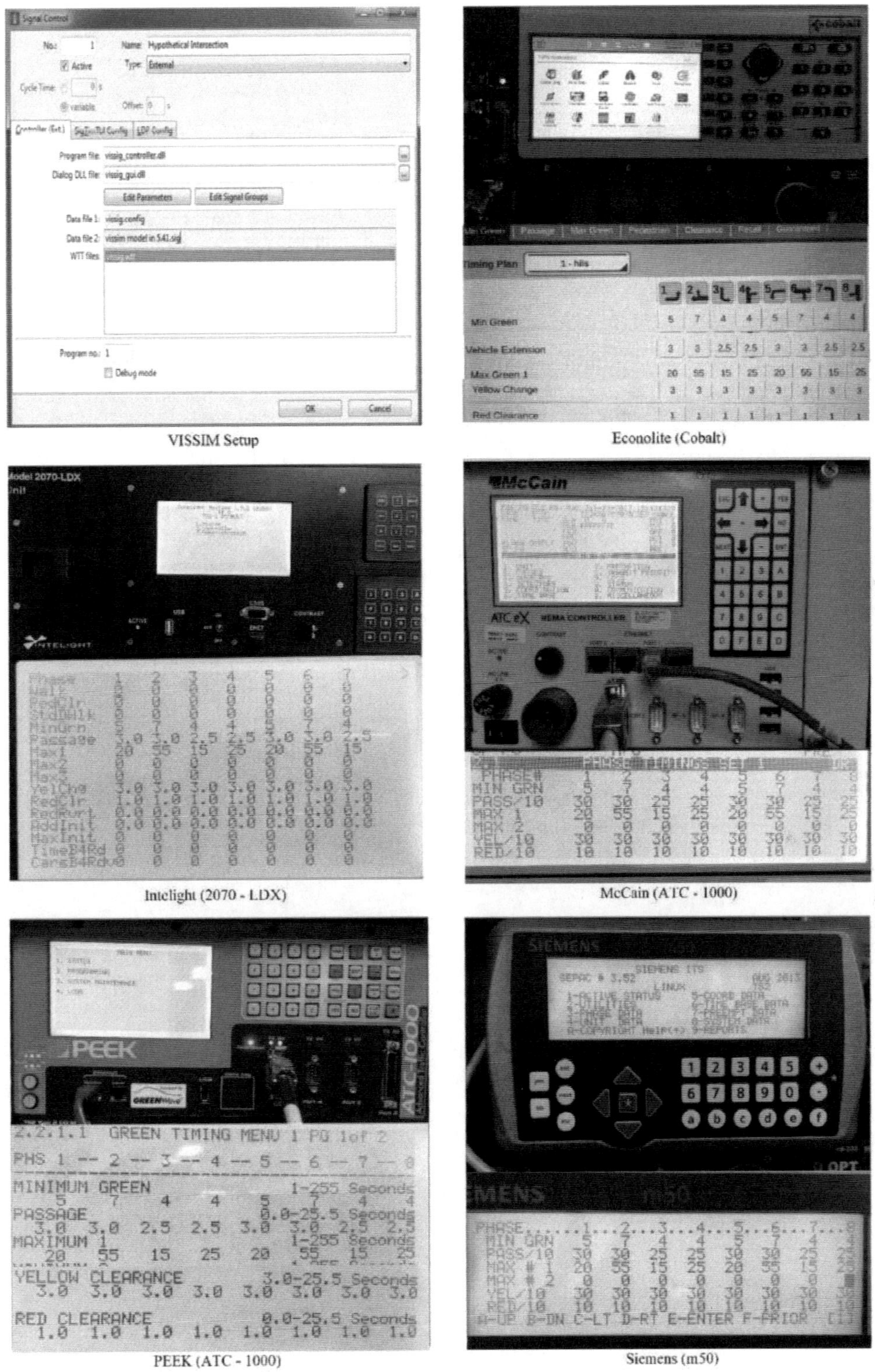

Figure 2. Screenshots of HILS controller with signal timings and VISSIM setup options.

is inherently part of each simulation run even when not designed; more information about this can be found elsewhere (*20*). All of the simulations were executed with 10 Hz simulation resolution, for 65 min: 1 h of evaluation time

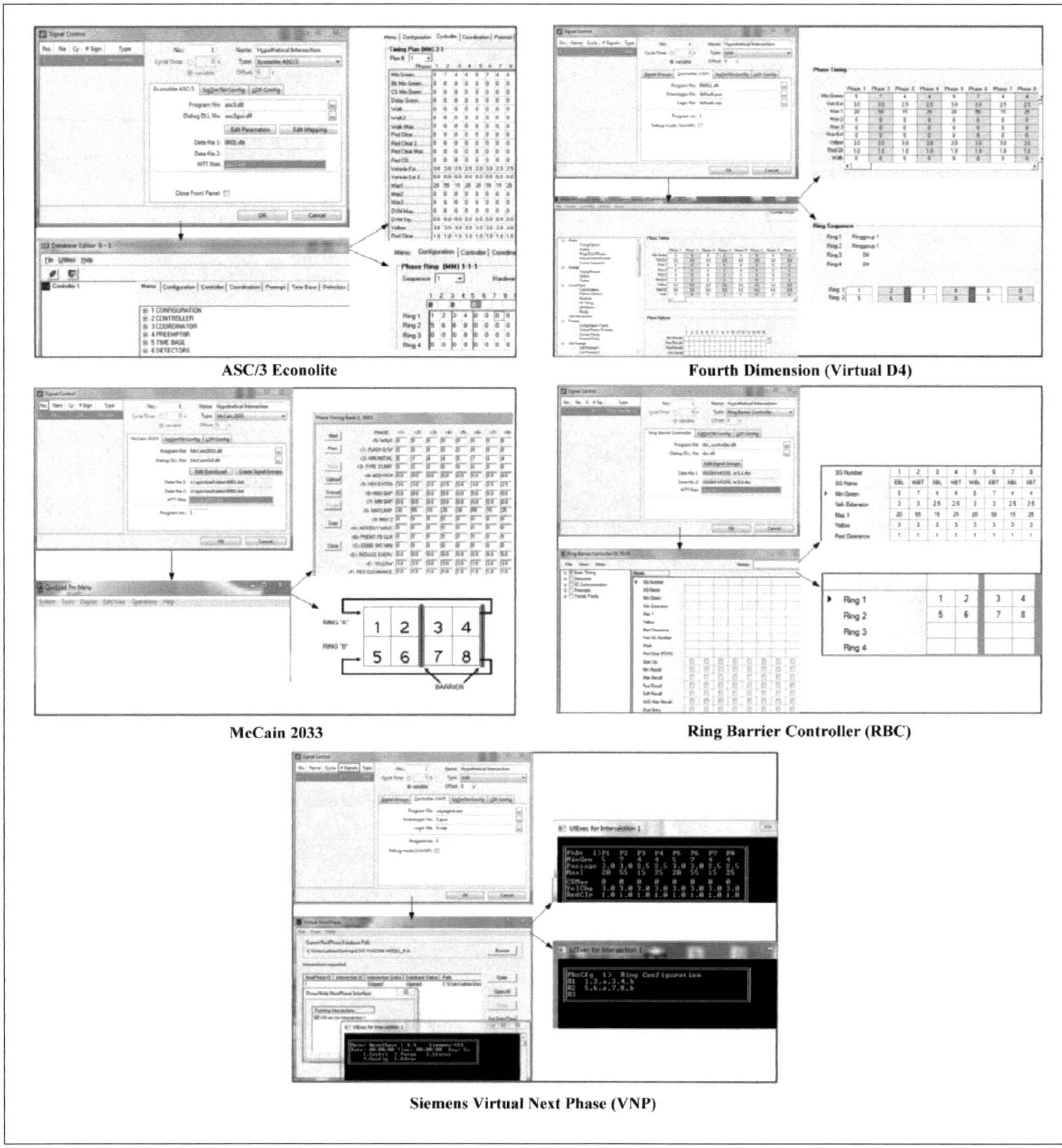

Figure 3. VISSIM setup for various SILS controllers with timing parameters.

and 5 min of warm-up time. To "stabilize" the operations of the HILS controllers and avoid unnecessary discrepancies caused by initialization, Phases 2 and 6 are set to "rest in green" before each simulation run. Various types of vehicular and signal performance measures were extracted from each of these experiments.

Initially the goal was to execute all of the HILS and SILS experiments using the same version of VISSIM, but most of the SILS platforms have not been updated to work with the newest version of VISSIM. Similarly, the interface between VISSIM and Trafficware CID has been enabled to work with VISSIM 7 and has not been updated yet to work with newer versions of VISSIM. Thus, VISSIM 5.4 was used to model SILS platforms and VISSIM 7 was used to model HILS controllers. To check consistency between two different VISSIM versions, the authors evaluated RBC performance measures, which was only SILS version that could be modeled in both VISSIM versions. Results show that the difference in performance measures generated by the two different VISSIM versions is significant, but remains within 10%. These differences are computed for each measure of effectiveness, (MOE) as:

$$Difference(\%) = \frac{MOE^{\max} - MOE^{\min}}{MOE^{\max}} \tag{1}$$

where $MOE^{\max}$ and $MOE^{\min}$ are the maximum and minimum value of a particular measure of effectiveness, respectively. The results also show that operational measures such as delay and (number of) stops are slightly higher (6–8%) if the experiments are performed in Version 5.4. However, the other parameters are quite similar but not identical. For this reason, the output from HILS and SILS are mainly compared separately in the rest of the paper.

To test the consistency among different SILS and HILS performances signal timing performance measures were examined. Figures 4 and 5 show averages and standard deviations of phase green times, which were recorded by VISSIM for each phase and each signal control scenario of each controller (base case, and 10% and 20% increase in traffic) for HILS and SILS setup, respectively. One can observe from Figures 4 and 5 that the results are quite consistent. The only exception are signal groups 2 and 6, which are assigned the longest green times to serve the through movements of the main street. In particular, the average green times of signal groups 2 and 6 are the most different for McCain HILS (see Figure 4) and ASC-3 SILS (see Figure 5). In spite of multiple checks it was impossible to detect reasons for these discrepancies. Figure 4 also shows, as expected, that the standard deviations of green time are comparatively larger for the major movement signal groups than for those servicing left-turn traffic. As shown in Figure 5, the standard deviations for SILS controllers are zero as expected—each simulation run with the same random seed gives identical values for various measures.

Table 1 presents the number of occurrences of each phase, during 1 h of simulation time, for the three experimental scenarios. This performance measure is utilized to check the consistency of the number of phase calls/actuations across various HILS and SILS platforms. The results from Table 1 show that number of phase activations varies significantly across different HILS and SILS platforms. It seems that such variations are larger for those signal groups that service higher traffic volumes. Although some of the results are more consistent (e.g., Intelight and Econolite HILS for SGs 2, 4, and 8) than others, there is no a single pair of HILS or SILS platforms where the number of activations, for all of the signal groups, is identical. These outputs may be associated with the randomness that exists in the internal controller logics, or caused by communications or inconsistent initialization (phase sequencing introduced at the beginning of simulation process where arrivals of the vehicles awake physical/virtual controllers in various states of operations). However, as the results in Figure 6 show, initialization seems to play much a smaller role than initially thought. Further investigation is needed to reveal true reasons for this randomness.

To understand the influence of the initialization process on the inconsistency of the results, the authors investigated when and how long each HILS and SILS controller runs their first phase (Phase 2) in the simulation. Figure 6 presents the time instant and duration of the first phase for each HILS and SILS controller. The results from Figure 6 show that all of the HILS controllers start their first Phase (# 2) at the same time (0.3 s). However, while most of the SILS controller starts at 0.2 seconds, Econolite (1.4 s) and McCain (3.3 s) take a longer time to start. In terms of the durations among HILS controllers, it seems that Intelight is consistent with Peek, and McCain is consistent with Siemens, whereas Econolite is not consistent with any others. On the SILS side, all of the controllers are consistent except McCain. The authors checked signal timings and available parameters multiple times; also, VISSIM's "ldp" files were checked to investigate the consistency of car arrivals under each scenario. None of those quality control checks revealed any reasons for these inconsistencies between various HILS platforms. However, one thing can be concluded—although inconsistent between themselves, each HILS platform performs self-consistently (Figure 6 presents repetition of ten runs, and standard deviations of Phase 2 durations are very small).

In addition to signal timing parameters, the authors compared common traffic performance measures, which were obtained by processing VISSIM network-based outputs. Tables 2 and 3 presents such network-based performance measures for all of the HILS and SILS alternatives in all three scenarios. The last row of each scenario in Table 2 and 3 shows the difference between minimum (italic) and maximum (bold) values of the particular MOE (see Equation 1). Table 2 shows that in terms of vehicular network performances, Econolite HILS produces the largest values for most of the performance measures whereas McCain HILS yields the smallest values. In the case of SILS platforms, Siemens SILS generates some of the highest performance measures whereas Econolite has opposite results (see Table 3). One can also observe that although the differences among various HILS and SILS options are not large in absolute aspects, their relative differences range from very small (e.g., throughput volumes) to considerable (e.g., delays). It is interesting to note that the differences in delay can be larger between various controllers within the same group (e.g., HILS or SILS) than between two different versions of VISSIM for the same type of traffic control.

Thus, there are enough reasons to speculate that inconsistencies in the operations of various controllers, caused most

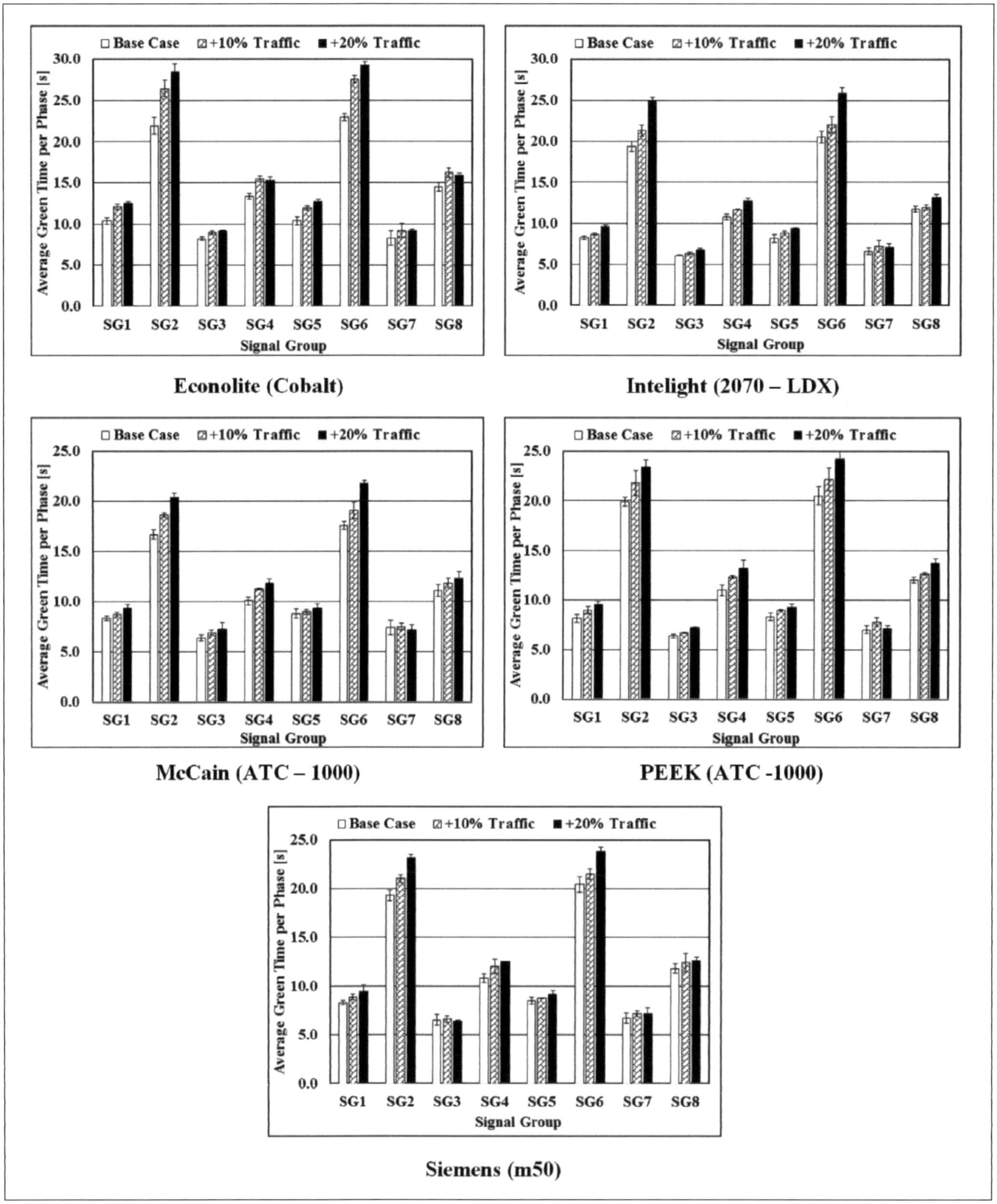

Figure 4. Average green time and standard deviation per phase (HILS).

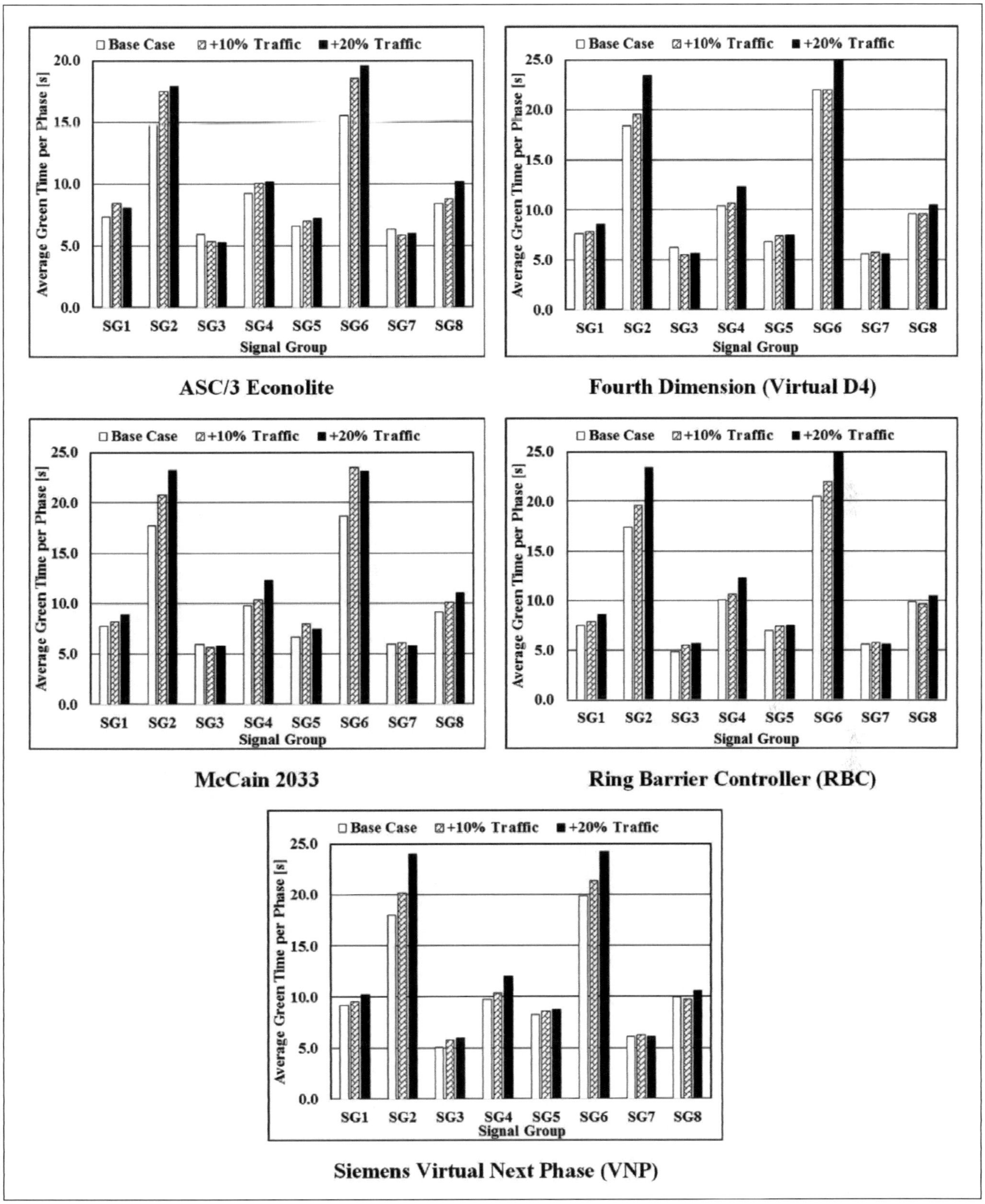

Figure 5. Average green time and standard deviation per phase (SILS).

Table 1. Average Number of Occurrences per Phase (HILS & SILS)

		HILS					SILS				
		Econolite	Intelight	McCain	Peek	Siemens	Econolite	D4	McCain	RBC	Siemens
SG1	Base	43.7	41.7	41.9	43.0	42.6	48.0	48.0	47.0	48.0	44.0
	10%	45.3	42.7	43.3	41.7	42.7	46.0	50.0	48.0	50.0	46.0
	20%	43.7	40.0	42.7	41.3	40.5	48.0	45.0	43.0	45.0	41.0
SG2	Base	61.9	59.0	64.6	57.8	58.5	69.0	60.0	61.0	63.0	59.0
	10%	56.0	54.3	58.0	52.3	53.7	61.0	56.0	54.0	56.0	52.0
	20%	54.0	48.0	54.3	48.7	50.5	59.0	50.0	50.0	50.0	46.0
SG3	Base	32.0	28.7	28.4	27.7	28.9	26.0	26.0	25.0	29.0	29.0
	10%	34.7	28.7	29.3	29.0	28.3	26.0	28.0	25.0	28.0	28.0
	20%	34.7	29.7	29.3	29.7	31.5	32.0	28.0	22.0	28.0	28.0
SG4	Base	60.9	58.8	61.0	58.2	57.6	65.0	54.0	59.0	57.0	58.0
	10%	55.7	53.3	55.7	52.0	53.7	57.0	53.0	54.0	53.0	54.0
	20%	53.7	47.7	52.7	48.7	49.5	57.0	48.0	50.0	48.0	49.0
SG5	Base	38.5	36.5	35.7	37.0	35.1	43.0	41.0	43.0	39.0	33.0
	10%	41.3	39.0	38.7	38.7	39.0	42.0	40.0	37.0	40.0	34.0
	20%	40.0	37.7	37.0	39.7	38.0	43.0	41.0	44.0	41.0	35.0
SG6	Base	62.5	58.9	64.0	59.1	59.1	71.0	57.0	63.0	60.0	57.0
	10%	56.0	54.3	58.3	53.0	54.3	63.0	56.0	54.0	56.0	53.0
	20%	54.3	48.0	54.3	48.7	51.0	59.0	50.0	52.0	50.0	47.0
SG7	Base	29.0	26.6	25.9	26.3	26.4	28.0	32.0	28.0	29.0	26.0
	10%	30.7	26.7	27.0	28.3	27.7	32.0	31.0	28.0	31.0	28.0
	20%	32.7	28.7	26.7	28.3	30.0	31.0	37.0	33.0	37.0	34.0
SG8	Base	58.6	54.4	56.5	53.1	55.0	62.0	55.0	60.0	58.0	57.0
	10%	55.7	51.7	54.3	49.7	50.3	59.0	53.0	52.0	53.0	52.0
	20%	53.7	46.3	51.0	48.0	48.5	56.0	48.0	48.0	48.0	47.0

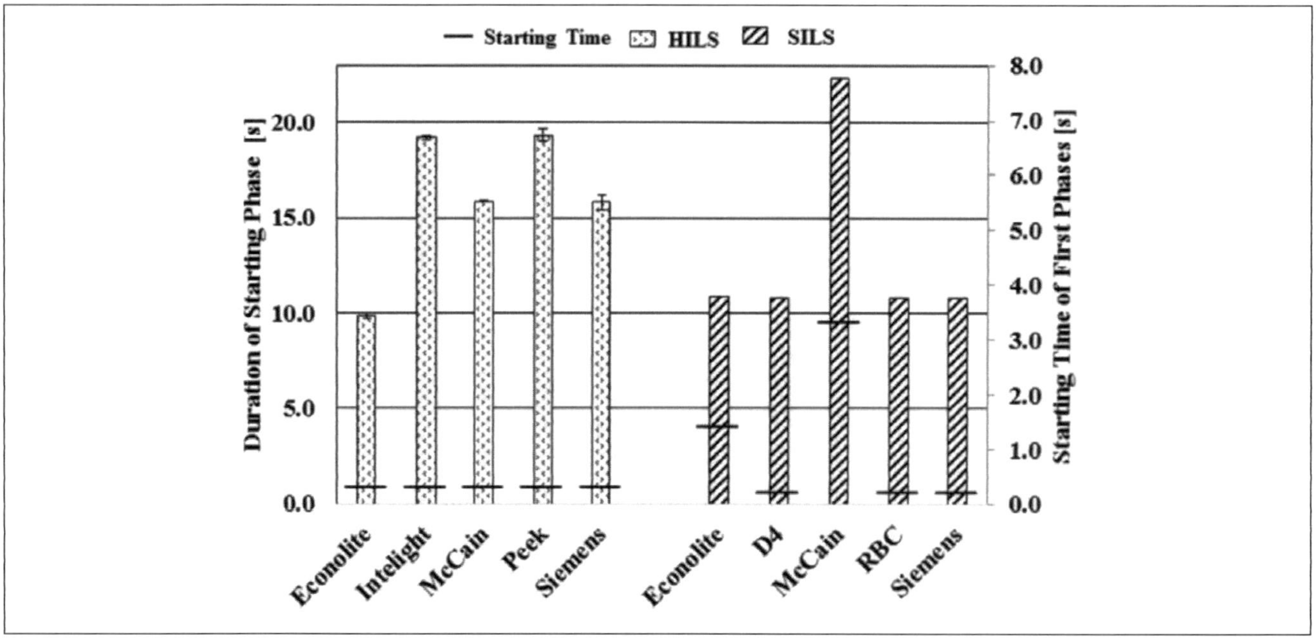

Figure 6. Initialization of starting phase 2 for HILS & SILS controllers.

Table 2. Operational MOEs of Various HILS Controllers

		Controllers in HILS environment						
		Total number of vehicles	Average delay/vehicle (s)	Total delay (h)	Average stopped delay/ vehicle (s)	Total stopped delay (h)	Average no. of stops/ vehicle	Total no. of stops
Base case scenario	Econolite	*2191.6*	**18.62**	**11.34**	**11.29**	**6.88**	**0.58**	**1280.32**
	Intelight	2193.6	17.08	10.4	9.89	6.03	0.57	*1254.09*
	McCain	2193.7	*16.04*	*9.77*	*8.82*	*5.37*	*0.57*	1245.85
	Peek	2192.8	17.34	10.56	10.28	6.26	0.58	1262.87
	Siemens	**2194**	16.75	10.21	9.61	5.86	0.57	1257.42
	Diff (%)	0.11%	13.86%	13.79%	21.93%	21.87%	2.78%	2.69%
10 % volume increase	Econolite	2418.33	**22.41**	**15.05**	**14.69**	**9.86**	**0.62**	**1506.87**
	Intelight	2416.67	18.23	12.24	10.79	7.24	*0.58*	*1407.53*
	McCain	2411.33	*17.72*	*11.87*	*10.17*	*6.81*	0.6	1439.59
	Peek	**2420**	18.74	12.6	11.23	7.55	0.6	1456.48
	Siemens	*2410*	18.67	12.5	11.17	7.48	0.6	1447.43
	Diff (%)	0.41%	20.94%	21.17%	30.78%	30.98%	6.53%	6.59%
20 % volume increase	Econolite	2630.33	**23.56**	**17.22**	**15.59**	**11.39**	**0.64**	**1686.57**
	Intelight	2635	19.99	14.63	12.3	9	0.61	1601.21
	McCain	*2629.67*	*18.77*	*13.71*	*11.24*	*8.21*	*0.61*	*1597.92*
	Peek	**2636.67**	20.06	14.69	12.27	8.99	0.62	1623.41
	Siemens	2634.5	20.29	14.85	12.45	9.11	0.64	1684.12
	Diff (%)	0.27%	20.34%	20.35%	27.92%	27.93%	5.25%	5.26%

Table 3. Operational MOEs of Various SILS Controllers

		Controllers in SILS environment						
Base case scenario	Econolite	2129	*16.05*	*9.49*	*8.22*	*4.86*	*0.58*	1226.3
	D4	**2134**	16.98	10.07	9.29	5.51	0.56	*1186.5*
	McCain	2129	17.49	10.34	9.37	5.54	0.59	1256.11
	RBC	2129	17.68	10.46	10.04	5.94	0.62	1319.98
	Siemens	*2126*	**19.17**	**11.32**	**11.14**	**6.58**	**0.66**	**1397.57**
	Diff (%)	0.37%	16.29%	16.17%	26.28%	26.17%	15.42%	15.10%
10 % volume increase	Econolite	*2328*	*17.23*	*11.14*	*9.21*	*5.96*	*0.58*	1345.58
	D4	**2334**	17.99	11.66	10.03	6.51	0.57	*1332.71*
	McCain	2333	18.94	12.27	10.74	6.96	0.59	1383.47
	RBC	2334	17.99	11.66	10.03	6.51	0.57	1332.71
	Siemens	2333	**19.48**	**12.62**	**11.48**	**7.44**	**0.65**	**1517.2**
	Diff (%)	0.26%	11.53%	11.72%	19.76%	19.94%	12.20%	12.16%
20 % volume increase	Econolite	**2567**	*17.97*	*12.82*	*9.7*	*6.92*	*0.58*	*1481.16*
	D4	2562	19.34	13.76	11.31	8.05	0.58	1493.65
	McCain	*2551*	20.08	14.23	11.57	8.2	0.61	1551.01
	RBC	2562	19.34	13.76	11.31	8.05	0.58	1493.65
	Siemens	2560	**20.15**	**14.33**	**12.15**	**8.64**	**0.66**	**1677.49**
	Diff (%)	0.62%	10.78%	10.54%	20.15%	19.93%	11.94%	11.70%

likely by internal controller logics and handling of the first actuations, may create more noise in the performance outputs than the systematic differences introduced by new versions of the traffic simulation models.

Two-tail *t*-tests were conducted to investigate if means of respective datasets (for all MOEs from Table 2) are significantly different at significance level of 0.05. Table 4 shows the *t*-stat values of intersection MOE for the controllers under HILS framework. If the *t*-stat values were considered significant they were denoted with *, **, *** for base case, 10% increase in traffic, and 20% increase in traffic, respectively. As observed from Table 4, there are significant

Table 4. Results of the *T*-tests for Various MOEs (HILS Controllers)

		Econolite			Intelight			McCain			Peek		
MOEs		Base	10%	20%	Base	10%	20%	Base	10%	20%	Base	10%	20%
Total number of vehicles	Intelight	1.2	2.8	0.4									
	McCain	0.9	3.4**	0.3	0	2.5	1.2						
	Peek	0.7	2.1	4.6***	0.6	1.8	1	0.6	0.9	2.1			
	Siemens	1.1	2.3	1.7	0.2	0.9	0.9	0.3	0.6	1.2	1.1	0.7	0.5
Average delay/ vehicle (s)	Intelight	3.1*	5.5**	12.5***									
	McCain	5.0*	17.8**	8.8***	9.1*	2.1	4.7***						
	Peek	2.7*	13.8**	11.6***	2.1	0.5	1.4	8.3*	4.8**	3.2***			
	Siemens	3.3*	11.3**	10.2***	1.8	0.7	2.1	4.1*	3.6**	3.7***	2.8*	0.1	1
Total delay (h)	Intelight	3.0*	5.7**	11.3***									
	McCain	4.9*	17.2**	8.4***	9.3*	2.1	4.9***						
	Peek	2.6*	15.0**	11.5***	2.1	0.6	1.4	8.4*	5.2**	3.2***			
	Siemens	3.2*	11.2**	10.0***	1.8	0.8	2.1	4.1*	3.4**	3.7***	2.7*	0.3	1.2
Average stopped delay/ vehicle (s)	Intelight	3.1*	6.6**	12.0***									
	McCain	5.4*	18.9**	8.9***	11.4*	2.4	1.6						
	Peek	2.7*	12.1**	21.8***	3.1*	0	1.2	10.6*	4.4**	2.6			
	Siemens	3.4*	10.1**	15.1***	1.6	0.5	1.8	5.9*	3.3**	3.6***	3.4*	1.2	1.1
Total stopped delay (h)	Intelight	3.1*	6.8**	11.8***									
	McCain	5.3*	18.3**	8.7***	11.9*	2.5	1.9						
	Peek	2.6*	13.2**	21.6***	3.1*	0.1	1.1	10.7*	4.6**	2.6			
	Siemens	3.3*	10.0**	14.9***	1.7	0.6	1.8	5.8*	3.2**	3.5***	3.4*	0.6	1.4
Average no. of stops/vehicle	Intelight	1.6	1.3	10.2***									
	McCain	2.1	1	2.1	1.5	1.2	0.8						
	Peek	0.8	4.5**	3.7***	0.8	0.3	2.8	1.1	1	0.8			
	Siemens	1.1	3.2**	0	0.4	0.5	4.4***	1.1	0.3	2.4	0.6	1.8	1.9
Total no. of stops	Intelight	1.5	1.7	9.9***									
	McCain	2	1.4	2.3	1.4	1.3	0.7						
	Peek	0.8	6.6**	3.6***	0.7	0.2	2.6	1.1	1.2	0.9			
	Siemens	1	3.3**	0	0.4	0.5	4.4***	1.1	0.8	2.3	0.5	1.4	1.9

*Statically significant difference for base case.
**Statically significant difference for 10% volume increase.
***Statically significant difference for 20% volume increase.

differences between certain pairs of HILS platforms. For most of the evaluated MOEs, there are significant differences between McCain and Econolite HILS.

Conclusions

This study investigated five HILS and five SILS methods of modeling basic traffic signal control operations and three different scenarios in a microsimulation environment. Similar studies have never been done before, and this research tried to investigate, basically, if these controllers can be reliably switched (if needed) without experiencing significant differences in resulting signal timings and vehicular traffic performance measures. A single intersection in the VISSIM model was used to test the basic free-actuated traffic signal operations. The experiments were done under limiting conditions dictated by the availability of the HILS and SILS platforms and microsimulation functionalities. The following conclusions have been reached in this study:

1. Differences between resulting signal timings of various HILS and SILS platforms are large enough that it cannot be claimed that use of any of these platforms will result in signal timings identical, or very similar, to another platform. Even when test conditions are very basic and limited, as in the case of a signalized uncoordinated intersection, the differences (of average green times and phase actuations) were large enough to question the repeatability of the results, if the experiments were done in two HILS/SILS platforms.
2. In terms of the magnitude and pattern of the signal timing errors/differences, the authors did not observe that a particular HILS or SILS behaved very erratically. Signal timing differences (green times and frequency of phase calls) are not very large, but they were all over the space, with two systems behaving slightly differently: McCain among HILS and Econolite among SILS. These two platforms had slightly shorter average times for main phases, but

also slightly higher frequency of those same phases, than the other platforms.

3. Signal timing differences inevitably caused variability of the operational performance measures, which were especially emphasized in the case of delays. Some delay measures from various HILS and SILS platforms differed so much that such a margin of difference can bring into question any experiments where two HILS or SILS platforms are compared in estimating traffic control strategies of a higher operational level. This means, for example, that an investigator who wants to compare adaptive operations of Centracs Adaptive (with local controllers modeled in ASC/3 SILS) and coordinated-actuated operations of RBC SILS cannot do such comparison without first investigating the underlying differences between Econolite ASC/3 and RBC SILS in coordinated-actuated modes. This finding puts a significant restriction on how various HILS and SILS platforms, either alone or in conjunction with other higher forms of traffic control strategies, can be used for similar comparative evaluations.
4. Although it originally seemed that the major cause of the observed discrepancies was an initialization process (when virtual traffic makes detector calls while the controllers are in various operational states/phases), further experiments showed that this was not a major problem. The authors were able to achieve relatively stable initialization, although some of the controllers behaved differently from the very beginning, showing that the cause of the discrepancies may exist within internal controller logic and the way they react to detector calls.
5. Inability to use the same version of VISSIM for all experiments additionally contributed to the potential stochasticity of the experiments. However, differences in the results from various VISSIM versions, when the same SILS platform was exposed to those, revealed a lower degree of significance than differences between HILS or SILS platforms tested in the same VISSIM version.

Future research should go in the direction of unifying all of the experiments under a single VISSIM version, when this becomes possible. Also, SILS controllers should be evaluated in a multi-intersection environment with more complex scenarios. Finally, researchers should attempt to investigate internal logics used by various controllers, to the extent possible.

Author Contributions

The authors confirm contribution to the paper as follows: study conception and design: S-E-S. Chowdhury, A. Stevanovic, and N. Mitrovic; data collection: S-E-S. Chowdhury; analysis and interpretation of results: S-E-S. Chowdhury, A. Stevanovic, and N. Mitrovic; draft manuscript preparation: S-E-S. Chowdhury, and N. Mitrovic. All authors reviewed the results and approved the final version of the manuscript.

References

1. PTV. *VISSIM 5.40 - User Manual*. PTV Planung Transport Verkehr AG, Karlsruhe, Germany, 2011.
2. Li, Z., M. Kyte, and B. Johnson. Hardware-in-the-Loop Real-time Simulation Interface Software Design. Proc., 7th International IEEE Conference on Intelligent Transportation Systems, IEEE, 2004, pp. 1012–1017.
3. Naztec Inc., *Naztec Operations Manual For TS2 and 2070 Test Box*, April 2003. http://naz.simpleapp2.net/sites/default/files/documents/manuals/testbox_manual.pdf. Accessed July 31, 2017.
4. Urbanik, T., M. Kyte, and D. Bullock. Software-in-the-Loop Simulation of Traffic Signal Systems. *SimSub Mid-year Newsletter*, 2006, pp. 15–19.
5. Engelbrecht, R., C. Poe, and K. Balke. Development of a Distributed Hardware-in-the-Loop Simulation System for Transportation Networks. Presented at 78th Annual Meeting of the Transportation Research Board, Washington, D.C., 1999.
6. Engelbrecht, R. Using Hardware-in-the-Loop Traffic Simulation to Evaluate Traffic Signal Controller Features. Proc., 27th Annual Conference of the Industrial Electronics Society, IECON'01, IEEE, Vol. 3, 2001, pp. 1920–1925.
7. Bullock, D., B. Johnson, R. B. Wells, M. Kyte, and Z. Li. Hardware-in-the-Loop Simulation. *Transportation Research Part C: Emerging Technologies*, Vol. 12, No. 1, 2004, pp. 73–89.
8. Byrne, N., P. Koonce, R. Bertini, C. Pangilinan, and M. Lasky. Using Hardware-in-the-Loop Simulation to Evaluate Signal Control Strategies for Transit Signal Priority. *Transportation Research Record: Journal of the Transportation Research Board*, 2005. 1925: 227–234.
9. Yun, I., M. Best, and B. Park. Evaluation of Adaptive Maximum Feature in Actuated Traffic Controller: Hardware-in-the-Loop Simulation. *Transportation Research Record: Journal of the Transportation Research Board*, 2007. 2035: 134–140.
10. Hunter, M., M. Roe, and S. Wu. Hardware-in-the-Loop Simulation Evaluation of Adaptive Signal Control. *Transportation Research Record: Journal of the Transportation Research Board*, 2010. 2192: 167–176.
11. Dickey, S., M. Li, J. Yee, M. Zennaro, H. Liu, W. Ma, and L. Li. *Development of Hardware in the Loop Simulation and Paramics/VS-PLUS Integration*. California PATH Program, Institute of Transportation Studies, University of California at Berkeley, 2008.
12. Abdelgawad, H., K. Rezaee, S. El-Tantawy, B. Abdulhai, and T. Abdulazim. Assessment of Adaptive Traffic Signal Control Using Hardware in the Loop Simulation. Proc., 2015 IEEE 18th International Conference on Intelligent Transportation Systems (ITSC), IEEE, 2015, pp. 1189–1195.
13. Zlatkovic, M., P. Martin, and A. Stevanovic. Predictive Priority for Light Rail Transit: University Light Rail Line in Salt Lake County, Utah. *Transportation Research Record: Journal of the Transportation Research Board*, 2011. 2259: 168–178.
14. Zlatkovic, M., P. T. Martin, and I. Tasic. Implementation of Transit Signal Priority and Predictive Priority Strategies in ASC/3 Software-in-the-loop Simulation. Proc., 2011 14th

International IEEE Conference on Intelligent Transportation Systems (ITSC), IEEE, 2011, pp. 2130–2135.

15.He, Q., K. L. Head, and J. Ding. *PAMSCOD*: Platoon-based arterial multi-modal signal control with online data. *Transportation, Research Part C: Emerging Technologies*, Vol. 20, No. 1, 2012, pp. 164–184.

16. Day, C., and D. Bullock. Design Guidelines and Conditions That Warrant Deployment of Fully Actuated Coordination. *Transportation Research Record: Journal of the Transportation Research Board*, 2014. 2439: 1–11.

17. Stevanovic, A., J. Stevanovic, and P. Martin. Optimizing Signal Timings from the Field: VISGAOST and VISSIM-ASC/3 Software-in-the-Loop Simulation. *Transportation Research Record: Journal of the Transportation Research Board*, 2009. 2128: 114–120.

18. Day, C., J. Ernst, T. Brennan Jr, C. S. Chou, A. Hainen, S. Remias, and D. Bullock. Performance Measures for Adaptive Signal Control: Case Study of System-in-the-Loop Simulation. *Transportation Research Record: Journal of the Transportation Research Board*, 2012. 2311: 1–15.

19. Mladenovic, M., M. Abbas, and M. Vodogaz. Implementation of the Software-in-the-Loop Simulation for Assessment of Operational Capabilities in the North American Advanced Transportation Controllers. *Journal of Road and Traffic Engineering*, Vol. 57, No. 2, 2011, pp. 27–32.

20Stevanovic, A., A. Abdel-Rahim, M. Zlatkovic, and E. Amin. Microscopic Modeling of Traffic Signal Operations: Comparative Evaluation of Hardware-in-the-Loop and Software-in-the-Loop Simulations. *Transportation Research Record: Journal of the Transportation Research Board*, 2009. 2128: 143–151.

21. *MaxTime Technical Reference Manual*, Intelight, Inc., MAN0012, Rev. B, 2010.

22. Econolite Group, Inc., "*Cobalt Programming Manual*", P/N 140-0903-001 Rev. 05, 2015.

23. McCain Inc., "*ATC eX NEMA CONTROLLER User manual*", Version: 1.0, 2012.

24. Siemens, *"SEPAC Activated Signal Control Software with ECOM and NTCIP User 44 Manual"*, PIM216-001, 2012.

25.Peek Traffic Corporation, "*GreenWave Basic Programming Guide*", p/n: 99-637, Rev 1, 40 2015.

26. PTV Group, "*Ring Barrier Controller (RBC) User Manual*", August 2014.

27. Econolite Control Products, Inc., *Advanced System Controllers ASC/3 Programming Manual*, P/N 100-0903-001 Rev. 03, 31 July 2007.

28. McCain Inc., *2033 Program Intersection Control Software for the 2070 Controller USER MANUAL Version: 3.3*, 2010.

29. Wright, M., *Virtual NextPhase (VNP): User Manual Including VISSIM Interface*, December 2004.

30. Siemens Energy & Automation, Inc., *NextPhase Operations Manual, Version 1.7.4*, Revised April 2008.

31. Fourth Dimension Traffic. *Virtual D4 Suite User's Manual.*

32.Highway Capacity Manual 2010. Transportation Research Board of the National Academies, Washington, D.C., 2010.

33. Stevanovic, A., I. Klanac, and D. Radivojevic. *Development of Minimum Standards for Event-Based Data Collection Loggers and Performance Measure Definitions for Signalized Intersections*. Department of Transportation, Florida, 2017.

The Standing Committee on Traffic Signal Systems (AHB25) peer-reviewed this paper (18-04109).

Article

TRR
JOURNAL OF THE TRANSPORTATION RESEARCH BOARD

Transportation Research Record
2018, Vol. 2672(18) 107–116

DOI: 10.1177/0361198118790638
journals.sagepub.com/home/trr
SAGE

Using Traffic Signal Control to Limit Speeding Opportunities on Bidirectional Urban Arterials

Peter G. Furth[1], Ahmed T. M. Halawani[1], Jin Li[1], Weimin (Jake) Hu[1], and Burak Cesme[2]

Abstract

Although controlling speed on urban arterials is important for safety, conventional traffic calming techniques cannot usually be applied on arterials, and many jurisdictions prohibit automated speed enforcement. Moreover, unlike unidirectional arterials, bidirectional arterials with short intersection spacing are not amenable to green waves that can remove the incentive to speed. This research explores the ways that traffic signal coordination creates – or limits – speeding opportunities on bidirectional arterials. Two measures of speeding opportunity are proposed: number of unconstrained vehicles, meaning vehicles arriving at a stopline on green and with no vehicle less than 5 s ahead of them, and number of speeders in a traffic microsimulation in which 20% of the vehicles have been assigned a desired speed in the "speeding" range. Theoretical analysis, confirmed by two case studies, show how speeding opportunities are related to degree of saturation, cycle length, specified progression speed (as in input to signal timing software), intersection spacing, and recall settings. The important role of clusters of intersections with near-simultaneous greens, a byproduct of bidirectional coordination with short intersection spacing, is examined. Clusters with many intersections are shown to create a strong speeding incentive, and cluster size can be reduced by lowering the cycle length and the progression speed. Case studies show it is sometimes possible to substantially reduce speeding opportunities with little or no increase in vehicular delay by lowering cycle length, lowering progression speed, dividing an arterial into smaller "coordination zones" with each zone having its own cycle length, and by abandoning coordination altogether.

In an effort to improve traffic safety and livability, many cities, often under the banner of Vision Zero (*1*), are paying increasing attention to speed control. According to NHTSA, 28% of the traffic fatalities in the United States between 2005 and 2014 were speeding related (*2*). Speeding on multimode roads (arterials, collectors, and locals) account for 83% of speeding-related fatalities (*3*). On urban roads, speeding is particularly dangerous due to the prevalence of vulnerable pedestrians and cyclists. In addition, by discouraging walking and cycling, speeding reduces livability and contributes to auto dependence with its negative effects on public health, congestion, energy resources, and climate.

The default method of speed control is setting and enforcing speed limits. Recently, for example, New York, Boston, and several other cities lowered their default speed limits from 30 to 25 mph. However, enforcement on urban roads is very difficult to accomplish by conventional methods. Speed cameras offer an effective solution if widely deployed, but they are politically controversial and are forbidden in many states. Without intense enforcement, drivers tend to ignore speed limits, choosing a speed at which they feel safe based on road geometry and other factors of the road environment such as intersection frequency (*4*).

Road geometry can be very effective at controlling speed, and is the basis for traffic calming devices such as speed humps, chicanes, and neighborhood traffic circles (*5*). However, these methods are unsuitable for arterials for several reasons including emergency vehicle response, bus service, and a desire to offer attractive speeds in order to discourage travel on local streets. And while the concept of "design speed" can be used to control speed on curvy roads, there is no such effect on most urban arterials because they have little curvature.

[1]Northeastern University, Boston, MA
[2]Kittelson & Associates, Orange, CA

Corresponding Author:
Address correspondence to Peter G. Furth: pfurth@coe.neu.edu

Where traffic volumes can be carried with a single lane per direction plus turning lanes, road diets have been highly effective at reducing speed (*6*), because with a single lane per direction, would-be speeders become impeded by vehicles ahead of them. But how can speed be controlled on multilane arterials?

It has long been recognized that on one-way arterials with close intersection spacing, traffic signals timed to offer a green wave can control speed by eliminating the incentive to go any faster than the progression speed. However, on two-way arterials, it is impossible to offer green waves in both directions when intersection spacing is short (*7*), as it is often on urban arterials. In fact, some believe that traffic signals contribute to the speeding problem by giving drivers an incentive to beat the red light when they see a signal that has been green for a while, or has just turned yellow (*8*). In practice, traffic signal timing is applied mainly to regulate conflicts, increase capacity, and decrease delay, with little or no attention given to speed control.

Research Objective and Measures of Speeding Opportunity

Though it may be impossible to provide green waves that positively control speed as easily on two-way arterials as on one-way roads, traffic signal timing may nevertheless play an important role in creating or limiting speeding opportunities. The objective of this research is to see how traffic signal timing on two-way urban arterials with short intersection spacing affects speeding opportunities, and to explore ways in which signal timing can reduce speeding opportunities without substantially increasing delay.

To speed, drivers must have both the desire and the opportunity. Absent constraining geometry or a strong threat of legal enforcement, it is natural for a substantial fraction of drivers to have the desire to speed, and so we focus on speeding opportunities. Stoplines are chosen as points of speed measurement, because intersections have the greatest potential for conflict with other road users (*9–11*). At a stopline, approaching drivers have an opportunity to speed if the signal is green and they are not impeded by a vehicle ahead of them.

Two measures of speeding opportunity are proposed:

- **Number (or fraction) of unconstrained vehicles:** A vehicle is considered unconstrained if it arrives at the stopline while the signal is green and its headway with respect to the vehicle ahead of it in the same lane is greater than 5 s. This quantity can be measured both in the field and using traffic simulation.
- **Number (or fraction) of speeding vehicles:** For traffic microsimulation, this quantity is heavily influenced by the use of a "desired speed" setting chosen by the user. To standardize this measure for microsimulation analysis, we propose assigning to 20% of the vehicles a desired speed within the range considered to be "speeding," and assigning to the remainder a desired speed not considered "speeding." In microsimulation, if the vehicles assigned a high desired speed have the opportunity to speed, they will.

The first measure does not account for speeding opportunities that arise when the vehicle ahead, though less than 5 s away, is speeding, nor does it account for speeders who may be decelerating as they close in on a slower vehicle. The second measure does not have these weaknesses, but suffers from having an arbitrary fraction of vehicles desiring to speed, and therefore cannot be expected to give a measurement that corresponds directly to a field measurement unless calibrated to match the fraction of motorists desiring to speed (a task we did not attempt).

Speeding Incentives and Opportunities with Two-Way Coordination

On a two-way arterial, ideal intersection spacing is when travel time between adjacent intersections equals half the cycle length, in which case two-way coordination can provide the same green waves as are possible with one-way coordination, and therefore they can provide a means of positive speed control. Offsets follow the "half-cycle alternate" pattern, with each intersection offset half a cycle from its neighbor, as shown in Figure 1*a*. Small adjustments to offsets can also be made to favor one direction over another. As travel time is segment length divided by progression speed, the progression speed for ideal two-way coordination is:

$$vProgression = \frac{S}{C/2} \qquad (1)$$

where

$vProgression$ = progression speed,
S = segment length or intersection spacing, and
C = cycle length.

Using lead-lag phasing, deviations from ideal spacing equal to half the split of a left-turn phase can also produce ideal bidirectional green waves (*7*), also shown in Figure 1*a*.

With ideal spacing and a progression speed at or below the target speed, there is no incentive or opportunity for the platoon leader to speed. In fact, it may be better to set the progression speed a bit lower than the

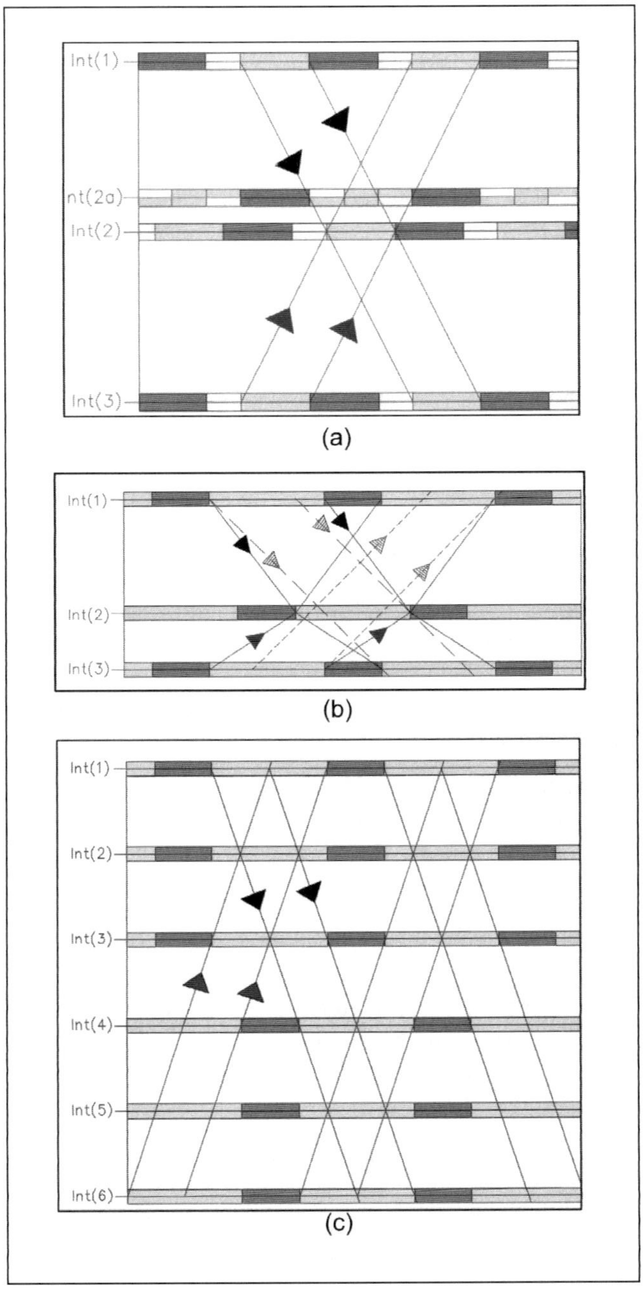

Figure 1. Coordination diagrams for a bidirectional arterial: (*a*) ideal two-way coordination, with lead-lag phasing at 2a. Yellow represents the main street's left-turn phases; (*b*) distorted progression envelope due to non-ideal intersection spacing; and (*c*) bidirectional coordination with clusters of three intersections.

target speed. Denney, Curtis, and Head (*12*) have shown how "holes" in the platoon form as vehicles turn off. A slightly depressed progression speed will slow the platoon leaders down enough to fill these holes, resulting in better capacity utilization; if progression speed is not depressed, vehicles following a hole may speed up to fill them.

Unfortunately, the conditions for ideal spacing cannot be met on most urban arterials. For example, for $C = 90$ s and *vProgression* = 40 ft/s or 27 mph, ideal segment length is 1,800 ft, far greater than signal spacing on many urban arterials. If *S* were actually 600 ft, then, holding *C* at 90 s, *vProgression* for ideal two-way progression would be 9 mph, a speed that is impractically low.

Degree of Saturation

When the degree of saturation is high, the platoon will fill almost the entire green period, and so nearly all vehicles will be constrained from speeding. With a low degree of saturation, the green interval will continue well beyond the time needed to clear the platoon, and vehicles approaching during that little-used part of the green will not be constrained.

In conventional practice, a low degree of saturation at many intersections is common, even during peak periods. One reason is the requirement of a common cycle length, typically set to meet the needs of an arterial's busiest or most complicated intersection. Intersections with less cross-street traffic or fewer phases end up with a cycle length far longer than they need, with long periods of unsaturated green that create speeding opportunities. Using smaller coordination zones could help diminish this phenomenon.

Non-Ideal Intersection Spacing

Where signal spacing is not ideal, as is the case for most urban arterials, optimal offsets still follow half-cycle synchronization, meaning every intersection's offset is either 0 or $C/2$, with offsets measured from the center of green averaged over the two directions (*13*). However, progression envelopes become distorted, with high progression speed on some segments and low progression speed on others, as illustrated in Figure 1*b*. Segments with high speed progression offer obvious speeding opportunities.

Unequal Green Intervals

Progression envelopes can likewise be distorted, creating associated speeding opportunities, when green intervals at successive intersections are unequal in length. Unequal green intervals are common; intersections where the cross-street demands are light typically give longer green periods to the arterial street. With coordinated-actuated control, arterial green intervals are random as slack time not needed by cross-street and left-turn phases is used to extend the arterial phases. Severe inequality can result where pedestrian phases, concurrent with the cross-street phase but requiring far more time, are pushbutton-actuated.

Short Segments and Intersection Clusters

On many urban arterials, intersection spacing is far too short to apply the ideal two-way progression paradigm, as discussed earlier. The standard solution, applied implicitly by signal timing software, is to cluster intersections together, with simultaneous green within each cluster, such that the travel time between adjacent clusters, measured between cluster centers, roughly equals $C/2$. Returning to the original example, if $C = 90$ s and $vProgression = 40$ ft/s, it was shown that ideal intersection spacing is 1,800 ft. If intersection spacing is actually 600 ft, then by forming clusters of three intersections, cluster spacing can be 1,800 ft, with two-way coordination as illustrated in Figure 1*c*.

If n = cluster size (i.e., number of intersections in a cluster), then cluster size for a given cycle length, progression speed, and intersection spacing is given by:

$$n = \frac{vProgression^{*}\,C/2}{S} \tag{2}$$

Common traffic signal timing software does not formally identify clusters, but clusters are apparent in their solutions. (Because of left-turn treatments and offset adjustments made to favor the peak direction, clusters with simultaneous green are not always obvious; to spot them, analysts should compare the middle of green, taking an average between the two directions.) Clusters of five or more intersections are not unusual; for example, if $C = 120$, $S = 600$ ft, and $vProgression = 50$ ft/s, n will equal 5.

In practice, Equation 2 will be rounded to an integer, because clusters must be made up of an integer number of intersections. The necessity of rounding means that small changes in *vProgression* may leave the optimal (rounded) cluster size unchanged, which in turn is likely to leave other signal timing parameters unchanged, because offsets are based on clusters and splits are largely independent of both clustering and progression speed. For example, if $C = 90$ s and $S = 600$ ft, changing *vProgression* from 30 mph (44 ft/s) to 25 mph (36.7 ft/s) when applying signal timing software is likely to leave the signal timing plan unchanged, as the ideal cluster size for the two cases (3.3 and 2.75) both round to 3.

The simultaneous green offered within a cluster (see Figure 1*c*) creates obvious speeding opportunities, especially with a large cluster size. Within a cluster, drivers may see several green lights ahead of them, giving them an incentive to go as fast as possible knowing that those green lights may not last long. (Lead-lag phasing can be used to smooth the transition between clusters, reducing—but not eliminating—this effect.)

Hypotheses

Based on the preceding analysis of the nature of two-way arterial coordination, the following hypotheses can be advanced:

H_1: Speeding opportunities tend to be greater with lower degrees of saturation, which involve longer periods of unsaturated green.

H_2: Large clusters of intersections with simultaneous green create many speeding opportunities, and arise when a combination of long cycle length and high progression speed make intersection spacing short, relative to the ideal.

H_3: Changes in progression speed that are too small to change rounded cluster size are likely to have little or no effect on optimal signal timing parameters and performance measures such as delay and speeding opportunities.

H_4: Shortening cycle lengths is particularly effective at limiting speeding opportunities because it both lowers the size of intersection clusters with simultaneous green and increases degree of saturation.

H_5: Compared with conventional arterial coordination, it may be possible to substantially reduce speeding opportunities with little or no increase in vehicular or pedestrian delay.

Study Site 1: Massachusetts Avenue

To test the effects of signal timing parameters on speeding opportunities, two corridors in Boston, Massachusetts were studied; both are sketched in Figure 2. The first is a 0.9 mi stretch of Massachusetts Avenue (Mass. Ave.), a 4-lane arterial, between St. Botolph Street and Melnea Cass Boulevard. This stretch involves nine traffic signals, with intersection spacing averaging 660 ft, with signal timing at the two extreme intersections held constant as a boundary condition. At the time of the study, the traffic signals ran coordinated-actuated with a common cycle of 120 s, except at the Southwest Corridor pedestrian crossing, 300 ft south of St. Botoloph Street, where the cycle length was 60 s. As a rule, the intersections have arterial left-turn phases.

A simulation model of the corridor was constructed using VISSIM, using its RBC module for signal control. The period studied was the a.m. peak hour, using traffic volume data supplied by the city of Boston in which the busiest northbound and southbound segments carry 1,317 and 877 vph, respectively. Each simulation run includes a 15-min warm up period, and reported results are averages of five simulation runs. Measures of performance included average network delay (average delay to all vehicles), corridor delay (average delay to vehicles running the full length of the corridor, averaged between the two directions), number of unconstrained arrivals summed over all of the Mass. Ave. stoplines, and average cycle length as a proxy for average pedestrian delay.

Three alternatives were evaluated: coordinated-actuated control (the scheme currently operated, with

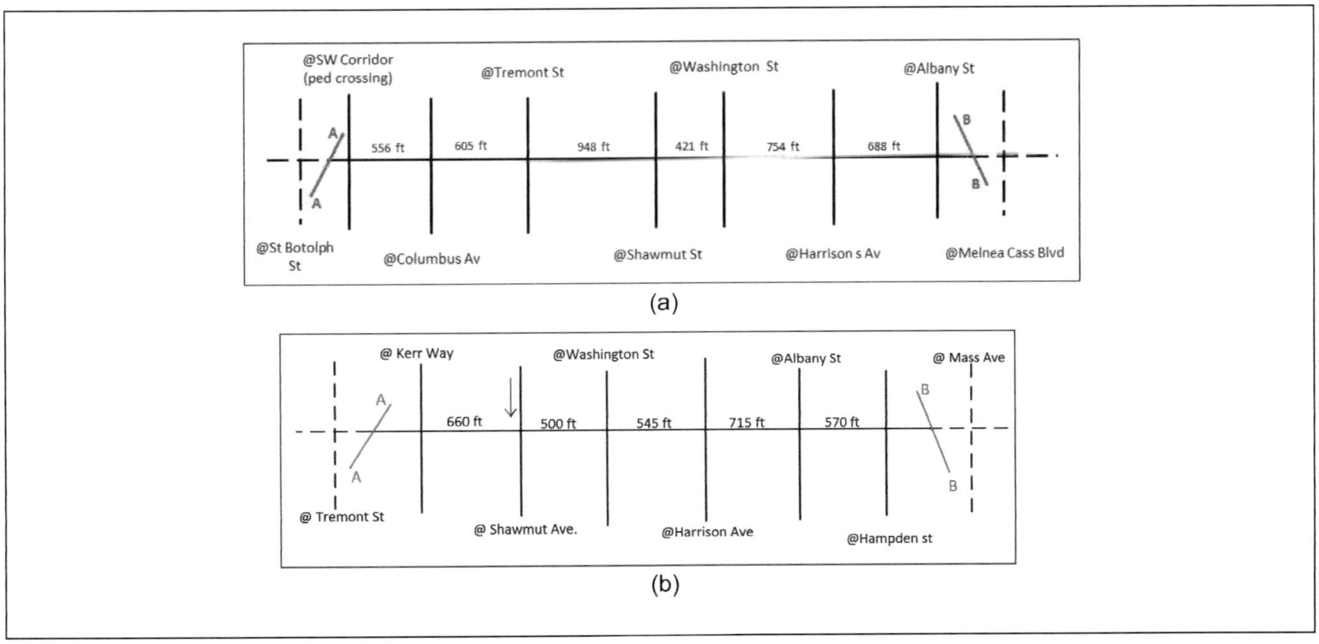

Figure 2. Study Corridors: (*a*) Massachusetts Avenue and (*b*) Melnea Cass Boulevard.

Table 1. Results for Three Signal Timing Plans (Massachusetts Avenue)

Signal timing plan	Average cycle length (s)	Corridor delay (s)	Average network delay (s)	Unconstrained arrivals per h
Coordinated-actuated	113	65	55	2,283
Actuated	78	135	61	798
Zonal coordinated	76	110	48	1,431

timings reoptimized), fully actuated control, and a "zonal coordination" plan. In the zonal coordination alternative, the Southwest Corridor pedestrian crossing ran fully actuated (with pedestrian recall), and the remaining six interior intersections were grouped into three zones of two intersections each, with coordinated-actuated control within each zone, but no coordination between them. Timing plans for the three zones were determined using Synchro, with small manual adjustments, which resulted in cycle lengths of 65 s for the middle zone and 80 s for the other two zones. Segments between zones were at least 750-ft long, long enough to avoid queue interactions.

Key results are shown in Table 1. Coordinated-actuated control, with its long signal cycle and long green periods, has the most speeding opportunities, while also offering the lowest corridor delay. Fully actuated control has the fewest speeding opportunities, but has the longest vehicular delays. The zonal coordination plan has an intermediate number of speeding opportunities, and although its corridor delay is greater than the unzoned coordination plan, it lowers delay so much for crossing traffic and turning traffic that it achieves the lowest average delay for all vehicles.

These results support both hypotheses H_4 and H_5. The two alternatives with substantially shorter cycle lengths allow substantially fewer speeding opportunities. And by placing less emphasis on corridor delay, one alternative (zonal coordination) was found that reduces speeding opportunities by 37% while simultaneously lowering average vehicular delay; another (full actuation) was found that reduces speeding opportunities by 65% while increasing average vehicular delay by only 11%, or 6 s per vehicle (albeit while increasing average corridor delay substantially).

Field Test Confirmation

A field test was undertaken to confirm the ability of the simulation software to model vehicular movements in a way that accurately represents unconstrained arrivals. The southbound approach to the intersection at Tremont Street was observed from 7:30 to 8:30 a.m. on a weekday, videoing the approach and the replaying the video to manually count vehicles and classify them as unconstrained or not. The total number of arriving vehicles was about 10% greater in the field study than in the simulation (807 versus 729).

The fraction of arrivals classified as unconstrained was 21.7% in the field study, versus 21.8% in the simulation study. This result confirms the validity of the simulation model for measuring unconstrained arrivals.

Study Site 2: Melnea Cass Boulevard

The second study site is a 0.87-mi stretch of Melnea Cass Boulevard (MCB), a 4-lane arterial, between Tremont Street and Mass. Ave. In the simulation model, control at the extreme intersections was left unchanged in order to provide consistent boundary conditions, leaving six interior intersections, as shown in Figure 2*b*. Intersection spacing averages 600 ft.

Currently, the traffic signals run coordinated-actuated with a common cycle of 120 s in the p.m. peak, with a cluster of five intersections (Kerr Way through Albany Street) whose green is essentially simultaneous. This large cluster means that drivers can see green signals for several intersections ahead of them, giving them a strong incentive to speed to try to get through as many intersections as possible before the green ends. Speeding is a common complaint.

The current layout has left-turn lanes on MCB for only some of the intersections, and has protected plus permitted left-turns throughout. In all of the alternatives studied including the base case, the missing left-turn lanes were added and left turns from MCB are protected only. These changes were made to permit a comparison for operation in the near future when the street is rebuilt with a full set of left-turn lanes and with protected lefts.

The corridor was modeled using VISSIM, using its RBC module for signal control. Two different volumes were assigned to the network: p.m. peak hour, using traffic volumes obtained from the city of Boston, and off-peak, defined as 50% of p.m. peak volumes. Each simulation run covers 60 min following a 15-min warm-up period, and results are averages from five simulation runs. Eighty percent of the vehicles were assigned to a class whose desired speed varies from 28 mph to 32 mph, and 20% to a class with desired speed between 38 mph and 42 mph. Performance measures included average vehicular delay, corridor travel time, unconstrained arrivals at all of MCB's interior stoplines, and number of speeders at all of MCB's stoplines. Delay is measured compared with desired speed, which had the same distribution in all alternatives. Speeders were defined as vehicles with speed exceeding 35 mph.

All of the control alternatives tested use coordinated-actuated control, with timing parameters determined using Synchro. The minimum split of through movements is sufficient for concurrent pedestrian crossings. An initial set of tests with and without lead-lag phasing allowed showed that lead-lag phasing resulted in significantly less delay, with little difference in speeding opportunities, and so all of the control alternatives allow lead-lag phasing.

Cycle Length, Progression Speed, and Cluster Size

As discussed earlier, two-way coordination creates a relationship between progression speed *vProgression*, cycle length C, and cluster size n, whose unrounded value is given by Equation 2. For MCB with its average intersection spacing of 600 ft, Table 2 shows the values of unrounded n that correspond to different choices of *vProgression* and C; bold formatting is used to indicate combinations expected to result in the same rounded value of n, assuming a bias in rounding in which fractional values above 0.4 are rounded up.

Impact of Volume, Cycle Length, Progression Speed, and Degree of Recall

Table 3 shows the performance measures for two sets of volumes (peak and off-peak), a variety of cycle lengths between 70 and 120 s, and a variety of progression speeds between 15 and 35 mph. Corridor travel time was measured but is not reported for conciseness because it was so strongly correlated with network delay (correlation coefficient 0.99 off-peak, 0.96 peak).

An important methodological finding is that the two measures of speeding opportunity proposed are strongly correlated, with correlation coefficient 0.97 off-peak and 0.98 peak. The overall average ratio of speeders to unconstrained vehicles is 0.23 off-peak and 0.19 in the peak, close to the specified fraction (0.20) of drivers whose desired speed lies in the "speeding" range. This finding confirms that unconstrained vehicles is a good measure of speeding opportunity; it also suggests that

Table 2. Implied Cluster Size as a Function of Cycle Length and Progression Speed When Intersection Spacing Is 600 ft

	Cycle length (s)			
Progression speed (mph)	70	80	100	120
15	1.3	**1.5**	**1.8**	**2.2**
20	**1.7**	**2.0**	2.4	2.9
25	**2.1**	2.4	3.1	**3.7**
30	2.6	**2.9**	**3.7**	**4.4**
35	3.0	**3.4**	**4.3**	5.1

Note: Bold indicates cells with a common rounded cluster size.

Table 3. Timing Plan Performance for Different Cycle Lengths and Progression Speeds

				Off-peak			Peak		
Cycle length (s)	Progression speed (mph)	Ideal cluster size	Pedestrian delay (s)	Vehicular delay (s)	% unconstrained vehicles	% speeders	Vehicular delay (s)	% unconstrained vehicles	% speeders
70	15	1.3	28	57	12%	2.7%			
	20	1.7	28	33	16%	4.1%			
	25	2.1	28	32	16%	4.1%			
	30	2.6	28	31	22%	5.5%			
	35	3.0	28	33	22%	5.4%			
80	15	1.5	33	54	16%	3.6%	80	9%	1.2%
	20	2.0	33	44	17%	3.7%	46	10%	1.8%
	25	2.4	33	36	22%	5.2%	45	10%	2.1%
	30	2.9	33	30	24%	5.7%	43	11%	2.3%
	35	3.4	33	41	24%	5.7%	50	14%	2.8%
100	15	1.8	43	46	23%	4.7%	57	12%	2.1%
	20	2.4	43	38	23%	4.7%	45	15%	2.9%
	25	3.1	43	37	25%	4.9%	43	15%	3.0%
	30	3.7	43	33	28%	6.0%	43	16%	3.3%
	35	4.3	43	35	32%	7.0%	58	16%	3.3%
120	15	2.2	53	62	23%	5.2%	74	12%	2.1%
	20	2.9	53	48	25%	5.5%	54	15%	3.2%
	25	3.7	53	35	35%	8.5%	54	17%	3.5%
	30	4.4	53	35	36%	8.6%	49	20%	4.0%
	35	5.1	53	35	36%	8.3%	49	20%	4.0%

either of the proposed measures of speeding opportunity can be used with similar fidelity.

Looking at differences in performance, one obvious result is that speeding opportunities are far greater with off-peak volumes than peak volumes, confirming hypothesis H_1.

Another is that speeding opportunities tend to increase with cycle length, illustrated in Figure 3 and confirming hypothesis H_4. Capacity analysis shows that the minimum common cycle that provides sufficient capacity and satisfied pedestrian minima is 70 s for off-peak and 80 s for peak volumes; increasing cycle length beyond that minimum, especially with the low degree of saturation prevalent in the off-peak, substantially increases speeding opportunity.

The impact of changes in progression speed tends to be stepped, as illustrated in Figure 4 for the case of off-peak volumes and $C = 70$ s. In most cases, it makes a difference only when the change in progression speed is enough to alter cluster size, confirming hypothesis H_3.

Figure 4 also shows how three recall options – no ped recall (pedestrians use a pushbutton), ped recall (pedestrian phase is called automatically, our default option), and maximum recall (same as pre-timed operation) – affect speeding opportunities. When not in pre-timed mode, fixed forceoff (as opposed to floating forceoff) is applied to non-coordinated phases, consistent with local practice. Weaker recall settings randomly start arterial phases early, creating additional speeding opportunities that are evident in the results. However, stronger recall settings tend to give more unsaturated green time to cross streets, which may create speeding opportunities there, and so it would be unwise to conclude anything about what are the best recall settings for speed control without studying speeding on the cross streets as well as the arterial.

With regard to hypothesis H_2, Table 3 shows a clear trend of speeding opportunities increasing with cluster size. For the off-peak, the case with the smallest values of n has only 12% unconstrained vehicles, versus 36% for the case with the largest n. This confirms the hypothesis that large clusters of intersections with roughly simultaneous green create speeding opportunities.

Table 3 also shows the vehicular delay, pedestrian delay, and corridor travel time for each signal timing alternative. Corridor travel time is so strongly correlated with vehicular delay in this case study that it is not further discussed. Pedestrian delay is likewise strongly correlated with cycle length. Note that vehicular delay was measured using VISSIM and is based on desired speed, which follows the same distribution with an average value of 32 mph in every alternative, regardless of the progression speed used to select signal timing settings.

Because average desired speed is 32 mph, one might expect that the least delay occurs when signal timing has been optimized for a speed of 30 or 35 mph. Indeed,

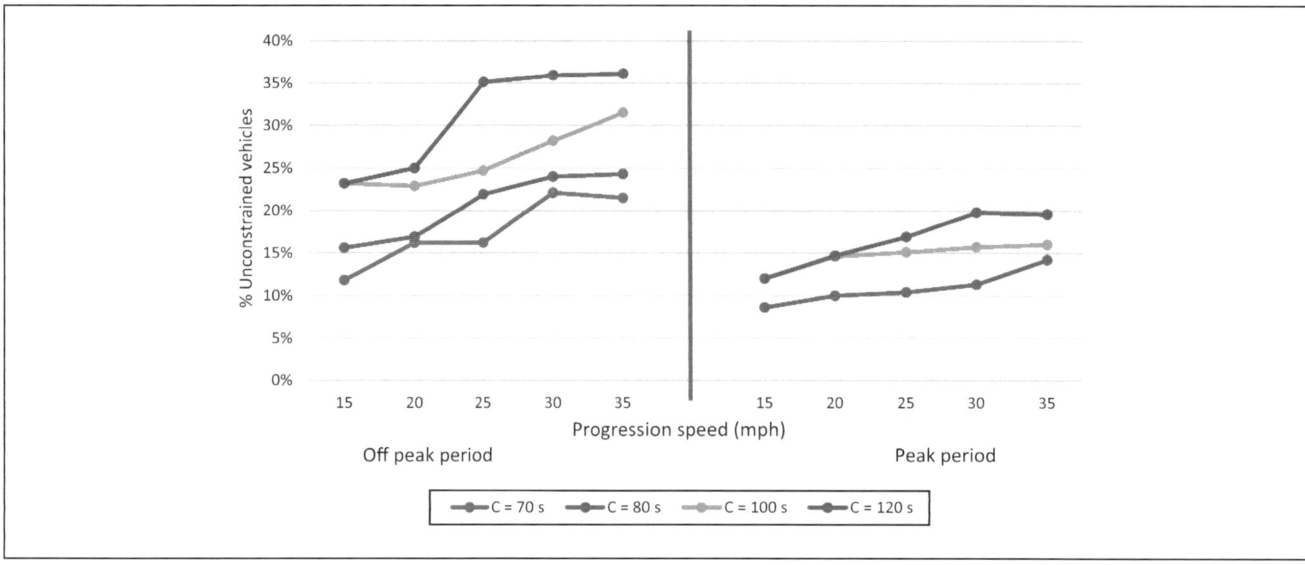

Figure 3. Unconstrained vehicles as a function of cycle length, specified progression speed, and traffic volume.

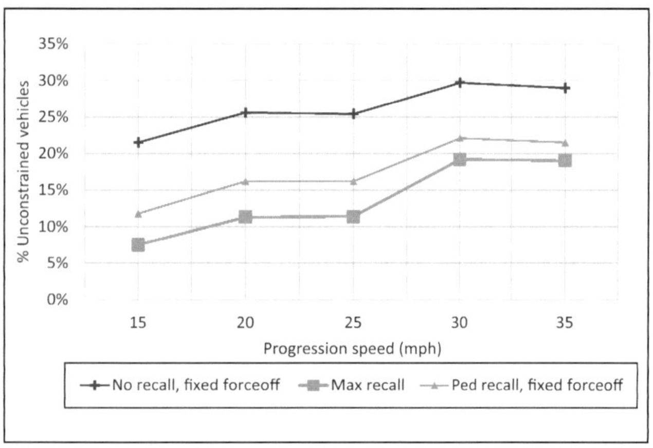

Figure 4. Unconstrained vehicles versus progression speed (off-peak volumes, cycle length = 70 s).

progression at 30 mph gives consistently low-delay results; in contrast, progression at 35 mph clearly tends toward greater delay. This confirms the beneficial effect mentioned earlier of a progression speed slightly below the average speed. Solutions with a progression speed of 15 mph have a large average delay, but most of those with progression speeds of 20 and 25 mph perform well with respect to vehicular delay.

Figure 5 illustrates the tradeoff between average vehicle delay and unconstrained vehicles for all the solutions in Table 3. One can see that for both off-peak and peak, Synchro's recommended timing plan based on default parameters performs well in relation to vehicular delay, but has considerably more speeding opportunities. Off-peak, reducing progression speed from 30 to 25 mph cuts speeding opportunities by 27% while increasing vehicular delay by only 5% (and leaving pedestrian delay unaffected); in the peak, reducing the cycle length from 100 to 80 s cuts speeding opportunities by 28% without any change in vehicular delay (and with a 23% decrease in pedestrian delay).

Conclusion and Further Research

An increasing number of cities recognize that controlling vehicle speeds in vital for improving safety and livability. This research has explored the potential for doing so using traffic signal settings on multilane, two-way urban arterials with close intersection spacing.

The number of unconstrained vehicles, measured at stoplines, was found to be a valid measure of speeding opportunities. A field measurement confirmed that this measure could be reliably measured by traffic microsimulation, and simulation study confirmed that it is strongly correlated with the number of speeding vehicles in a setting in which 20% of all vehicles were specified as having a desired speed in the "speeding" range.

Both case studies found that compared with conventional arterial coordination, it was possible to substantially reduce speeding opportunities with little or no increase in vehicular delay. Case study experiments confirmed that with standard arterial coordination, speeding opportunities increased with longer cycles, lower degree of saturation, and closer intersection spacing. Speeding opportunities are related to cluster size, that is, the number of consecutive intersections with simultaneous offsets, which is inversely proportional to effective intersection spacing (travel time between neighboring intersections measured in number of cycles). Shortening cycle length is particularly effective at limiting speeding

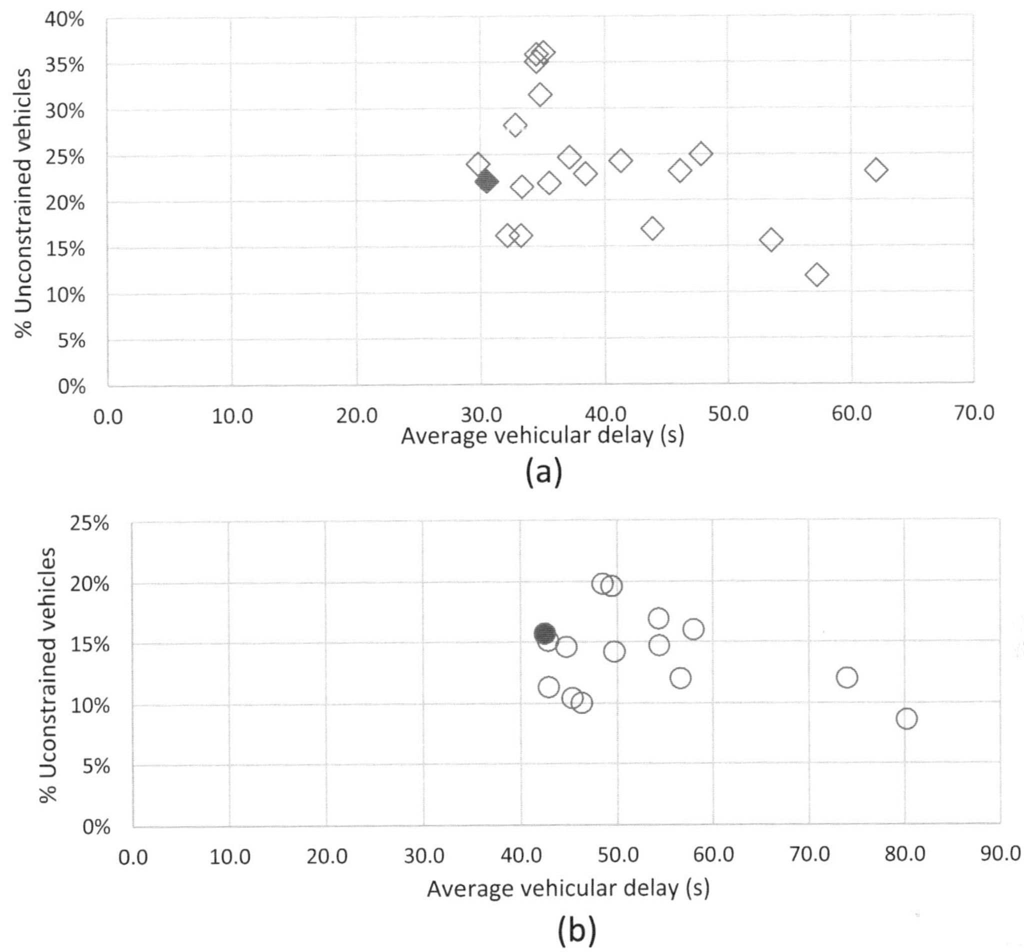

Figure 5. Delay vs. speeding opportunities for timing plans with different cycle length and progression speed. Solid markers indicate timing plans recommended by Synchro: (*a*) off-peak; and (*b*) peak.

opportunities because it both increases effective intersection spacing and reduces periods of unsaturated green; it also has the side benefit of lowering pedestrian delay.

Lowering progression speed – an input to signal timing software – can also reduce speeding opportunities, because it increases effective intersection spacing. However, small changes in progression speed can have no effect at all if they do not lead to a decrease in cluster size. Recall and minimum green parameters that make the length of the arterial through phases less variable also help reduce speeding opportunities.

Abandoning coordination altogether has the strongest effect on reducing speeding opportunities, but the delays and queue interactions this could lead to may be unacceptable. Short of that, breaking an arterial into small coordination zones, with coordination within each zone but each zone free to have its own cycle length, was found in one case study to sharply reduce speeding opportunities without increasing network delay. Adaptive control methods that achieve a degree of coordination without imposing a common cycle, such as self-organizing control (*14*), may also be effective at limiting speeding opportunities while maintaining a good level of service; that question is left for further research.

Author Contributions

The authors confirm contribution to the paper as follows: study conception and design: PF, AH; data collection: AH, JL, WH; analysis and interpretation of results: PF, AH, BC; draft manuscript preparation: PF, AH, JL, WH. All authors reviewed the results and approved the final version of the manuscript.

References

1. Vision Zero Network. http://visionzeronetwork.org/resources/vision-zero-cities/.

2. National Center for Statistics and Analysis. *Speeding: 2014 Data.* (Traffic Safety Facts. Report No. DOT HS 812 265). National Highway Traffic Safety Administration, Washington, D.C, 2016.
3. Elvik, R., P. Christensen, and A. Amundsen. *Speed and Road Accidents. An Evaluation of the Power Model.* TØI Report 740/2004. Institute of Transport Economics, TØI, Oslo, Norway, 2004.
4. Debnath, A., H. Narelle, and A. Rakotonirainy. Driver Beliefs Regarding the Benefits of Reduced Speeds. *Journal of Transportation Safety & Security*, Vol. 9, No. 4, 2017.
5. Ewing, R. *Traffic Calming: State of the Practice.* Institute of Transportation Engineers, Washington, D.C., 1999.
6. Knapp, K., B. Chandler, J. Atkinson, T. Welch, H. Rigdon, R. Retting, S. Meekins, E. Widstrand, and R. J. Porter. *Road Diet Informational Guide.* FHWA-SA-14-028, Federal Highway Administration, Washington, D.C., 2014.
7. Urbanik, T., A. Tanaka, B. Lozner, E. Lindstrom, K. Lee, S. Quayle, S. Beaird, S. Tsoi, P. Ryus, D. Gettman, and S. Sunkari. *NCHRP Report 812: Signal Timing Manual* - Second Edition. Transportation Research Board, of the National Academies, Washington, D.C. 2015.
8. Persaud, B., E. Hauer, R. A. Retting, R. Vallurupalli, and K. Mucsi. Crash Reductions Related to Traffic Signal Removal in Philadelphia. *Accident Analysis and Prevention*, Vol. 29, No. 6, 1997, pp. 803–810.
9. Stutts, J. C., W. W. Hunter, and W. E. Pein. Pedestrian Crash Types: 1990s Update. *Transportation Research Record: Journal of the Transportation Research Board*, 1996. 1538: 68–74.
10. Choi, E. *Crash Factors in Intersection-related Crashes: An On-scene Perspective.* Report DOT HS 811 366. National 38 Highway Traffic Safety Administration (NHTSA), Washington, D.C., 2010.
11. Bhesania, R. P. Using Crash Statistics and Characteristics to Improve Safety. *ITE Journal*, Vol. 61, No. 3, 1991.
12. Denney, R. W., Jr, E. Curtis, and L. Head. Long Green Times and Cycles at Congested Traffic Signals. *Transportation Research Record: Journal of the Transportation Research Board*, 2009. 2128: 1–10.
13. Morgan, J. T., and J. D. C. Little. Synchronizing Traffic Signals for Maximal Bandwidth. *Operations Research*, Vol. 12, No. 6, 1964, pp. 896–912.
14. Cesme, B., and P. G. Furth. Self-organizing Traffic Signals Using Secondary Extension and Dynamic Coordination. *Transportation Research Part C: Emerging Technologies*, Vol. 48, 2014, pp. 1–15.

The Standing Committee on Traffic Signal Systems (AHB25) peer-reviewed this paper (18-04160).

TRR
JOURNAL OF THE TRANSPORTATION RESEARCH BOARD

Article

Transportation Research Record
2018, Vol. 2672(18) 117–127

DOI: 10.1177/0361198118786840
journals.sagepub.com/home/trr
SAGE

A Simulation Study on Max Pressure Control of Signalized Intersections

Xiaotong Sun[1] **and Yafeng Yin**[1]

Abstract

In this paper, a decentralized traffic signal control strategy named max pressure control is reviewed and examined. This control strategy aims at optimizing overall network throughputs, but applies a distributed approach that only requires local information to generate timing plans for each intersection. A Vissim simulation study is conducted to compare existing max pressure schemes. The results show that a recently proposed cyclic-based approach performs more poorly than the original non-cyclic approach. Further, to address two issues that hinder existing schemes: frequent changes of phase and queue spillover/blockage, two modifications are suggested. The simulation reveals that network performance can be improved after modifications.

Max pressure signal control (MP) is a distributed traffic signal control strategy that has attracted much attention in recent years (*1–4*). Unlike traditional centralized control strategies that involve high computational complexity, the essence of MP is a decentralized algorithm that requires only local information to perform a flexible control for each intersection. MP holds great promise for real-time adaptive control of traffic signals, particularly when robust communication is made available with the deployment of connected vehicle technologies. The control is highly scalable and can be readily implemented for large-scale urban traffic networks. More importantly, MP has been theoretically proven to bound the queue length under largest possible arrival demands, a global optimality that has not yet been established for some other decentralized approaches (*1–3*, *5*).

Since the global optimality of MP is theoretically guaranteed by assuming point or vertical queues (*2–4*), simulation studies are necessary to validate the effectiveness of MP signal control. However, previous simulation studies either adopted macroscopic models that do not well represent driving behaviors and traffic dynamics, or considered overly simplified network settings in which MP control cannot be adequately examined. For example, Kouvelas et al. (*6*) developed an event-based simulator to compare two variations of MP, but this macroscopic simulation study neglects considerations of car following and lane changing behaviors. Other studies applied Simulation of Urban Mobility (SUMO) as their simulation platform (*4*, *7*), which, again, is weak in replicating realistic driving behaviors. Although some other studies applied microscopic traffic simulators, the scenario settings are oversimplified, for example, isolated intersection (*8*) or two-phase timing plans (*9*). These studies cannot fully examine the distributed nature of MP signal control and thus impede fair validation of its performance. As a conclusion, a simulation study that employs a widely-accepted microscopic traffic simulator and well-defined traffic network becomes necessary.

In this paper, we review the potential benefits and drawbacks of MP policies through a comprehensive simulation study based on a real traffic network. Unlike many of the previous simulation studies (*4*, *7*, *10*), Vissim is applied in this study for better replication of traffic network topology and flow dynamics. The original non-cyclic MP, cyclic phase MP, and the currently-implemented actuated signal control are compared in terms of travel time, delay, throughput, and queueing dynamics. Possible modifications of MP to overcome policy deficiencies are further discussed.

The remainder of this paper is organized as follows. A literature review on MP control is presented first, followed by mathematical formulations of two existing MP policies as well as modifications of the original non-cyclic MP policy. The next two sections introduce the simulation settings, and then analyze and compare the

[1]Department of Civil and Environmental Engineering, University of Michigan, Ann Arbor, MI

Corresponding Author:
Address correspondence to Xiaotong Sun: xtsun@umich.edu

performances of the control policies in the simulated traffic network. The last section presents the conclusions of the paper.

Literature Review

Network-based signal control strategies can be classified into centralized and decentralized controls depending on their system architectures, while the latter is also referred to as distributed control. Centralized control systems usually possess a central computing unit where all control decisions take place. In practice, centralized controls normally have limited scalability since they are subject to increasing computational complexity and the lack of reliable communication links. In contrast, fully distributed signal control systems have no central computing units so that control operations rely on local intersection controllers. In some classic multi-hierarchy control strategies, such as OPAC (*11*) and RHODES (*12*), distributed controls are implemented for individual intersections using local information from detection and prediction. In recent years, research interest in distributed signal control has been aroused worldwide because of its scalability and flexibility (*13–16*). MP control policy, which is also referred to as back-pressure control policy, is one such system.

The idea of MP was first proposed for packet scheduling problems in communication networks (*5*). The control policy was constructed as a linear programming to maximize the total network throughput with queueing dynamics being described as a Markov process while a new terminology of pressure was introduced to denote the difference of upstream and downstream queues. An efficient algorithm was designed and has been successfully implemented in wireless network controls. Several adjustments were made when applying MP to traffic signal control. For example, since in wireless network control the destinations and routing information of packets are known, control units directly control flow rates. In contrast, hard to obtain destinations and routing information of vehicles are in signalized traffic networks: MP signal control determines green time durations instead of flow rates (*1*). It is also worth noting that MP signal control differs from another pressure-based signal control proposed by Lämmer and Helbing earlier (*17*), which calculates pressure with information at each isolated intersection while neglecting dependency and coordination between adjacent intersections.

To prove that MP achieves maximum throughput for traffic networks, previous studies considered point or vertical queues and assumed that storage capacities are infinite (*1*, *2*). A store-and-forward network is often used to describe queueing dynamics (*18*). However, performing store-and-forward model in microscopic simulation studies may lead to traffic conservation failures and congestion propagation (*8*). Several efforts have been made to relax this assumption. A concept called weighted queues was introduced and tested by an event-based macroscopic simulation model (*2*, *6*) and showed significant improvement of total throughput as compared with the original MP control. Gregoire et al. reformulated pressure and proposed a variation of MP named "capacity-aware backpressure control." This capacity-aware control was also validated by a simulation study and proven to be stabilized in a grid network under high demand scenarios where original MP fails. Another denotative study was focused on MP control in a switching network with limited buffer size (*19*). Although not considering traffic networks, this pioneering study proved that when the control strategy is still modeled as a linear programming problem, the maximum throughput cannot be achieved: the actual throughput achieved is regulated by the buffer size and number of nodes in the network.

While achieving maximum throughput, MP produces unpredictable phase sequence. A straightforward remedy is to fix cycle length and phase sequence before determining the phase durations based on pressure. Kouvelas et al. suggested determining phase durations proportionally based on the pressure magnitudes (*6*). The suggestion fails when pressures of some phases become non-positive. An enhanced version named "cyclic phase backpressure control policy" was later proposed in which an exponential function of pressure is introduced to determine the green splits, and was proved to stabilize the traffic network for a certain demand region (*4*). This policy was further extended to allow different priorities for different phases to be activated. However, this extended cyclic phase MP maintains network stability only under mild traffic conditions where the maximum queue length does not exceed half of the maximum storage capacity (*20*, *21*).

Another relevant stream of research is to integrate MP control with dynamic traffic routing due to its simplicity and adaptivity. For example, Zaidi et al. (*22*) applied MP in a multi-commodity network, where vehicles adopt fixed or adaptive routing respectively. The comparison study showed that adaptive routing performed better when traffic demand is high. Cai et al. (*10*) incorporated another variation of MP with dynamic traffic routing. This MP variation not only considers the difference between in-link and out-link queue lengths, but also includes upstream demand during service time as a part of pressure calculation. However, the authors did not prove or demonstrate the stability of the proposed MP variation.

Max Pressure-Based Signal Control

In this section, the control of a general traffic network with signalized intersections is studied. Traffic related variables as well as control related variables are introduced first. Control policies stated by Vairaya (2) are then reviewed. Mathematical models are established to describe the vehicle arrivals and departures and queueing process to facilitate a better understanding of the stability property of MP. Two modifications on MP and another category of cyclic control proposed by Le et al. (3) are also presented.

Notations and Terminologies

The notations used are listed below. As for the important terminology "phase," the Manual on Uniform Traffic Control Devices (MUTCD) defines it as the right-of-way, yellow change and red clearance intervals in a cycle that are assigned to an independent traffic movement (23). In the NEMA standard, eight phases corresponding with through and left turning movements from four approaches in a typical intersection are numbered from 1 to 8. Meanwhile, the Signal Timing Manual defines a "phase" as a controller timing unit associated with the control of one or more movements (24). Both manuals use the term "phase pair" to describe a couple of phases that can be activated simultaneously in an actuated signal scheme. In this paper, we use the term "phase" consistent with the usage in the Signal Timing Manual.

Max Pressure Control—Varaiya's Model

Varaiya (2, 3) described the MP signal control policy (referred to hence as non-cyclic MP) as follows.

Considering the *weight* associated with each movement from link l to link m as the difference between the queue and the average queue from downstream intersections:

$$w_{l,m}(t) = q_{l,m}(t) - \sum_{n \in O_m} p_{m,n}(t) q_{m,n}(t) \tag{1}$$

The associated *pressure* for each phase is thus given by:

$$\gamma_\sigma(t) = \sum_{(l,m) \in \sigma} s_{l,m}(t) w_{l,m}(t) \tag{2}$$

in which $(l, m) \in \sigma$ means that phase σ serves movement from l to m.

Traffic-related notations	
$G(\mathcal{N}, \mathcal{A})$	General traffic network
$\mathcal{N}$	The set of all nodes or intersections
N	Total number of intersections
$\mathcal{A}$	The set of all directed links
$\mathcal{A}_I$	The set of all entry links, through which traffic flows enter the network
I_l	The set of all links whose ending nodes are the starting node of link l. For, $l \in \mathcal{A}_I$, $I_l = \emptyset$
O_l	The set of all links whose starting nodes is the ending node of link l
f_l	Average flow on link l
d_l	Average hourly traffic demand entering from link l, $l \in \mathcal{A}_I$
$r_{l,m}$	Average proportion of flow on link i that enters link j
$N_{l,m}$	Number of lanes for movement from link l to link m
$L_{l,m}$	Length of lanes for movement from link l to link m
$q_{l,m}(t)$	Queue length of movement from link l to link m at time t
$p_{l,m}(t)$	Proportion of flow on link l that enters link m at time t
$a_l(t)$	Exogenous vehicle arrival rate at time t, for $l \in \mathcal{A}_I$
Control-related notations	
Φ_i	Phase sequence predetermined for each cycle at intersection i
Π_i	The set of phases at intersection i
T_0	Time horizon for signal control that is discretized into a number of time slots
Δ	Time interval of each time slot
$w_{l,m}(t)$	Weight associated with movement from link i to link j
$\gamma_\sigma(t)$	Pressure associated with phase σ at time t
$s_{l,m}(t)$	Saturation flow rate on movement from l to m
$I^\sigma_{l,m}(t)$	A time-dependent binary variable, which equals 1 if phase σ serving movement from link l to link m is actuated at time t; otherwise 0.
ST	Time horizon for signal control in simulation study
Δs	Simulation second
o_i	Offset of intersection i from intersection $i-1$, which equals 0 for intersection 1
r	Mandatory red time set in signal control used in simulation study
a	Mandatory yellow time set in signal control used in simulation study
g_{min}	Universe minimum green time used in simulation study

The phase with the maximum pressure is activated in $t+1$th time slot:

$$\sigma^*(t+1) = \text{argmax}\{\gamma_\sigma(t) | \sigma \in \Pi_i\} \tag{3}$$

Equation 3 is a decomposition of the global maximization objective

$$\boldsymbol{\sigma}^*(t+1) = \text{argmax} \sum_i^N \{\gamma_\sigma(t) | \sigma \in \Pi_i\} \tag{4}$$

where $\boldsymbol{\sigma}^*(t)$ is the vector of the activated phases of all intersections, and N is the total number of intersections. Notice that the phase sequence determined by this rule will always be changeable, which makes settings on cycle length and offsets meaningless.

With regard to queueing process, the store-and-forward modeling paradigm (*18*) is applied to describe it in a discrete manner. Equation 5 describes how the queue length of movement from link l to link $m(m \in O_l)$ progresses from time t to $t+1$. In this equation, $s_{l,m}(t)I^\sigma_{l,m}(t) \wedge q_{l,m}(t)$ is equal to the minimum value of $s_{l,m}(t)I^\sigma_{l,m}(t)$ and $q_{l,m}(t)$, measuring the maximum number of vehicles that could leave link l to link m under the infinite storage capacity assumption. $\left(\sum_{k \in I_l} s_{k,l}(t)I^\sigma_{k,l}(t) \wedge q_{k,l}(t)\right)p_{l,m}(t)$ is the endogenous arrivals from upstream of l. a_l is the exogenous arrival rates at time t, which is known during the controling process. If $l \notin \mathcal{A}_l$, then $a_l(t)$ is zero for all t.

$$q_{l,m}(t+1) = q_{l,m}(t) - s_{l,m}(t)I^\sigma_{l,m}(t) \wedge q_{l,m}(t) + a_l(t) + \left(\sum_{k \in I_l} s_{k,l}(t)I^\sigma_{k,l}(t) \wedge q_{k,l}(t)\right)p_{l,m}(t) \tag{5}$$

MP control policy defined in this way is proved to guarantee maximum throughput. To elaborate, let us introduce the concept of the traffic demand region. Consider that in the controling horizon T_0, the average demand and turning proportions are defined in Equations 6 and 7:

$$d_l = \frac{1}{T_0}\sum_{t=1}^{T_0} a_l(t) \tag{6}$$

$$r_{l,m} = \frac{1}{T_0}\sum_{t=1}^{T_0} p_{l,m}(t) \tag{7}$$

With the notations listed above, we can express the flow conservation conditions as follows:

$$f_i = d_i, i \in \mathcal{A}_I \tag{8}$$

$$f_j = \sum_{i \in I_j} r_{i,j} f_i, j \in \mathcal{A} \backslash \mathcal{A}_I \tag{9}$$

For the overall network, vector $\mathbf{f} = \{f_l, f_m, f_n, \ldots\}$ denotes all link flows; demand vector $\mathbf{d}$ denotes traffic demands, and the combinations of turning ratios can be defined by a proportion matrix $\mathbf{P}$. We thus have:

$$\mathbf{f} = \mathbf{dP} \tag{10}$$

For the supply side, define

$$\Sigma_{l,m} = \frac{1}{T_0}\sum_{t=0}^{T_0} I^\sigma_{k,m}(t) \tag{11}$$

Evidently, $\Sigma_{l,m}$ is the average amount of time in each time slot that movement (l, m) is activated. The average number of vehicles driven from link l to m thus will be bounded by $\Sigma_{l,m}s_{l,m}$. Consequently, we have

$$f_l r_{l,m} \le \Sigma_{l,m} s_{l,m} \tag{12}$$

Let D be the set of demand vectors $\mathbf{d}$ that follow conservation Equation 10 and capacity constraint (Equation) 12; clearly, the demand region D is a convex set. Varaiya (*2*) demonstrated that all queue lengths in the network are bounded under MP if traffic demand is in the demand region D. It has been also proved that any demand region larger than D cannot be stabilized by any control policy, meaning that MP achieves the maximum throughput. More detailed proofs and analyses about stability can be found in Varaiya (*2*) and Tassiulas and Ephremides (*5*).

Modifications on Non-Cyclic MP

As demonstrated in the simulation study below, although setting a smaller time interval utilizes the most up-to-date traffic information, it results in frequent phase switches and relative short green durations for each phase. This limitation has also been identified in a previous test experiment on MP using a discrete event simulator, and a modified rule is established to reduce the number of phase switches (*25*). As a remedy, we propose to introduce a minimum green time to the MP algorithm. To let the control algorithm be aware of the finiteness of the storage capacity, we further employ a normalized queue instead of the absolute queue length in the pressure calculation. The idea of normalized queue is similar to weighted queue proposed by Varaiya (*2*), but differs in *weight* calculation. The modified normalized MP is shown as follows. The *weight* is defined as

$$w_{l,m}(t) = \frac{q_{l,m}(t)}{L_{l,m}N_{l,m}} - \sum_{n \in O_m} \frac{p_{m,n}(t)q_{m,n}(t)}{L_{m,n}N_{m,n}} \tag{13}$$

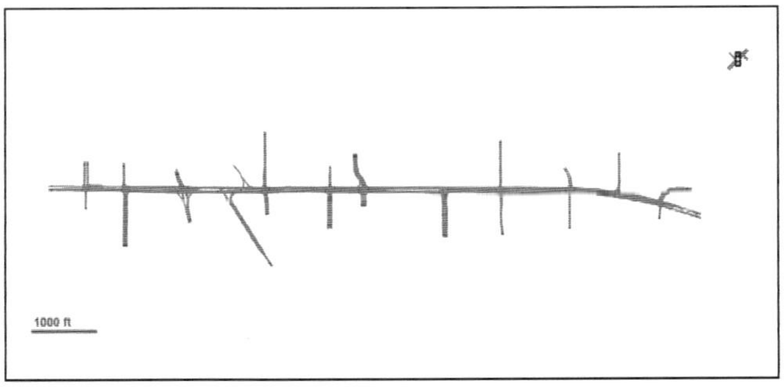

Figure 1. Vissim network model of Newberry Road in Gainesville, FL.

The associated pressure is still calculated as per Equation 2, and the right-of-way is assigned according to Equation 3.

Cyclic Phase Max Pressure Control—Le's Model

The cyclic phase MP control policy (referred to as cyclic phase MP) requires setting fixed cycle lengths, offsets, and phase sequence in the first place. The timing plan is determined by the following procedure: *weight* associated with each movement and *pressure* associated with each phase are still determined by Equations 1 and 2. In contrast, green splits for intersection i are given by the following equation:

$$P^i_\sigma(t+1) = \frac{exp\{\eta\gamma_\sigma(t)\}}{\sum_{\psi\in\Phi_i}\exp\{\eta\gamma_\psi(t)\}} \quad (14)$$

Equation 14 yields phase proportions in a cycle, and further determines phase durations when cycle length is known. Here, Φ_i is the predetermined phase sequence and η is a parameter to help control the distribution of green time. If η is relatively small, the proportion of each group will tend to be uniform, albeit Le et al. (*4*) did not elaborate how to determine the value of η. Although this MP scheme provides fixed phase sequence and right of way for each phase in a cycle, tests of its stability property indicate that this control policy cannot achieve the maximum throughput (*4, 20, 21*)

Simulation Settings

For the simulation study, we selected a 1.5-mile-long corridor with 12 signalized intersections located in Gainesville, FL (Figure 1). Each intersection normally has three or four phases. The origin-destination demand matrix was obtained from a field survey conducted in the evening peak hours (5 p.m. to 6 p.m.) in April 2015. The data can be provided to interested readers on request.

The simulation study was operated using Vissim, which allows external programs to control traffic signals in the simulation via the COM Interface. In this study, we mainly used C# to implement MP control algorithms and interact with the COM Interface. The detailed algorithms are attached in the appendix.

Figure 2 shows the simulation procedure when a COM Interface is involved in the control study. In cyclic phase MP, phase and phase sequence are predetermined. An initial timing plan is provided to the simulator at the beginning to activate the system. Once activated, Vissim transmits the required data to the COM Interface at every customized time step. Static data including network topology, signal controlers, vehicle inputs, and other fixed parameters are transferred only at the beginning of the simulation, while queueing information is updated dynamically. The COM Interface thus performs control algorithms based on those data and outputs signal status to Vissim for the next step's simulation. In simulation, the observation of queueing process and pressure calculation are performed for each lane. For safety concerns, mandatory yellow and red clearance times are also inserted in the simulation settings.

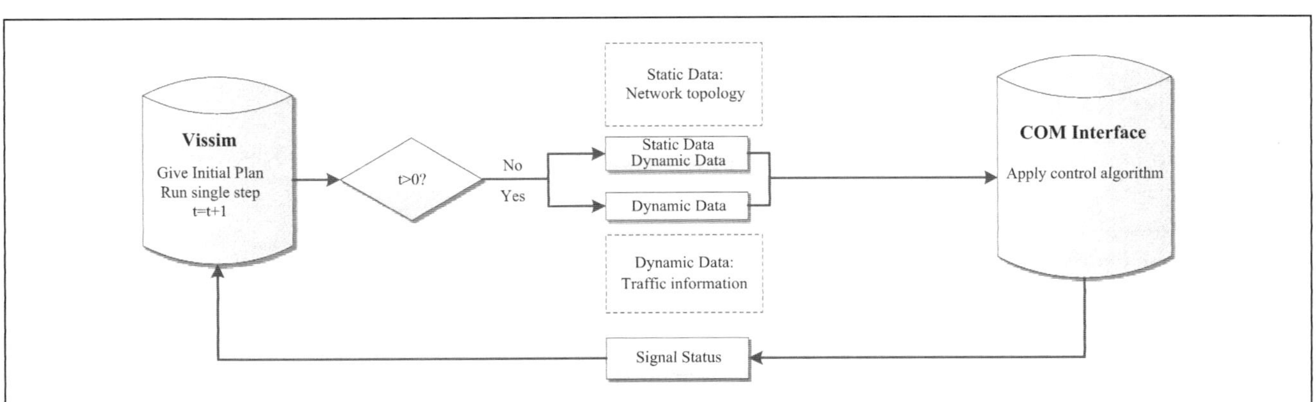

Figure 2. Simulation procedure.

Simulation Results

This section mainly compares the simulation performance of non-cyclic MP, cyclic phase MP, and a benchmark control strategy—a coordinated actuated signal control scheme implemented through the Ring Barrier Control (RBC) module in Vissim (*26*). The actuated control scheme adopts timing plans used in practice. However, since the signal control system in the field location is SynchroGreen (*27*), whose control logic is not available in Vissim, the benchmark simulation study did not replicate the real-world control logic, but used actuated signal control provided in the RBC module. For fair comparisons among different control strategies, basic timing parameters from practice for signal coordination purpose, including cyclic length, offsets, and green splits, are further optimized and adjusted by using TRANSYT-7F before implementation. Parameters in cyclic phase MP are fine-tuned at first by examining different cycle lengths and two types of queueing information. After that, a comparison is made among the cyclic phase MP settings with greatest throughput, non-cyclic MP, and actuated control. The performance of modified MPs is demonstrated at the end.

Simulation Performance of Cyclic Phase MP

We first determine the cycle length for the corridor. According to the Traffic Signal Timing Manual, two commonly accepted methods are applied (*23*). The optimal cycle length of the critical intersection, in this case, 162 seconds, is chosen when using the critical intersection method. Since field data implies that flow rates vary significantly among all the intersections, the network approach considering minimum cycles for all intersections is applied; as a result, another cycle length of 90 seconds is considered. Optimal offsets are calculated accordingly. Regarding the fact that time slot is aborted in cyclic phase MP, Le et al. suggested that the queueing information is obtained at the beginning of each cycle, with *pressures* calculated once per cycle (*4*). Such an instantaneous measurement of queues greatly reduces communication costs and computation time, but fails to capture the queueing dynamics of the whole cycle. In contrast, calculating average queue lengths based on multiple measurements in a cycle can address this weakness.

Based on variations of cycle lengths and queueing measurements, we conduct four scenarios of cyclic phase MP. The cycle lengths are 162 and 90 seconds, respectively; queue lengths are measured instantaneously and calculated on average, respectively. For scenarios using average queue lengths, queueing information is read from the simulator every 5 seconds. The control parameter η is determined via a heuristic approach. The average delay and travel time of four scenarios are compared in Table 1*a*. Here, cyclic_162i indicates the scenario that applies cyclic phase MP with 162 seconds as cycle length and using the instantaneous queueing lengths for calculation; while cyclic_90a indicates the scenario which applies cyclic MP with 90 seconds as cycle length and average queueing lengths. The simulation period is one simulation hour, and each scenario was run five times due to possible randomness.

Figure 1*a* represents the re-scaled cumulative vehicle departure curves $N(t) - q_0(t)$. For each simulation scenario, $N(t)$ is the network overall cumulative departures up to time t, q_0 is a background departure rate. A suitable choice of q_0 is chosen to highlight the visual identification of change on departure rates from re-scaled curves (*28*). At the beginning of the simulation, the departure rates of all the scenarios are nearly the same. Divergences begin when the number of vehicles in the network increases. Figure 3*b* shows queued vehicles at each moment. Correspondingly, a scenario with a higher average departure rate generates a smaller number of queued vehicles. For this network, hypothesis tests show that the average throughput under the scenario with 90-seconds cycle length and applying average queue lengths (cyclic_90a) is statistically higher than that under other scenarios with a level of significance of 0.01.

Comparison of Non-Cyclic MP, Cyclic Phase MP, and Actuated Control

Non-cyclic MP, cyclic phase MP and the benchmark actuated signal control scheme are compared in this subsection. All the simulation scenarios are performed for three simulation hours to fully examine the stability of these control policies. Considering flow variations, the peak hour demand (5 p.m. to 6 p.m.) is scaled 2.5 times and serves as the vehicle input for the three-hour simulation. The settings' best performance in the last section is applied for cyclic phase MP. For non-cyclic MP, time interval is selected as 5 seconds after testing several different values, meaning that the control algorithm is called every 5 seconds. The high frequency of calls is set to guarantee that the most recent traffic information is obtained and applied. Though costly in traditional communication settings, high frequency is expected to be achievable in a connected vehicle environment.

Table 1*b* displays the average delay and travel time under three control strategies. Once again, re-scaled cumulative vehicle departure curves are drawn to highlight the departure rate divergence (Figure 4*a*). Both non-cyclic MP and cyclic phase MP results reveal that MP-based control outperforms the optimized actuated control in terms of average travel time, average delay, and departure rate. The unstable departure rates under

Table 1. Network Performance Comparison

Average travel time and delay of cyclic phase MPs (one hour)	cyclic_162i	cyclic_162a	cyclic_90i	cyclic_90a
Average delay (min)	11.66	9.92	9.71	8.59
Average travel time (min)	18.97	15.42	14.84	13.08
Average travel time and delay of non-cyclic MP, cyclic phase MP and actuated control (one hour)	**cyclic_90a**	**non-cyclic**	**AC**	
Average delay (min)	11.22	7.15	29.47	
Average travel time (min)	14.19	9.68	38.76	
Average travel time and delay of modified MPs (three hours)	**non-cyclic**	**non-cyclic_nor**	**non-cyclic_min**	**non-cyclic_nor_min**
Average delay (min)	6.80	8.33	6.01	7.86
Average travel time (min)	8.76	10.23	7.95	9.74
Overall decrease in number of stops by movements	**Through movements**	**Left movements**	**Movements on minor streets**	
Reduction of stops	3,255	15,297	5,829	
Percentage of reduced stops	2.35%	37.67%	23.02%	

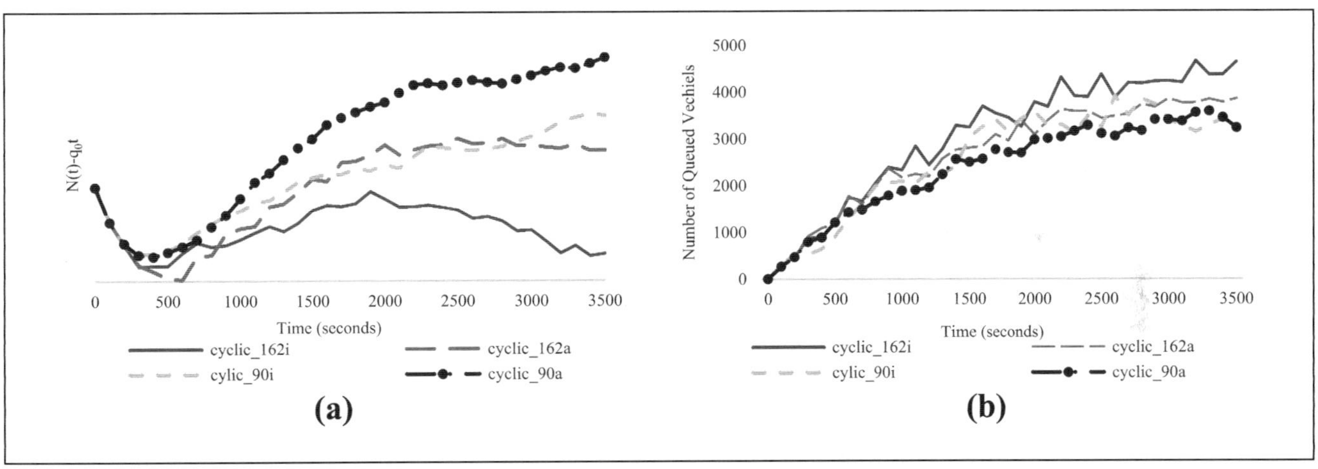

Figure 3. (*a*) Re-scaled cumulative vehicle departure curves of cyclic phase MPs. (*b*) Total number of queued vehicles of cyclic phase MPs.

actuated control result in significantly poorer network performance, nevertheless both departure rates under non-cyclic MP and cyclic phase MP are quite stable. Non-cyclic MP and cyclic phase MP are further compared in terms of the total number of vehicles queued in the network (Figure 4*b*). The scenario under non-cyclic MP creates a less congested network, which corresponds to the fact that it achieves a significantly higher departure rate. In a conclusion, non-cyclic MP is preferable in terms of stability.

Simulation Performance of Modified MP

In this sub-section, four scenarios based on the modifications previously suggested are compared. The basic simulation settings remain unchanged, the control algorithm is still called every 5 seconds. Minimum green time is tuned to be equal to 20 seconds, considering the traffic demand and the average number of detectable vehicles stored before stop bars during red time (*29*). Table 1*c* compares the average travel time and delay of the four scenarios. For the notations, "nor" means normalized queue is applied, "min" means minimum green time is applied.

Overall comparison indicates that introducing the minimum green time helps to reduce travel time and delay as well as increasing vehicle departure rate (Figure 5*a*), which yields the increase in throughput. Our hypothesis tests confirm that these improvements are statistically significant, with a level of significance of 0.01. Given that

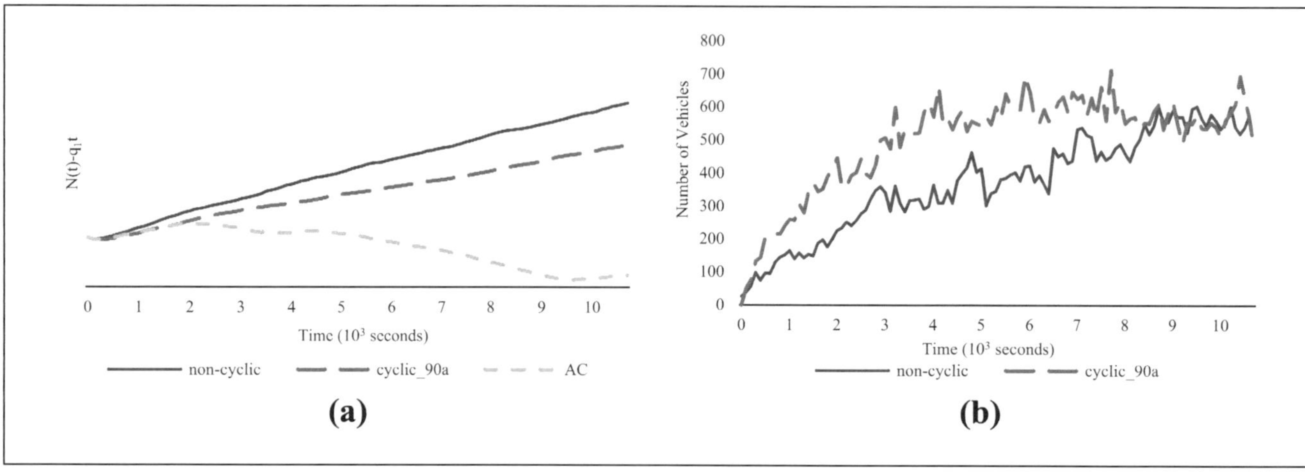

Figure 4. (*a*) Re-scaled cumulative vehicle departure curves of MP, cyclic phase MP and actuated control. (*b*) Total number of queued vehicles of MP and cyclic phase MP.

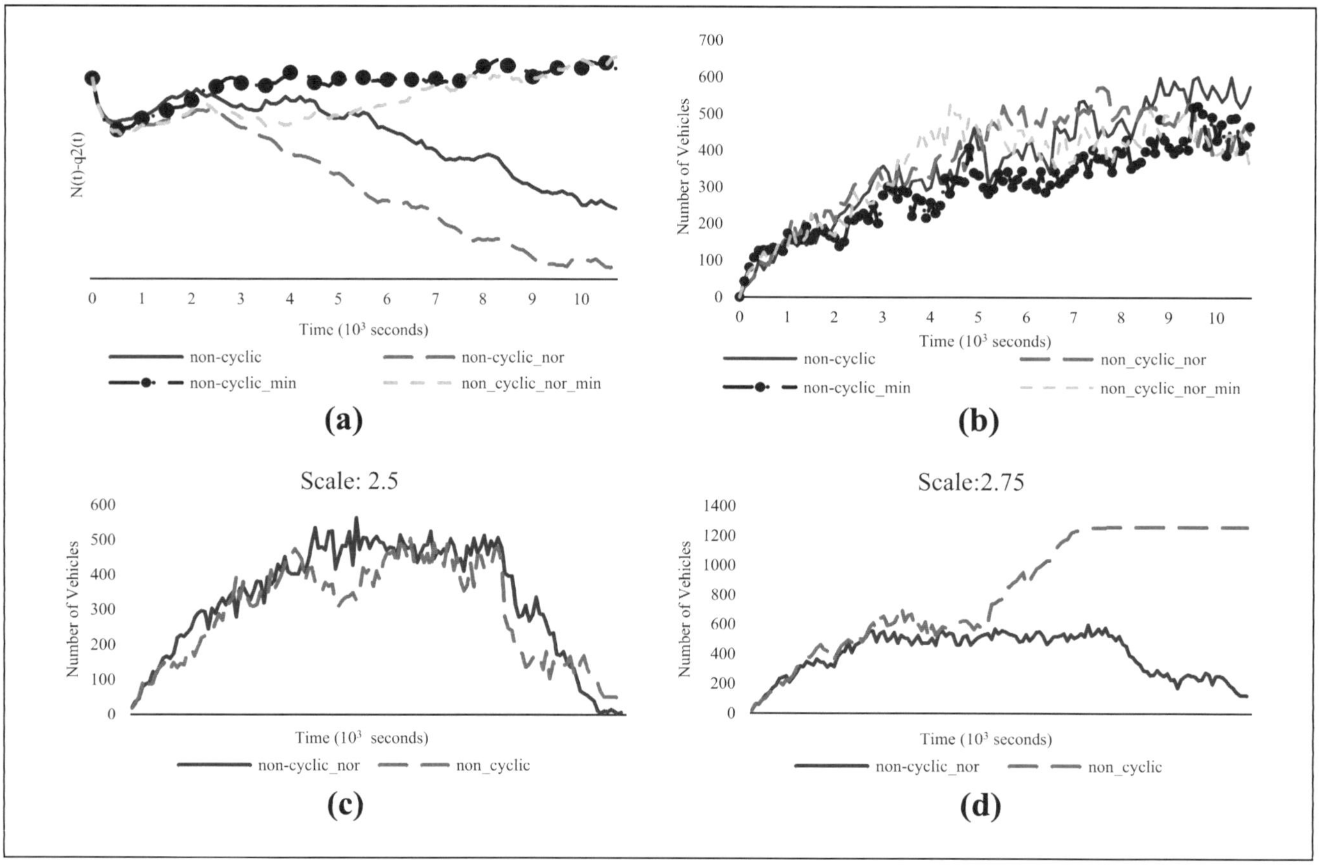

Figure 5. (*a*) Re-scaled accumulative vehicle departure curves of modified MPs. (*b*) Total number of queued vehicles of modified MPs. (*c*) Number of queue vehicles with scale parameter = 2.5. (*d*) Number of queue vehicles with scale parameter = 2.75.

the start-up lost time usually lasts for 3 to 4 seconds and is rather stable (*23*), increasing the minimum green time from 5 seconds (under the original non-cyclic MP control) to 20 seconds (under the modified MP control) will greatly increase the effective green ratio. An observation of the number of stops at stop lines demonstrates this point. Table 2*d* indicates that the number of stops greatly decreases after minimum green time is introduced, especially for left movements on the main street and movements on the minor streets.

In this three-hour simulation study, traffic operation did not apparently benefit from using normalized queues. We therefore extended the simulation period to four simulation hours and conducted two sets of comparison studies. When the vehicle input is the same as that in the three-hour study, the queueing processes perform similarly and can be cleared under both controls (Figure 5*c*). In view of the fact that the demand is relatively small, so that the queues lengths are far less than storage capacity, total demand is increased to 2.75 times peak-hour demand in the second set of comparison studies. The results show that queues are bounded and cleared at the end of the simulation period under normalized non-cyclic MP, while under non-cyclic MP queues reach storage capacity, suggesting that the network is gridlocked (Figure 5*d*). A similar phenomenon was also observed in a previous study when a capacity-aware MP algorithm was tested, although the queues were normalized differently (*8*).

Conclusion

This paper firstly reviewed previous work on MP control policies. A comprehensive simulation study on the Vissim platform was then conducted. To mitigate the instability created by the infinite storage capacity assumption and frequent phase switches, modifications considering the minimum green time and normalized queues, respectively, were introduced to the non-cyclic MP control scheme. Our simulation results showed that non-cyclic MP outperformed cyclic phase MP and both MP control policies performed better than the actuated control scheme which adopts the timing-plans being implemented in the field. Under control by the modified MP scheme with a minimum green time, the average travel time was reduced and the network throughput was significantly improved. The results also showed that MP using normalized queue was more capable of handling relatively larger demands and prevented the network from being blocked in nearly saturated conditions. These results have proved that the modifications on MP are helpful.

Our future work will explicitly consider the physical queueing dynamics and then theoretically compare the performance of MP and MP with normalized queues. Also, since this study mainly focuses on fixed routings, MP control under adaptive routing scenario or dynamic traffic assignment would also be an interesting topic.

Appendix

The control algorithms for non-cyclic MP, cyclic phase MP, and MP with minimum green time, respectively, are presented below.

Algorithm 1. Non-cyclic MP

Initial settings: $t = 0$, $T = 0$
For each intersection i
  $u_i^*(t) = argmax\{\gamma_\sigma(t)|\sigma \in \Pi_i\}$
  $g_i(t) = 0$
End For
Loop $T = T + \Delta s$
  If mod $(T, \Delta) = 0$
    $t = t + 1$
    For each intersection i
      $u_i^*(t) = argmax\{\gamma_\sigma(t)|\sigma \in \Pi_i\}$
      If $u_i^*(t) \equiv u_i^*(t-1)$
        $g_i(t) = g_i(t-1) + \Delta$
      Else
        $g_i(t) = 0$
        stop green light and **activate** yellow light for $u_i^*(t-1)$
      End if
    End for
  End If
  If mod $(T, \Delta) = a$
    Activate red light for $u_i^*(t)$
  End If
  If mod $(T, \Delta) = a + r$
    Activate green light for $u_i^*(t)$
  End If
Until $T = ST$

Algorithm 2. Cyclic MP

Initial settings, $t = 0$, $T = 0$, deciding phase sequence.
For each intersection i
  Apply one actuated control timing plan
End for
Loop $T = T + \Delta s$
  If mod $(T, \Delta) = o_i$
    $t = t + 1$
    For each phase k
      $P_k^i(t) = \frac{\exp\{\eta\gamma_k(t-1)\}}{\sum_{\psi\in\Phi_i}\exp\{\eta\gamma_\psi(t-1)\}}$
    End for
  End if
  If mod $(T, \Delta) = o_i + C\sum_{j=1}^{k-1} P_k^i(t) - a - r$
    Activate yellow light for phase σ_{k-1}^i
  End if
  If mod $(T, \Delta) = o_i + C\sum_{j=1}^{k-1} P_k^i(t) - a$
    Activate red light for phase σ_{k-1}^i
  End if
  If mod $(T, \Delta) = o_i + \sum_{j=1}^{k-1} P_k^i(t)C$
    Activate green light for phase σ_k^i
  End if
Until $T = ST$

Algorithm 3. MP with minimum green time

Initial settings, $t = 0$, $T = 0$
For each intersection i
  $g_i(t) = 0$
  $w_i^*(t) = argmax\{\gamma_\sigma(t)|\sigma \in \Pi_i\}$
Loop $T = T + \Delta s$
  If mod (T, Δ) **=** 0
    $t = t + 1$
    For each intersection i
      $u_i^*(t) = argmax\{\gamma_\sigma(t)|\sigma \in \Pi_i\}$
      If $g_i(t) < g_{min}$
        $g_i(t) = g_i(t-1) + \Delta$

      Else
        If $u_i^*(t) = w_i^*(t-1)$
          $g_i(t) = g_i(t-1) + \Delta$
          $w_i^*(t) = w_i^*(t-1)$
        Else
          $g_i(t) = 0$
          activate yellow light for $w_i^*(t-1)$
          $w_i^*(t) = u_i^*(t)$
        End if
      End if
    End for
  End If
  If mod (T, Δ) **=** a
    Activate red light for $w_i^*(t-1)$
  End If
  If mod (T, Δ) **=** $a + r$
    For each intersection i
      If $g_i(t) = 0$
        Activate green light for $w_i^*(t-1)$
      End if
    End for
  End If
Until $T = ST$

Acknowledgments

The authors acknowledge the support of Lloyd's Register Foundation (LRF). LRF helps to protect life and property by supporting engineering-related education, public engagement, and the application of research.

Author Contributions

The authors confirm contribution to the paper as follows: study conception and design: Yafeng Yin, Xiaotong Sun; data collection: Xiaotong Sun; analysis and interpretation of results: Xiaotong Sun, Yafeng Yin; draft manuscript preparation: Xiaotong Sun. Both authors reviewed the results and approved the final version of the manuscript.

References

1. Wongpiromsarn, T., T. Uthaicharoenpong, Y. Wang, E. Frazzoli, and D. Wang. Distributed Traffic Signal Control for Maximum Network Throughput. *Proc., International 15th IEEE Conference on Intelligent Transportation Systems, ITSC*, Anchorage, AK, 2012, pp. 588–595. https://doi.org/10.1109/ITSC.2012.6338817
2. Varaiya, P. Max Pressure Control of a Network of Signalized Intersections. *Transportation Research Part C*, Vol. 36, 2013, pp. 177–195. https://doi.org/10.1016/j.trc.2013.08.014
3. Varaiya, P. The Max-Pressure Controller for Arbitrary Networks of Signalized Intersections. *Advances in Dynamic Network Modeling in Complex Transportation Systems*, Springer, New York, NY, 2013, pp. 27–66.
4. Le, T., P. Kovács, N. Walton, H. L. Vu, L. L. Andrew, and S. S. Hoogendoorn. Decentralized Signal Control for Urban Road Networks. *Transportation Research Part C: Emerging Technologies*, Vol. 58, 2015, pp. 431–450. https://doi.org/10.1016/j.trc.2014.11.009
5. Tassiulas, L., and A. Ephremides. Stability Properties of Constrained Queueing Systems and Scheduling Policies for Maximum Throughput in Multihop Radio Networks. *Proc., 29th IEEE Conference on Decision and Control*, Honolulu, HI, Vol. 37, No.12, 1990, pp. 2130–2032. https://doi.org/10.1109/CDC.1990.204000
6. Kouvelas, A., J. Lioris, S. Fayazi, and P. Varaiya, Maximum Pressure Controller for Stabilizing Queues in Signalized Arterial Networks. *Transportation Research Record: Journal of the Transportation Research Board*, 2014. 2421: 133–141.
7. Gregoire, J., X. Qian, E. Frazzoli, A. De La Fortelle, and T. Wongpiromsarn. Capacity-Aware Backpressure Traffic Signal Control. *IEEE Transactions on Control of Network Systems*, Vol 2, No. 2, 2015, pp. 164–173. https://doi.org/10.1109/TCNS.2014.2378871
8. Lee, S., S. C. Wong, and P. Varaiya. Stage-Based Max-Pressure Adaptive Traffic Control Policy at Isolated Signalized Junctions. Presented at 95th Annual Meeting of the Transportation Research Board, Washington D.C., 2016.
9. Hosseini, P., and K. Savla. A Comparison Study Between Proportionally Fair and Max Pressure Controllers for Signalized Arterial Networks. Presented at 95th Annual Meeting of the Transportation Research Board, Washington D.C., 2016.
10. Chai, H., H. M. Zhang, C. N. Chuah, D. Ghosal, and D. Smith. Dynamic Traffic Routing in a Network with Adaptive Signal Control. *Transportation Research Part C: Emerging Technologies*, Vol 85, 2017, pp. 64–85. https://doi.org/10.1016/j.trc.2017.08.017
11. Gartner, N. H., F. J. Pooran, and C. M. Andrews. Implementation of the OPAC Adaptive Control Strategy in a Traffic Signal Network. *2001 IEEE Intelligent Transportation Systems. Proceedings (Cat. No.01TH8585)*, Oakland, CA, 2001, pp. 195–200. DOI: 10.1109/ITSC.2001.948655
12. Pitu Mirchandani, L. H. A Real-Time Traffic Signal Control System: Architecture, Algorithms, and Analysis. *Transportation Research Part C: Emerging Technologies*, Vol. 9, No. 6, 2001, pp. 415–432. https://doi.org/10.1016/S0968-090X(00)00047-4
13. Gershenson, C. *Self-Organizing Traffic Lights*, 2004. http://arxiv.org/abs/nlin/0411066 Accessed May 30, 2015.
14. Vasirani, M., and S. Ossowski. A Computational Market for Distributed Control of Urban Road Traffic Systems. *IEEE Transactions on Intelligent Transportation Systems*, Vol. 12, No. 2, 2011, pp. 313–321. https://doi.org/10.1109/TITS.2011.2104956

15. El-Tantawy, S., B. Abdulhai, and H. Abdelgawad. Multiagent Reinforcement Learning for Integrated Network of Adaptive Traffic Signal Controllers (MARLIN-ATSC): Methodology and Large-scale Application on Downtown Toronto. *IEEE Transactions on Intelligent Transportation Systems*, Vol. 14, No. 3, 2013, pp. 1140–1150. https://doi.org/10.1109/TITS.2013.2255286
16. Timotheou, S., C. G. Panayiotou, and M. M. Polycarpou. Distributed Traffic Signal Control Using the Cell Transmission Model via the Alternating Direction Method of Multipliers. *IEEE Transactions on Intelligent Transportation Systems*, Vol. 16, No. 2, 2015, pp. 919–933. https://doi.org/10.1109/TITS.2014.2347300
17. Lämmer, S., and D. Helbing. Self-Control of Traffic Lights and Vehicle Flows in Urban Road Networks. *Journal of Statistical Mechanics: Theory and Experiment*, Vol. 2008, No. 4, 2008, pp. P04019. https://stacks.iop.org/1752-5468/2008/i = 04/a = P04019
18. Aboudolas, K., M. Papageorgiou, and E. Kosmatopoulos. Store-and-Forward Based Methods for the Signal Control Problem in Large-Scale Congested Urban Road Networks. *Transportation Research Part C: Emerging Technologies*, Vol. 17, No. 2, 2009, pp. 163–174. https://doi.org/10.1016/j.trc.2008.10.002
19. Giaccone, P., E. Leonardi, and D. Shah. On the Maximal Throughput of Networks with Finite Buffers and Its Application to Buffered Crossbars. *Proc., IEEE 24th Annual Joint Conference of the IEEE Computer and Communication Societies*, Miami, FL, Vol. 2, 2005, pp. 971–980. https://doi.org/10.1109/INFCOM.2005.1498326
20. Xiao, N., E. Frazzoli, Y. Li, Y. Luo, Y. Wang, and D. Wang. Further Study on Extended Back-Pressure Traffic Signal Control Algorithm. *Proc., 54th IEEE Conference on Decision and Control (CDC)*, Osaka, Japan, 2015, pp. 2169–2174. https://doi.org/10.1109/CDC.2015.7402528
21. Xiao, N., E. Frazzoli, Y. Li, Y. Luo, Y. Wang, and D. Wang. Throughput Optimality of Extended Back-Pressure Traffic Signal Control Algorithm. *23rd Mediterranean Conference on Control and Automation (MED)*, Torremolinos, Spain, 2015, pp. 1059–1064. https://doi.org/10.1109/MED.2015.7158897
22. Zaidi, A. A., B. Kulcsár, and H. Wymeersch. Back-Pressure Traffic Signal Control with Fixed and Adaptive Routing for Urban Vehicular Networks. *IEEE Transactions on Intelligent Transportation Systems*, Vol. 17, No. 8, 2016, pp. 2134–2143. https://doi.org/10.1109/TITS.2016.2521424
23. Federal Highway Administration. *The Manual on Uniform Traffic Control Devices (2009 Edition with Revision Numbers 1 and 2 incorporated)*. Publication FHWA. U.S. Department of Transportation, Washington D.C., 2012, pp. 60.
24. Koonce, P., L. Rodegerdts, K. Lee, S. Quayle, S. Beaird, C. Braud, and T. Urbanik. *Traffic Signal Timing Manual*. Publication FHWA-HOP-08-024. Federal Highway Administration, U.S. Department of Transportation, 2008.
25. Lioris, J., A. Kurzhanskiy, D. Triantafyllos, and P. Varaiya. Control Experiments for a Network of Signalized Intersections Using the ".Q" Simulator. *Proc., 12th IFAC-IEEE International Workshop on Discrete Event Systems (WODES)*, 2014, pp. 332–337.
26. PTV Group. *Ring Barrier Controller (RBC) User Manual*, 2014, 7–39.
27. Trafficware. SyncrhoGreen: Real-time Adaptive Control System. *Technical White Paper*. Sugar Land, TX, 2012. http://www.trafficware.com/uploads/2/2/2/5/22256874/synchrogreen_technical_white_paper_1.pdf
28. Cassidy, M. J., and R. L. Bertini. Some Traffic Features at Freeway Bottlenecks. *Transportation Research Part B: Methodological*, Vol. 33, No. 1, 1999, pp. 25–42. https://doi.org/10.1016/S0191-2615(98)00023-X
29. Henry, R. D. *Signal Timing on a Shoestring*. Publication FHWA-HOP-07-. Federal Highway Administration, U.S. Department of Transportation, 2005.

The Standing Committee on Traffic Signal Systems (AHB25) peer-reviewed this paper (18-01319).

Article

TRR
JOURNAL OF THE TRANSPORTATION RESEARCH BOARD

Transportation Research Record
2018, Vol. 2672(18) 128–142

DOI: 10.1177/0361198118794546
journals.sagepub.com/home/trr
SAGE

Evaluating the Performance of Coordinated Signal Timing: Comparison of Common Data Types with Automated Vehicle Location Data

Stephen M. Remias[1], Christopher M. Day[2], Jonathan M. Waddell[1], Jenna N. Kirsch[1], and Ted Trepanier[3]

Abstract

Performance measures are essential for managing transportation systems, including signalized corridors. Coordination is an essential element of signal timing, enabling reliable progression of traffic along corridors. Improved progression leads to less user delay, which leads to user cost savings and lower vehicle emissions. This paper presents a comparative study of signal coordination assessment using four different technologies. These technologies include detector-based high-resolution controller data, Bluetooth/Wi-Fi sensors, segment-based probe vehicle data, and automated vehicle location data consisting of GPS-based vehicle trajectories, representing the data anticipated from emerging connected vehicle technologies. The data were compiled for a 4.2-mi corridor in Holland, Michigan. The results show that all of the data sources were able to identify, at some level, where coordination issues existed. Detector-based controller data and GPS-based vehicle trajectory data were capable of showing greater detail, and could be used to make offset adjustments. The paper concludes by demonstrating the identification of signal coordination issues with the use of visual performance metrics incorporating automated vehicle location (AVL) trajectory data.

Coordination is an essential aspect of signal operation. One basic way to evaluate the quality of progression is to measure the travel time of vehicles along the corridor. Historically, this has been done with floating car studies. GPS instruments increased the fidelity of floating car studies (*1*), but data collection remained labor intensive. Automatic vehicle identification (AVI) methods (*2–5*), such as using Bluetooth/Wi-Fi technologies, have greatly increased the sample sizes in travel time studies, but require additional infrastructure and do not provide the trajectory data.

With the advent of ubiquitous mobile GPS devices such as smartphones and commercial navigation devices, opportunities to obtain automated vehicle location (AVL) data at low cost have emerged. Initial data products using this strategy were timestamped average segment speeds, which have been applied to arterials (*6–8*). The advantage of this type of data is that no infrastructure is needed, but aggregation to 1-min or longer intervals loses a considerable amount of operational detail present in the vehicle trajectories. Statistical comparisons of the AVI and AVL data sets show a comparable accuracy for speed assessment over time (*9*, *10*).

Recently, AVL data has been explored for such purposes, either in anticipation of dedicated short range communications–based (DSRC) connected vehicle data in future, or using crowdsourced vehicle trajectories from commercial data providers. A critical issue with AVL data is the penetration rate of the vehicle fleet; at present, very few vehicles are equipped with the onboard units necessary to communicate with roadside infrastructure, which is itself nascent. Previous studies on use of AVL data have mainly focused on opportunities to improve signal control by integration of AVL data into the control system (*11–14*). Such methods have required relatively high penetration rates of perhaps 25% or more, which is much higher than present rates. A few studies have focused on performance evaluation, for which much lower penetration rates can provide useful results, as has

[1]Wayne State University, Detroit, MI
[2]Iowa State University, Ames, IA
[3]INRIX, Kirkland, WA

Corresponding Author:
Address correspondence to Stephen M. Remias: sremias@wayne.edu

been shown in previous studies (*15–19*). It is possible to compensate for low penetration rates by extending the sample period; this is possible by aligning similar time-of-day and day-of-week patterns. This paper examines potential uses of AVL data to evaluate signal coordination by aggregating trajectory data for similar time periods to build a large enough sample of data to determine characteristics of vehicle coordination.

The paper uses a signalized corridor in Holland, Michigan to investigate four different approaches for identifying and characterizing coordination issues. Bluetooth/Wi-Fi media access control (MAC) address devices, segment-based probe vehicle data, detector-based signal performance measures, and AVL trajectory data are used to evaluate the same corridor. The paper summarizes the strengths and weaknesses of each of the technologies and provides a perspective on the future of AVL trajectory data.

Performance Measurement with AVL Data

A few previous studies have examined the use of AVL data for the purpose of developing signal performance measures. Argote-Cabañero et al. considered an application for estimating *Highway Capacity Manual* metrics for 15-min analysis periods (*15*). They suggested that a penetration rate of 15% would likely be needed for accurate results for that time frame.

In a proof of concept study for measuring cyclic arrival patterns, Day and Bullock examined the ability of such patterns to feed an offset optimization process at very low penetration rates, considering sample periods of 3 h or 15 min (*16*). For these applications, the results suggested that penetration rates of 5% would work for 15 min and 1% would work for 3 h (e.g., for generating cyclic flow profiles for a time-of-day plan). Layering data from multiple days could potentially lower the penetration rate to 0.1%. This study was followed up with a field study using AVL data from a private sector provider. With penetration rates varying between approximately 0.1% and 0.5%, and data aggregated over a 6-week period, cyclic flow profiles were generated and used for offset optimization; they performed as well as optimal offsets generated from detector data on the same corridors.

A study by Wünsch et al. used AVL data to estimate movement delays at signalized intersections in Bavaria (*18*). This study applied a geofencing technique to identify zones of influence around intersections, and an automated process to determine the paths of vehicular movements at 2,300 intersections. This work was accomplished using the available penetration rates of navigation devices at the time of the study.

Zheng and Liu developed a method for estimating movement volumes using AVL data at low penetration rates (*19*). This method included aggregating AVL data over numerous days. The method was applied on corridors in the Ann Arbor, Michigan connected vehicle testbed and with commercial data in China. The method was found to estimate volumes with mean absolute percentage errors of 9% to 12%.

US-31 Corridor in Holland, Michigan

The US-31 corridor in Holland, Michigan is a 4.2-mi north–south corridor with an annual average daily traffic (AADT) of between 30,000 and 40,000 vehicles (see Figure 1). The corridor currently operates in fixed time with seven signals in the southbound (SB) direction and six signals in the northbound (NB) direction. As the

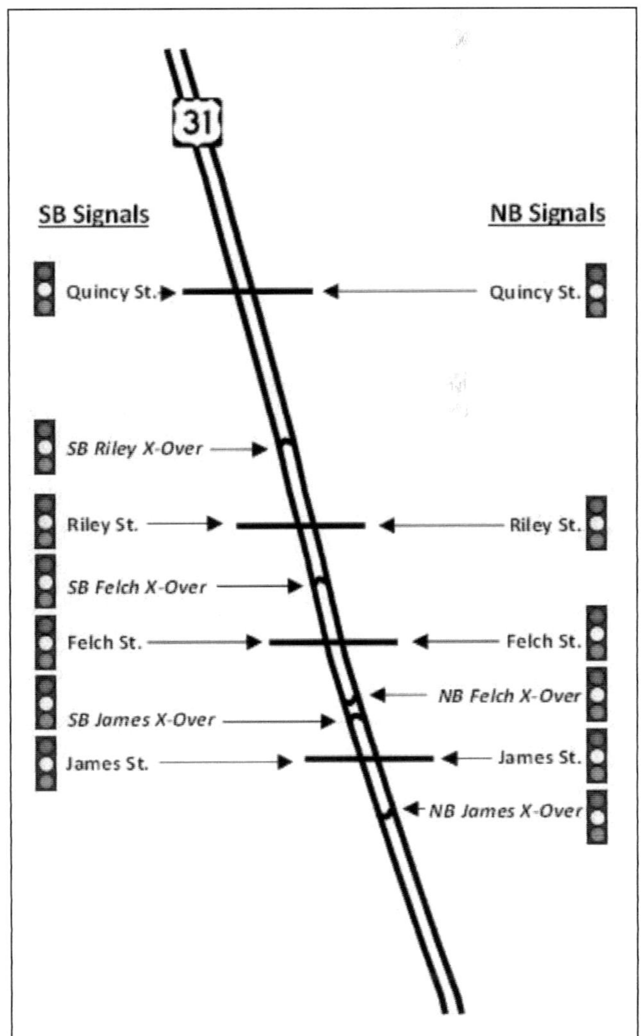

Figure 1. US-31 corridor in Holland, Michigan.

corridor was recently reconstructed, it consistently operates below capacity. The corridor is located just east of Holland, which is a 30,000-resident town on Michigan's western lake shore. The Michigan Department of Transportation is currently installing and testing a detector-based signal performance measure system on this corridor; however, no adjustments have been made on the corridor's signal timings throughout this data collection procedure. As part of this project, data is being collected to determine the baseline travel behavior on this corridor. The data being collected includes Bluetooth/Wi-Fi probe vehicle data, crowdsourced GPS probe vehicle data, and detector and phasing events from the signal controller. In addition, 2 weeks of AVL trajectory data were provided for comparison by INRIX.

Methodology

Four technologies were assessed for corridor performance measurement. It was desired to use the data to determine whether coordination issues exist and identify necessary adjustments. The four technologies are explained in depth below.

Signal Performance Measures

Detector-based signal performance measures based on high-resolution data are becoming a standard tool for determining performance of a signalized corridor. The Purdue Coordination Diagram, or PCD, allows the visualization of arrivals on green using phase and detector data from the signal controller. This visualization was constructed for each signalized intersection of the Holland corridor for a single day (Wednesday, July 12th, 2017). The corridor has various coordination issues as highlighted below.

Observations for NB US-31:

- Figure 2*i*—Random arrivals at the South James crossover;
- Figure 2*ii*—Poor offset during the 05:00–09:00 timing plan at James Street;
- Figure 2*iii*—Early arrivals at the South Felch crossover during the 05:00–09:00 timing plan;
- Figure 2*iv*—Late arrivals at the South Felch crossover in the 14:00–21:00 timing plan;
- Figure 2*v*—Late arrivals at the Felch intersection in the 14:00–21:00 timing plan;
- Figure 2*vi*—Platoon clipping at Riley Street; and
- Figure 2*vii*—Heavy platoon dispersion on the approach to Quincy Street.

Observations for SB US-31:

- Figure 2*viii*—Random arrivals at the Quincy intersection;
- Figure 2*ix*—Numerous turning vehicles from Riley arriving on red; and
- Figure 2*x*—Slightly early arrivals in the 14:00–21:00 timing plan at James Street.

These graphics suggest that the SB direction is better coordinated during all of the timing plans throughout the day. The percentages of arrivals on green were calculated for each of the intersections in each direction. The callouts in Figure 2 are placed in Table 1 at the corresponding approach and time-of-day. Over the entire day, a majority of the intersections seem to be performing well. The worst performing non-random arrival approach is NB at Quincy, with arrivals of 65.8% over the course of the day. These issues identified by the detector data will help determine the ability of other data sets to also identify such issues.

Bluetooth Data

Bluetooth and Wi-Fi sensors have become a prominent method of detecting congestion issues on various roadways. These sensors read MAC addresses from vehicles passing the roadside devices. These MAC addresses are then matched between different sensor locations and a travel time is derived. Historically, these sensors have been used to monitor overall travel time of corridors, but they have also been used to monitor signal performance (*6*).

Five sensors were used on the US-31 corridor in Holland. The placement of these sensors relative to the signalized intersections can be seen in Figure 3*a*. Using these devices, a week's travel time data were collected between June 19th and June 23rd, 2017. The 15th percentile travel time was used as a baseline travel time, and cumulative distribution functions were created as a measure of control delay for each timing plan and direction. The 15th percentile was chosen to eliminate outliers that may have represented cyclists, pedestrians, or vehicles stopping on the corridor. These results are visualized in Figure 3. The numbers of Bluetooth/Wi-Fi devices used in each of the cumulative distribution function (CDF) lines are provided in the legends.

The signal timing issues previously observed in Figure 2 are again pointed out in the CDFs in Figure 3. In CDFs, a more vertical line represents higher reliability and a more horizontal line represents less reliable travel times. For example, in Figure 3, callout "*i*" shows random arrivals at the South James crossover, which manifests as a uniform control delay distribution, represented by a sloped CDF. Additionally, the offset issues at Felch

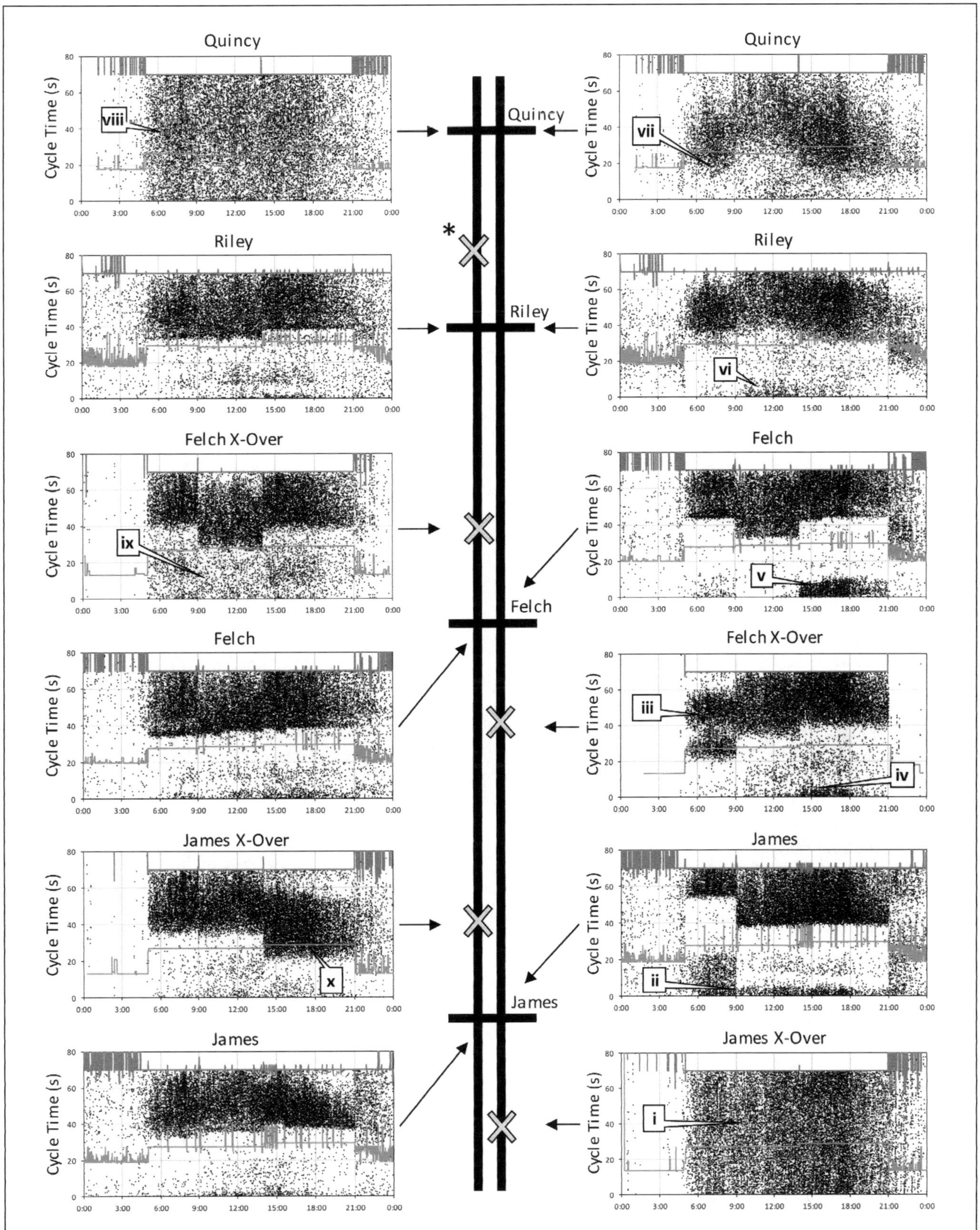

Figure 2. Purdue Coordination Diagrams for the US-31 Holland corridor (7/12/2017). The North Riley crossover (marked *) is signalized, but is currently without signal performance measures.

Table 1. Percentage Arrivals on Green Using Signal Performance Measures

			Percentage On Green (POG)/Percentage Green Time (PGT)						
			05:00–09:00		09:00–14:00		14:00–21:00		All Day
Order	Direction	Intersection	POG	PGT	POG	PGT	POG	PGT	POG
1	NB	South James X-Over	64.1%[i]	62.9%	63.3%[i]	62.9%	61.7%[i]	60.0%	64.8%
2	NB	James Street	67.3%[ii]	62.9%	92.1%	62.9%	89.4%	60.0%	82.4%
3	NB	South Felch X-Over	86.5%[iii]	62.9%	89.4%[iv]	61.4%	86.2%[iv]	60.0%	87.7%
4	NB	Felch Street	95.2%	62.9%	93.6%	61.4%	77.7%[v]	60.0%	87.2%
5	NB	Riley Street	94.9%	60.0%	88.0%[vi]	61.4%	87.0%[vi]	57.1%	86.7%
6	NB	Quincy Street	71.8%[vii]	64.3%	78.7%[vii]	65.7%	50.3%[vii]	60.0%	65.8%
1	SB	Quincy Street	64.0%[viii]	64.3%	65.1%[viii]	65.7%	59.3%[viii]	60.0%	64.4%
2	SB	Riley Street	93.3%	60.0%	87.6%	61.4%	85.0%	57.1%	87.3%
3	SB	North Felch X-Over	82.7%[ix]	62.9%	84.6%[ix]	61.4%	83.4%[ix]	60.0%	84.6%
4	SB	Felch Street	95.0%	62.9%	91.8%	61.4%	87.8%	60.0%	90.6%
5	SB	North James X-Over	95.6%	62.9%	89.0%	62.9%	80.9%[x]	60.0%	88.0%
6	SB	James Street	96.9%	62.9%	93.7%	62.9%	94.0%	60.0%	94.2%

Note: NB = northbound; SB= southbound; X-Over = crossover as in Figure 2. Footnotes (i–x) correspond to callouts in Figure 2.

and the South Felch crossover induce characteristic variations in control delay in Segment 3 (Figure 3, callouts "*iv*" and "*iii*"). The SB direction has little disparity amongst the control delay distributions. However, the slight variation in Segment 3 (Figure 3, callout "*ix*") represents the left-turning vehicles from Riley that are not arriving on green.

Overall, the probe travel data collected via Bluetooth/Wi-Fi sensors can be used to identify significant deficiencies in coordinated signal timing. But although the data can locate issues, they are less useful for suggesting potential offset adjustments. The data have some inherent limitations. Vehicle re-identification does not guarantee that the device is taking the most direct approach to the next sensor, or that the device represents a motor vehicle (i.e., it could potentially be a pedestrian or bike). The sensors have the advantage of being mobile; therefore, the segments can be customized for each corridor. However, non-midblock installations could potentially skew the data by detecting queued vehicles on the side streets.

Segment-Based Probe Vehicle Data

Commercial probe vehicle data are amongst the most widely used sources for mobility and travel time data. In this study, segment-based travel time data were used to create cumulative distribution function graphics of control delay to determine whether signal coordination issues could be discovered. Data from the week of June 19, 2017 were used. The traffic message channel (TMC) segmentation scheme was used. This scheme included two segments encompassing the signals on the corridor of interest. Because the segment-based data contain fewer outliers as a result of their aggregation to the nearest minute, the 5th percentile travel time was used as the baseline for the control delay diagrams in Figure 4. This differs from the 15th percentile used in the Bluetooth/Wi-Fi control delay diagrams because the segment-based data has eliminated a majority of outliers in the initial data aggregation. The numbers of minutes used in each of the CDF lines are provided in the legends.

The entire signalized portion of the corridor is represented in two segments. Having only two segments for a span of several intersections makes locating specific signal issues difficult. However, the data do reveal coordination problems. In the NB direction, during the a.m. and p.m. timing plans, Segment 2 has noticeably higher delay than the midday plan. This corresponds to offset deficiencies revealed by the PCDs (Figure 2). In the SB direction, random arrivals at Quincy are clearly visible in all three timing plans. The p.m. plan in the SB direction also shows a more severe issue in Segment 2, corresponding to the offset deficiency at the North James crossover.

In summary, segment-based probe data can be used to identify problems with signal coordination, but because of the broad spatial and temporal aggregation of the data, it is difficult to determine the intersection where the deficiencies are occurring, or the specific deficiency that is taking place.

Automated Vehicle Location Trajectory Data

The final approach used is a relatively new data source. The same commercial providers of the segment-based data are now providing raw data that allow for the creation of individual vehicle trajectories. The AVL trajectory data set contains anonymized unique vehicle trip

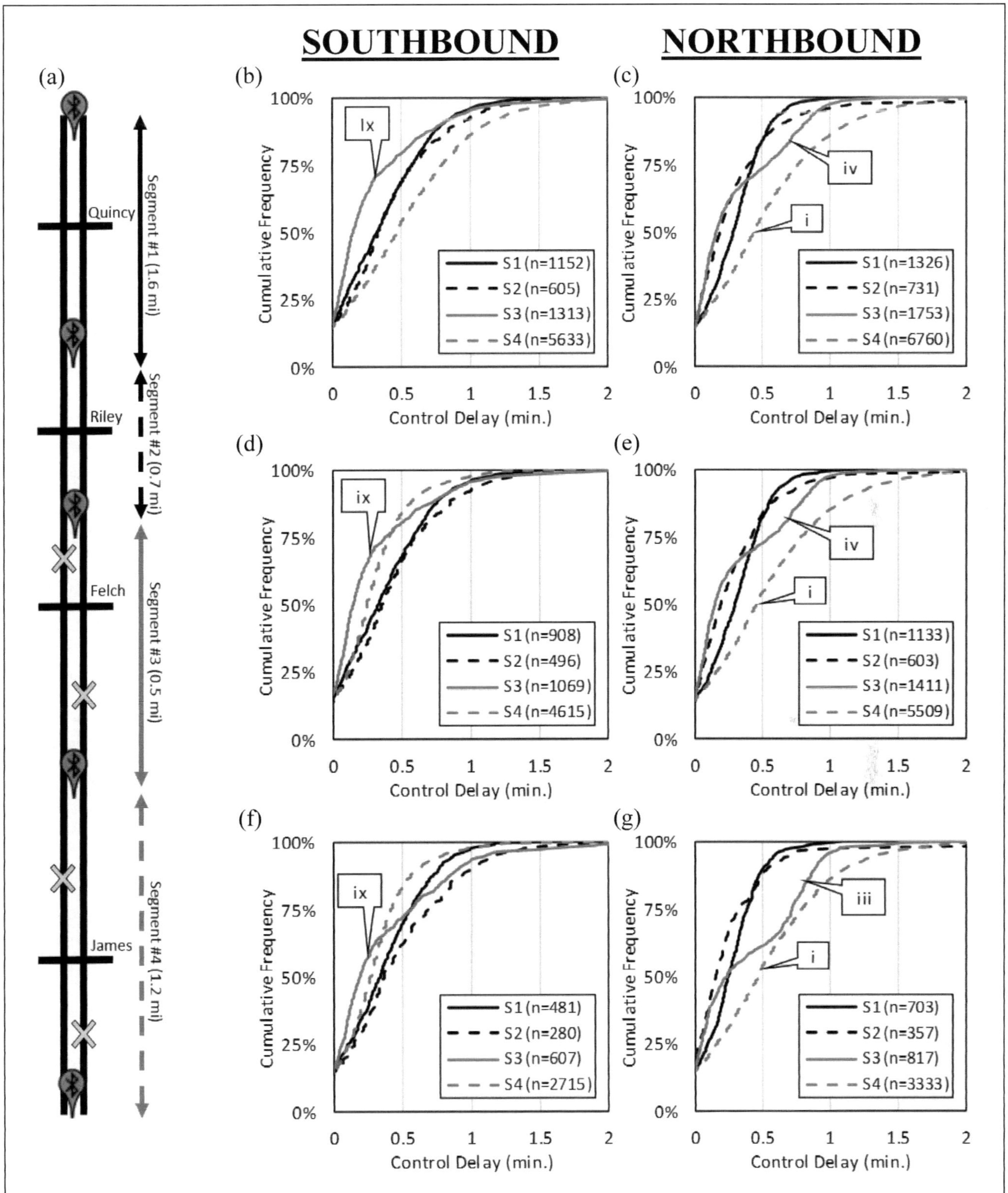

Figure 3. Control delay of probe segments using Bluetooth/Wi-Fi sensors, showing (*a*) map of the US-31 corridor, and results for (*b*) SB 05:00–09:00, (*c*) NB 05:00–09:00, (*d*) SB 09:00–14:00, (*e*) NB 09:00–14:00, (*f*) SB 14:00–21:00, and (*g*) NB 14:00–21:00.

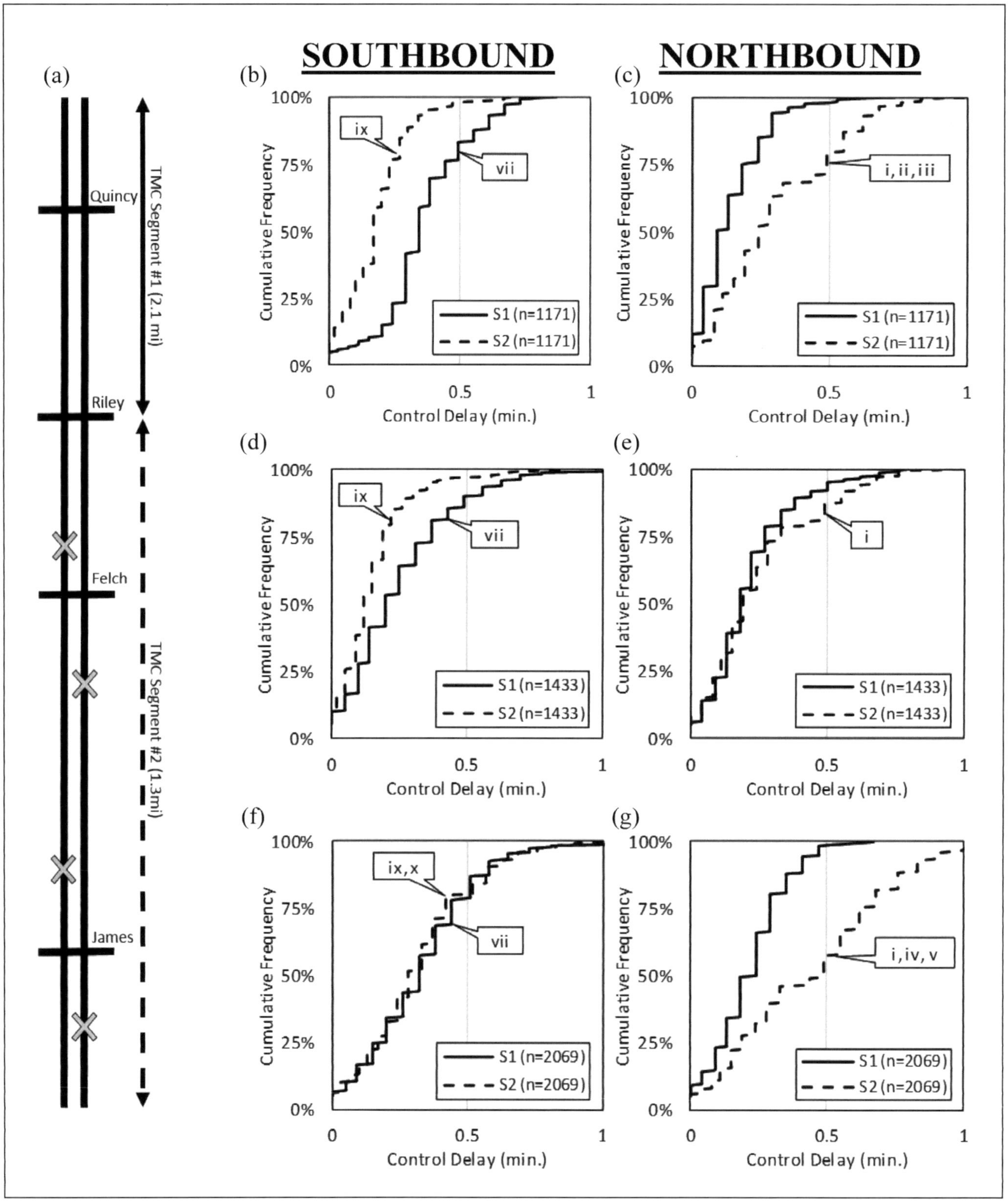

Figure 4. Control delay of probe segments using crowdsourced probe data, showing (*a*) map of the US-31 corridor, and results for (*b*) SB 05:00–09:00, (*c*) NB 05:00–09:00, (*d*) SB 09:00–14:00, (*e*) NB 09:00–14:00, (*f*) SB 14:00–21:00, and (*g*) NB 14:00–21:00.

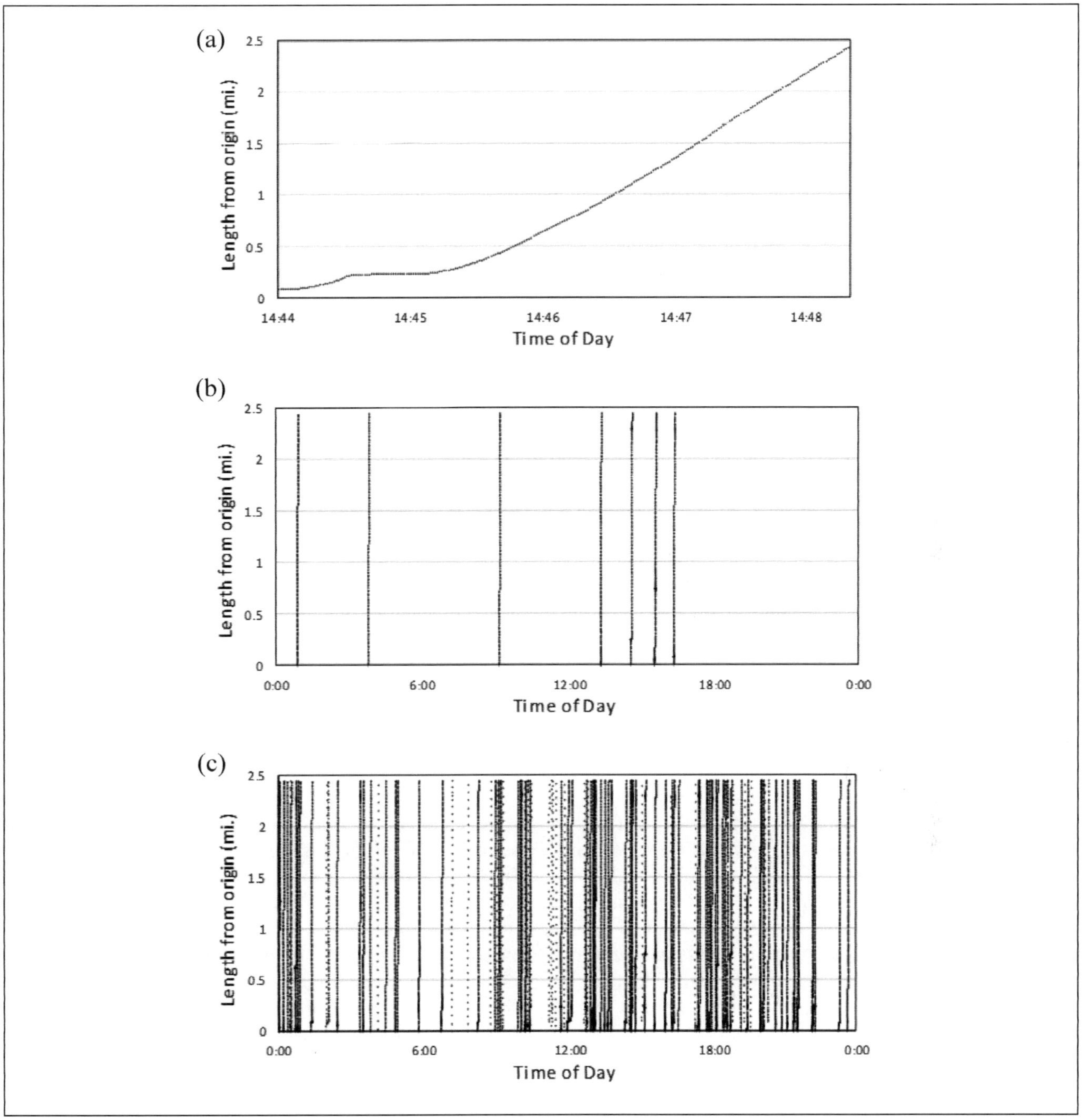

Figure 5. Vehicle trajectory aggregation on NB US-31, showing (*a*) single AVL trajectory, (*b*) single day of AVL trajectories (7/3/2017), and (c) 2 weeks of aggregated trajectories (6/30–7/14/2017).

identifiers, timestamps, waypoints, latitudes, and longitudes. Over a 2-week period 10,466 unique vehicle IDs were recorded with a total of 109,879 unique pings. Assuming the highest AADT as the entering volume of the corridor, approximately 1.8% of vehicles were recognized. A Structured Query Language (SQL) server was used to create a spatial reference point to organize the geospatial data. This reference is referred to as the origin. Vehicles traveling the entire stretch of road were selected and grouped together into a single data set. From this, 702 unique vehicle IDs with 45,539 pings were found to travel the entire corridor. The remaining data were processed using MATLAB to calculate the distance from the origin along the defined roadway and the

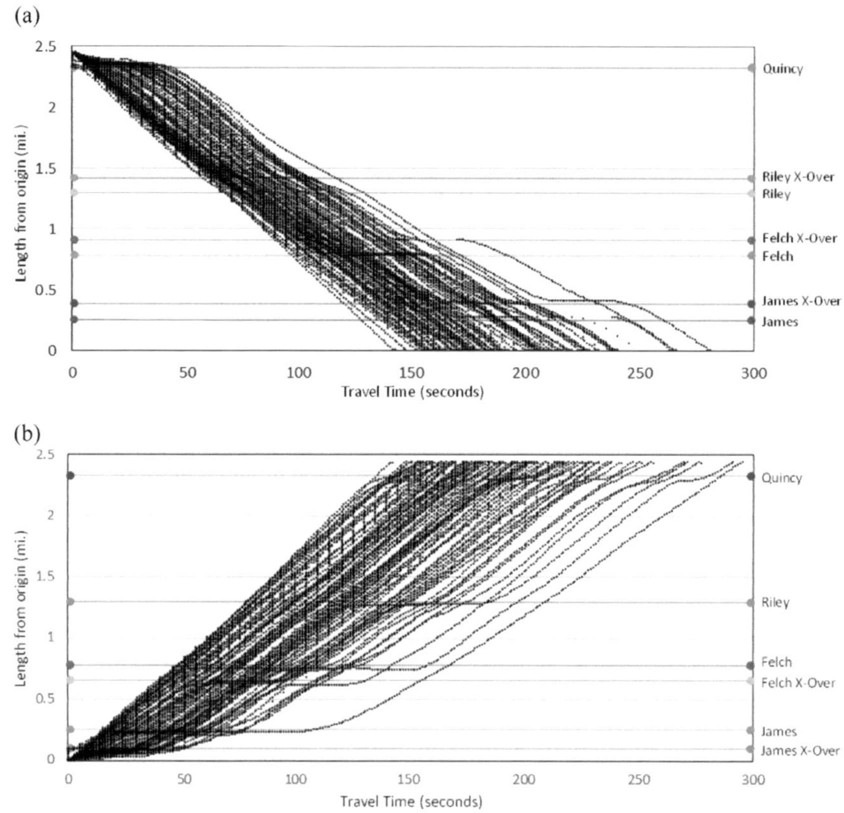

Figure 6. Aggregated corridor level vehicle trajectories, showing (*a*) SB trajectory data (2 weeks) and (*b*) NB trajectory data (2 weeks).

elapsed trip time. This allowed for both straight and curved roadways to be assessed. The MATLAB script removed trips with fewer than 10 pinged locations, and pings greater than 5 s apart. The filtering of infrequent pings resulted in 235 unique vehicle trips with 34,663 pings over a 2-week period. These points were then plotted as vehicle trajectories (Figure 5*a*). This data source is rich for use in transportation engineering and operations. However, the high volume of data makes it challenging to manage.

For this analysis, only probe vehicles that traversed the entire corridor were included to measure coordination from start to finish. Shorter segments will be used later to assess individual intersection performance. In addition, any trips for which the time between pings exceeded 5 s were excluded. Consequently, relatively few individual probe vehicle trip records were retained for any given time period. To counter this, the data were aggregated over a longer time period. Figure 5*b* shows a single day of probes traveling the corridor, and Figure 5*c* shows 2 weeks (6/30–7/14/2017) of trajectories overlaid onto a single day. These trajectories were then aggregated on a single point on the corridor and plotted to create a visualization of corridor progression. Figure 6*a* shows the SB direction and Figure 6*b* shows the NB direction. As was previously observed, progression in the SB direction is obviously superior to the NB direction because more NB trajectories exhibit control delay represented by horizontal segments on the trajectory. Additionally, the distribution of trajectory points exhibiting optimal progression outnumbers the trajectory points exhibiting poor progression, which may cause a misinterpretation of coordinated corridor performance. Supplemental numerical representation may be required to truly quantify performance.

To make better sense of it, the trajectory data can be broken down by timing plan (Figure 7). The a.m. timing plan in the NB direction (Figure 7*a*) had few probes to visualize, but one of the three did stop at both the South James crossover and James, which could reflect the issue shown in the PCD (Figure 2, callout "*ii*"). The midday timing plan in the NB direction appears not to have any issues, using the trajectory data in Figure 7*b*. This is also in agreement with the PCDs in Figure 2. Figure 7*c* shows the p.m. timing plan for the NB direction. There were clearly issues in the PCD at the Felch crossover/Felch

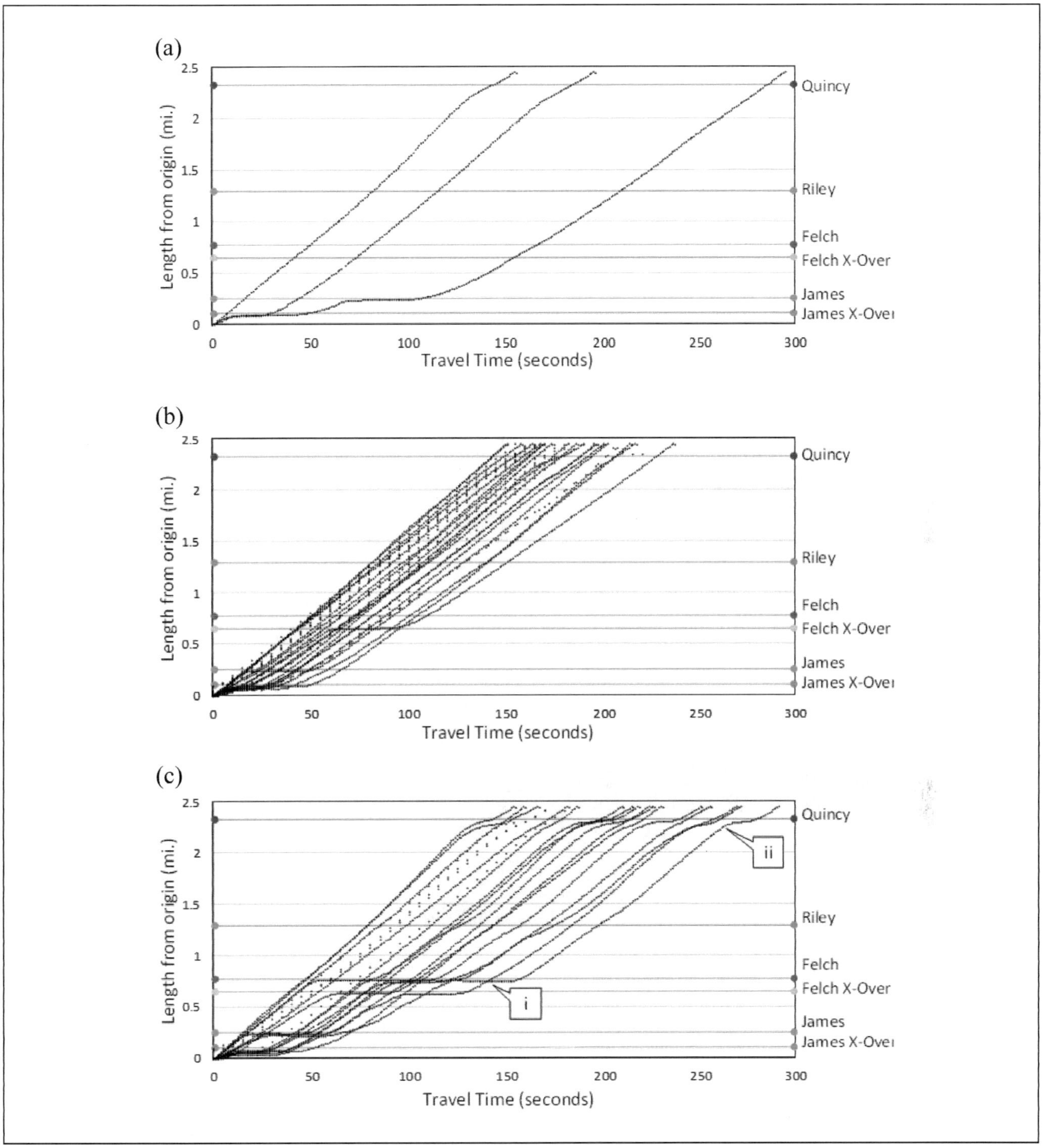

Figure 7. Vehicle trajectories by timing plan, showing (*a*) Timing Plan 1 (05:00–09:00), (*b*) Timing Plan 2 (09:00–14:00), and (*c*) Timing Plan 3 (14:00–21:00).

(Figure 2 callouts "*iv*" and "*v*") and at Quincy Street (Figure 2 callout "*vii*"). These issues are also observable in Figure 7*c*, callouts "*i*" and "*ii*", respectively. Unsurprisingly, the trajectory data is a useful source to monitor corridor progression. Such data is frequently used in floating car studies in current practice. However, by crowdsourcing the data collection, a much broader time period can be captured.

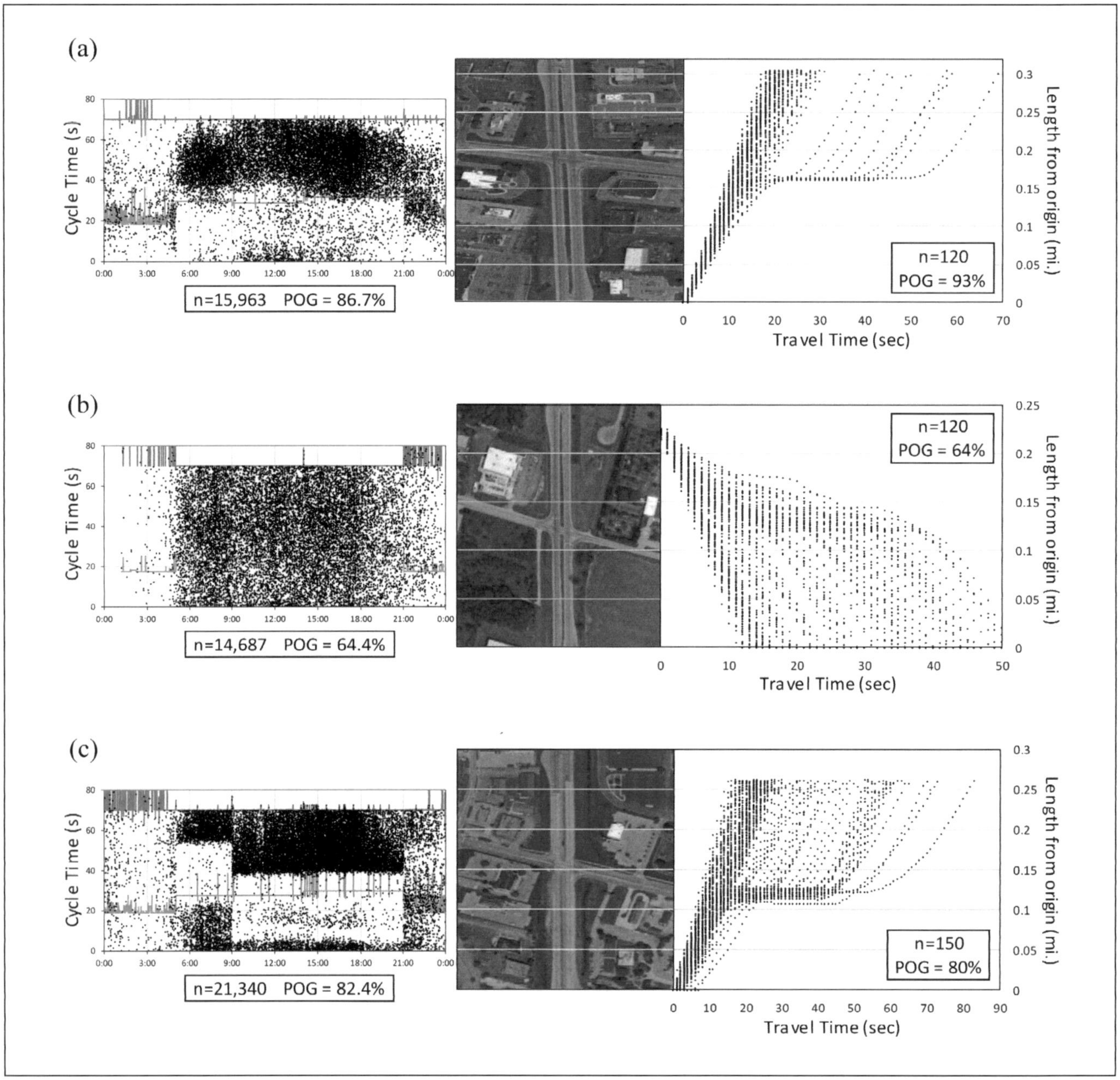

Figure 8. Arrival pattern comparison between PCD and trajectory data, showing (*a*) NB Riley Street, (*b*) NB James Street, and (*c*) SB Quincy Street.

The trajectory data can also be used to estimate arrivals on green. A similar approach to the corridor trajectory strategy was used to create intersection-based graphics. Figure 8 shows three intersections on the US-31 corridor with the corresponding PCD. Figure 8*a* shows the NB Riley intersection. This approach operates well, with a daily percentage on green (POG) of 86.7% according to the detector data. Using trajectory data, a POG of 93% is calculated. Figure 8*b* shows the NB James intersection, which has an offset issue in the AM period. The detector-based POG is 82.4%, whereas trajectory data yields a POG of 80%. Figure 8*c* shows the SB Quincy intersection, which has random arrivals as it is on the end of the corridor. The detector POG is 64.4%, whereas the trajectory POG is 60%.

There is some variance between these two POG numbers, but the former represents 1 day of detections whereas the latter comprises 2 weeks of trajectories. Nevertheless, the figures appear to correlate in this initial exploration. There is likely some promise for use of

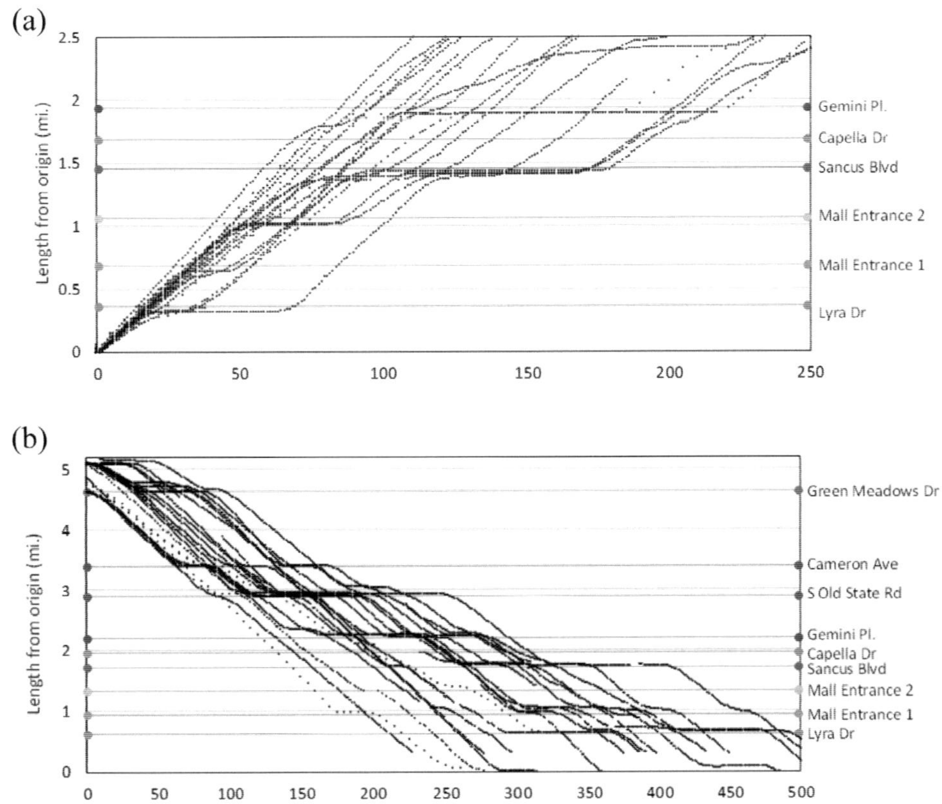

Figure 9. OH-750 signalized corridor in Columbus, Ohio, showing (*a*) portion of WB OH-750 and (*b*) portion of EB OH-750.

trajectory data to create infrastructure-free performance measures. Figure 9 further demonstrates the scalability of an analysis strategy based on trajectory data. A corridor in Columbus, Ohio was analyzed using trajectory data. The authors had no knowledge of the corridor before acquiring the data. Trajectory diagrams were created for WB OH-750 and EB OH-750. Aerial imagery was used to determine the signal locations for these charts. However, even if those locations were not explicitly shown in the charts, coordination issues would be apparent. The westbound (WB) direction (Figure 9*a*) has fewer, but rather long stops; the eastbound (EB) direction (Figure 9*b*) shows many numerous stops on practically every trajectory.

In addition, these trajectory diagrams can be fused with controller data to create composite time space diagrams. Such diagrams were created for the Holland corridor, as shown in Figure 10. The SB direction (Figure 10, *a–c*) shows stops at Quincy (callout "*i*"), Felch (callout "*ii*"), and James (callout "*iii*"). The NB direction shows stops at James crossover (callout "*iv*"), James (callout "*vi*"), and the Felch crossover (callout "*v*"). This information could potentially be used to validate software-based timing plans immediately after field implementation, rather than waiting for user complaints.

Similar data fusion strategies have been executed with floating car data (*1*); however, crowdsourcing such data can greatly reduce the labor required to obtain the data. These charts demonstrate that it is possible to do so with commercial trajectory data. It had been speculated that clock synchronization would present a formidable challenge in such data fusion, but the data in Figure 10 do not exhibit the types of anomalies that were anticipated, such as vehicles stopping on green or proceeding on red. This agrees with the previous comparison of detector and trajectory arrival profiles (*17*), suggesting that as long as the controller data is updated using a common time server, clock synchronization is unlikely to prove an Achilles' heel for data fusion between controller and vehicle data.

Conclusions

The assessment of signal coordination is a demanding effort that has historically relied on traditional measures, such as user complaints and individual travel runs.

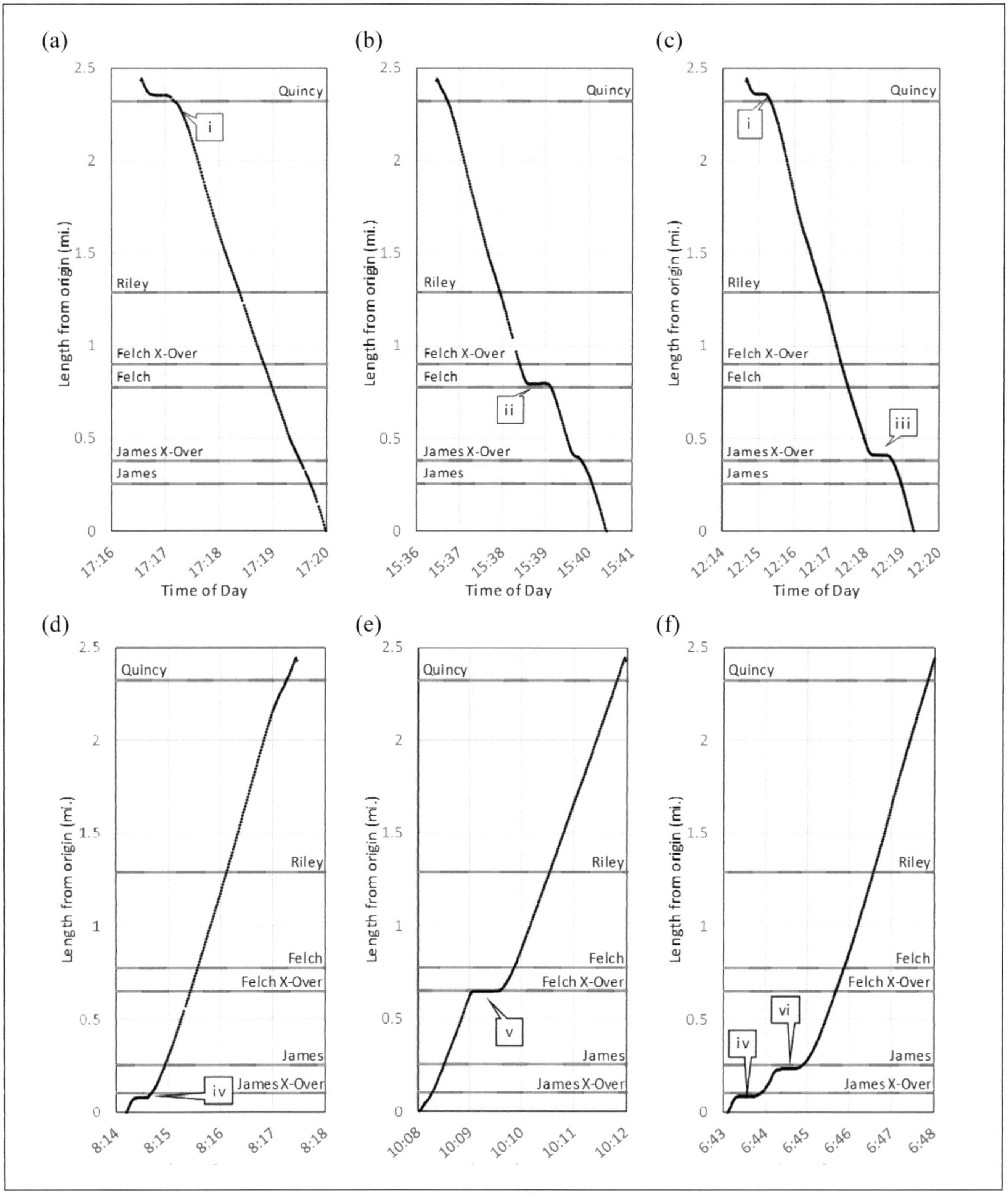

Figure 10. Fusion of high-resolution controller data and probe trajectory data, showing (*a*) SB Trajectory 1 (7/11), (*b*) SB Trajectory 2 (7/11), (*c*) SB Trajectory 3 (7/12), (*d*) NB Trajectory 1 (7/13), (e) NB Trajectory 2 (7/12), and (*f*) NB Trajectory 3 (7/10).

Recently, numerous technologies have been developed that can assist with the assessment of coordination. Both segment-based probe data and Bluetooth/Wi-Fi sensors can be used to detect if there are any issues; however, the ability of that data to suggest changes is limited. Bluetooth/Wi-Fi requires the purchase of infrastructure, but the corridors are customizable. Segment-based probe data requires no permanent infrastructure, but the segments are predefined, and may span numerous signals. As more agencies have purchased segment-based probe data for evaluating traffic operations, the cost of historical data has come down. Bluetooth sensors still require infrastructure and maintenance costs, generally requiring more of an investment than segment-based probe data. Segment-based analysis with Bluetooth/Wi-Fi sensors or probe data may be useful in problem identification with more intensive and costly evaluation, such as trajectory based methods, to be used in development of refined signal timing approaches.

Detector-based signal performance measures are excellent tools for measuring signal performance and adjusting signal timings. The data can identify issues and can be used to optimize coordination on a corridor (*3*). However, if there is no existing advanced detection on a corridor, a substantial investment is needed to acquire this data. If detection is present (as would be common for actuated-coordinated operation), the costs may be reasonable, depending on the abilities of the traffic controller and communication between the cabinet and an operations center. At the time of writing, six different controller manufacturers had at least one controller model with the capability of logging high-resolution data (*20*). In addition, several third-party products have emerged to add that capability to locations with legacy equipment.

Crowdsourced trajectory data can also be used to identify coordination issues, and adjust offsets to address those issues. No permanent infrastructure is required to collect this information; however, the temporal coverage is currently limited. This requires data aggregation over time to develop useful performance measures. The spatial coverages of these vehicle trajectories are limitless. At present, because this is a new data source, the costs are high. This is an economy-of-scale issue. As more agencies purchase the data, and the population of probe vehicles increases, the cost per data point will decrease. In the future, these data sources can be used to create aggregate performance metrics, such as average control delay, and to compare various corridors.

In future research, larger periods of time will be used for each of the data sets to truly compare data accuracy. These data can then be used to develop aggregate corridor performance metrics for agency use to prioritize signal improvements. Connected vehicle data based on DSRC may provide another path to the same type of data. However, because of privacy concerns, such data may only be available within close proximity of individual intersections; connecting trajectories along corridors might not be feasible. In summary, the future of vehicle trajectory data is very promising, with potential use cases for planning, operations, maintenance, and rapid validation of the models used in those endeavors.

Acknowledgments

The vehicle trajectory data provided in this paper were provided by INRIX. Signal performance measure data for this paper were provided by an ongoing research project with the Michigan Department of Transportation.

Author Contributions

The authors confirm contribution to the paper as follows: study conception and design: Remias; data collection: Wadell, Kirsch, Trepanier; analysis and interpretation of results: Remias, Day, Wadell, Kirsch; draft manuscript preparation: All. All authors reviewed the results and approved the final version of the manuscript.

References

1. Quiroga, C. A., and D. Bullock. Measuring Control Delay at Signalized Intersections. *Journal of Transportation Engineering*, Vol. 125, 1999, pp. 271–280.
2. Quayle, S. M., P. Koonce, D. DePencier, and D. M. Bullock. Arterial Performance Measures with Media Access Control Readers: Portland, Oregon, Pilot Study. *Transportation Research Record: Journal of the Transportation Research Board*, 2010. 2192: 185–193.
3. Day, C. M., R. Haseman, H. Premachandra, T. M. Brennan, J. S. Wasson, J. R. Sturdevant, and D. M. Bullock. Evaluation of Arterial Signal Coordination: Methodologies for Visualizing High-Resolution Event Data and Measuring Travel Time. *Transportation Research Record: Journal of the Transportation Research Board*, 2010. 2192: 37–49.
4. Moghaddam, S., and B. Hellinga. Real-Time Prediction of Arterial Roadway Travel Times Using Data Collected by Bluetooth Detectors. *Transportation Research Record: Journal of the Transportation Research Board*, 2014. 2442: 117–128.
5. Hainen, A. M., J. S. Wasson, S. M. L. Hubbard, S. M. Remias, G. D. Farnsworth, and D. M. Bullock. Estimating Route Choice and Travel Time Reliability with Field Observations of Bluetooth Probe Vehicles. *Transportation Research Record: Journal of the Transportation Research Board*, 2011. 2256: 43–50.
6. Remias, S. M., A. M. Hainen, C. M. Day, T. M. Brennan, Jr, H. Li, E. Rivera-Hernandez, J. R. Sturdevant, S. E. Young, and D. M. Bullock. Performance Characterization of Arterial Traffic Flow Using Probe Vehicle Data. *Transportation Research Record: Journal of the Transportation Research Board*, 2013. 2380: 10–21.

7. Day, C. M., S. M. Remias, H. Li, M. M. Mekker, M. L. McNamara, E. D. Cox, and D. M. Bullock. Performance Ranking of Arterial Corridors Using Travel Time and Travel Time Reliability Metrics. *Transportation Research Record: Journal of the Transportation Research Board*, 2015. 2487: 44–54.
8. Young, S. Use of Probe and Bluetooth Data for Arterial Performance Measures in the I-95 Corridor Coalition. *Proc., NATMEC Workshop on Arterial Data and Performance Monitoring,* Chicago, Ill., 2014.
9. Kim, S., and B. Coifman Comparing INRIX Speed Data against Concurrent Loop Detector Stations over Several Months. *Transportation Research Part C: Emerging Technologies* Vol. 49, 2014, pp. 59–72. 2014.
10. Center for Transportation Research and Education. *Evaluation of INRIX Probe Data*. Trans Project 13-480. August 2015.
11. He, Q., K. L. Head, and J. Ding. PAMSCOD: Platoon-Based Arterial Multi-Modal Signal Control with Online Data. *Transportation Research Part C*, Vol. 20, 2012, pp. 164–184.
12. Lee, J., B. Park, and I. Yun. Cumulative Travel-Time Responsive Real-Time Intersection Control Algorithm in the Connected Vehicle Environment. *Journal of Transportation Engineering*, Vol. 139, 2013, pp. 1020–1029.
13. Goodall, N. J., B. L. Smith, and B. Park. Traffic Signal Control with Connected Vehicles. *Transportation Research Record: Journal of the Transportation Research Board*, 2013. 2381: 65–72.
14. Feng, Y., K. L. Head, S. Khoshmagham, and M. Zamanipour. A Real-Time Adaptive Signal Control in a Connected Vehicle Environment. *Transportation Research Part C*, Vol. 55, 2015, pp. 460–473.
15. Argote-Cabañero, J., E. Christofa, and A. Skabardonis. Connected Vehicle Penetration Rate for Estimation of Arterial Measures of Effectiveness. *Transportation Research Part C*, Vol. 60, 2015, pp. 298–312.
16. Day, C. M., and D. M. Bullock. Opportunities for Detector-Free Signal Optimization with Limited Connected Vehicle Market Penetration: A Proof-of-Concept Study. *Transportation Research Record: Journal of the Transportation Research Board*, 2016. 2558: 54–65.
17. Day, C. M., H. Li, L. Richardson, J. Howard, T. Platte, and D. M. Bullock. Detector-Free Optimization of Traffic Signal Offset with Connected Vehicle Data. *Transportation Research Record: Journal of the Transportation Research Board*, 2017. 2620: 54–68.
18. Wünsch, G., F. Bölling, A. von Dobschütz, and P. Mieth. Bavarian Road Administration Uses Probe Data for Large-Scale Traffic Signal Evaluation Support. *Transportation Research Record: Journal of the Transportation Research Board*, 2015. 2487: 88–95.
19. Zheng, J., and H. X. Liu. Estimating Traffic Volumes for Signalized Intersections Using Connected Vehicle Data. *Transportation Research Part C*, Vol. 79, 2017, pp. 347–362.
20. Day, C. M., M. Taylor, J. Mackey, R. Clayton, S. Patel, G. Xie, H. Li, J. R. Sturdevant, and D. M. Bullock. Implementation of Automated Traffic Signal Performance Measures. *ITE Journal*, Vol. 86, No. 8, 2016, pp. 26–34.

The Standing Committee on Traffic Signal Systems (AHB25) peer-reviewed this paper (18-03728).

Article

TRR
JOURNAL OF THE TRANSPORTATION RESEARCH BOARD

Transportation Research Record
2018, Vol. 2672(18) 143–153

DOI: 10.1177/0361198118798457
journals.sagepub.com/home/trr
SAGE

An Improved Inductive Loop Detector Design for Efficient Traffic Signal Operations and Leaner Space Requirements

Goli Koti Veera Yogesh[1], Anuj Sharma[2], and Lelitha Vanajakshi[1]

Abstract

Inductive loop detectors (ILDs) are one of the most widely deployed traffic sensors. At present, for lane-by-lane detection, ILDs require separate connecting cables for each loop (each lane) and separate data acquisition systems or detector channels to process them. This becomes problematic with limited conduit and cabinet space. In most cases, transportation agencies use ILDs connected in series to avoid these constraints, in which case the lane-by-lane information is lost. However, research has shown that lane-by-lane detection can lead to safer and more efficient operations at signalized intersections. In order to ease the application of lane-by-lane detection, the current study proposes a solution that uses electronic circuit modification to convert the existing serially connected loops to carry out lane-by-lane detection. This system achieved 100% accuracy of lane-by-lane detection in test runs. The paper also proposes an improved loop design, for future installations, that can be used for vehicle classification and wrong way detection. The study implemented machine-learning algorithms for vehicle classification and direction determination with an accuracy of 99.6% and 78.57%, respectively, using single loop configuration.

Inductive loop detectors (ILDs) are one of the most extensively used traffic sensors because of their high accuracy and cost effectiveness. However, two main limitations affecting usage of ILD sensors are complexity in connections to collect data from multiple lanes and their inability to identify the direction of vehicular movement. For lane-by-lane detection, ILDs require separate connecting cables for each loop (each lane) and separate data acquisition systems (DAQs) to process them. This becomes especially problematic when the conduit space to pull these cables is limited or the cabinet space is filled up and is not able to accommodate additional DAQs, in addition to making it more expensive. In such cases, traffic signal operators make a trade-off between operational efficiency and space utilization. They connect the sensors detecting same movement in series and only pull one cable pair to the cabinet. The detector connected in series across multiple lanes will be able to detect the presence of vehicles in either of the two lanes but cannot differentiate in which lane the vehicle is present. On the other hand, only a single DAQ is required to process the data coming from multiple sensors connected in series, thus reducing the space needed in the cabinet. The trade-off of reducing space requirements with no lane differentiation increases the probability of maxing out the phase significantly.

This study proposes an algorithm that can be used to convert existing loop detectors, connected in series, to achieve lane-by-lane detection by adding a few more electronic components and providing an advanced signal processing unit. This will also reduce the space needed in conduit, as all the lanes on an approach, left, through, and right, can be brought back to the cabinet over a single-wire pair and one DAQ can be used per approach. Thus, in extreme cases all the sensors for the whole intersection can be serially connected and can be handled by a single advance DAQ unit. This implies that the number of sensors being used at an intersection would not increase the number of DAQs needed, thus eliminating the need for a bigger detector channel space. The advance

[1]Department of Civil Engineering, Indian Institute of Technology Madras, Chennai, India
[2]Department of Civil Construction and Environmental Engineering, Iowa State University of Science and Technology, Ames, IA

Corresponding Author:
Address correspondence to Anuj Sharma: anujs@iastate.edu

DAQ can act as a detector card that can report lane-by-lane actuations to the controller on different detector channels. The study is an extended application of the multiple loop system proposed by Ali et al. (*1*). The basic principle used is to input a superimposed wave with the resonant frequencies of all loops, which can be separated out lane-wise using suitable band-pass filters.

This study also proposes changes to include the functionality of vehicle classification and wrong way detection. In the field, advance loop detectors often used for dilemma zone protection or minimum green extension are tied in series and follow simultaneous gap-out logic. These loops are not able to detect in which lane the vehicle is present and what is the class of the vehicle. The proposed design, with additional asymmetrically placed inner loop and additional electronics, will be able carry out lane-by-lane detection, vehicle classification, and could be connected in series across multiple lanes. Vehicle classification will be especially useful for applications such as dilemma zone protection and minimum gap extension, as more extension time could be provided for trucks or other special classes of vehicle. An added feature of this loop design is that it can also provide wrong way detection, which can be very useful for counting entries and exits and vehicle types at parking lots and rest areas. Another limitation of single ILDs in their present form is their inability to identify the direction of vehicular movement, which is important for applications such as parking counts, wrong way detection in bi-directional traffic systems, and so forth. Currently, such applications use either two in-ground sensors or a video detector (*2–5*), which can be replaced with this proposed loop system.

Literature Review

Reported advantages in using ILDs are the ability to collect many basic traffic parameters such as volume, occupancy, speed, headway, and gap, and robustness to extreme weather conditions such as rain, fog, and snow (*5–9*). In addition to the above-mentioned conventional parameters, Cherrett et al. (*10*) proposed methods to determine vehicle flow, time duration a loop is occupied, average loop-occupancy time per vehicle, average headway time between vehicles, and proportion of long and short vehicles for a fixed period using an existing SCOOT (split cycle and offset optimization technique) type single inductive loop.

Several studies (*11–13*) showed significant advantages of lane-by-lane detection over serially connected simultaneous gap-out logic. Smaglik et al. (*11*) studied traditional vehicle detection, in which all lanes were connected to an actuated controller and lane-by-lane detection. Results showed a 13% increase in efficiency of actuated controllers for lane-by-lane detection strategy. Tian and Urbanik (*13*) studied two detection schemes, single-channel detection, and emerging lane-by-lane detection, over a wide range of traffic scenarios, comparing the difference in green extension. Results showed a superior performance achieved by lane-by-lane detection for high-volume traffic and significant differences between the two schemes did exist for certain signal cycles. In addition to an improvement in capacity, lane-by-lane advance detection at a signal high-speed intersection can significantly reduce the probability of max-out and thus increase the safety of the high-speed intersection (*14–16*).

ILD performance has been widely investigated and improved. Researchers who have investigated loop shape parameters for improved sensing abilities (*17*, *18*), have proposed better de-noising algorithms (*19*) and compared the performance of ILD with other detectors under different traffic scenarios (*20–22*). Over the years, several researchers have proposed methodologies for performing vehicle classification using single loop (*23–31*). The overall accuracy of classification hovers around 90% with some classes of vehicles achieving higher accuracies than others. FHWA Class 1 vehicles (motorcycles), and Class 3 vehicles (four-tire single-unit trucks) had accuracies ranging in the 80s. Many of these studies focused on improving the signal processing algorithms for improving vehicle classification accuracies. In this paper, we propose an improved design of loop detector to generate signatures that can achieve higher classification accuracies. The approach then builds on signal processing algorithm–based pattern recognition proposed in the field of ILD-based vehicle classification as well as other areas (*32–34*). The vehicle classification accuracies in a controlled test were closer to 98% for Class 1 and Class 3 vehicles. In addition to the vehicle classification, the improved ILD design is also able to detect the direction of travel, which is useful in applications such as parking space monitoring.

Methodology

Improved Shape Design

Different loop shapes, including square, rhombus, octagonal, and circular designs were reported for applications such as vehicle classification, speed, and headway measurements (*5*). However, there were no reported studies about the determination of directional information of traffic flow using a single inductive loop system. The only reported work to determine direction was using a double loop system (*5*). This system contained two adjacently placed loops on a single lane. Each loop creates a signature when a vehicle is passed, with a time lag. The order of activation of loop determines the direction of vehicles. The present study develops a new loop shape for

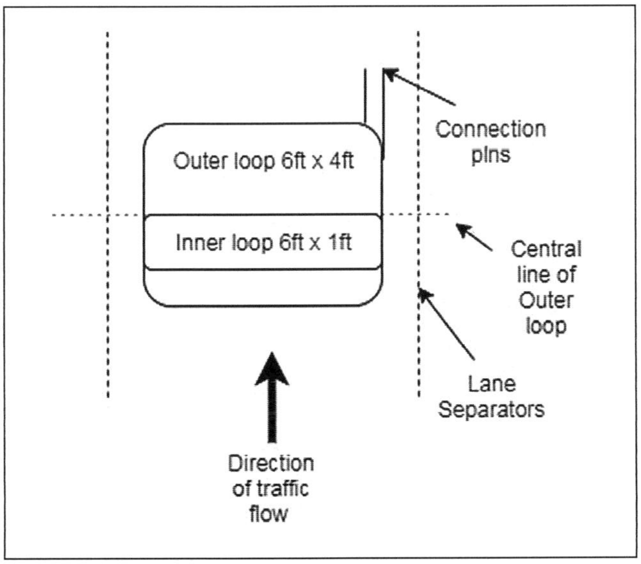

Figure 1. Loop design for direction determination.

determining direction of travel along with classification of vehicles using a single loop. The underlying concept for this single loop direction detection system is to create an asymmetry in the loop structure, thereby creating asymmetry in the vehicle signatures. Figure 1 shows the new loop structure that was designed based on this.

Here, the loop contains an inner smaller loop and an outer bigger loop. The smaller loop has dimensions 6 ft × 1 ft and the outer bigger loop has dimensions 6 ft × 4 ft. The inner loop is arranged with a shift of a half foot from the center of the bigger loop. This shift causes asymmetry in the design as well as in the signature generated, which enables the determination of the direction of movement of vehicles and also provides additional features in the signature for improved vehicle classification. Sample signatures formed before and after the inner loop shift are shown in Figure 2. Figure 2*a* shows a sample car signature without any shift in inner smaller loop, and 2*b* shows one with the newly designed loop. The marked part in Figure 2*b* shows the variation in signature, caused by the asymmetry in the loop design, which helps in determining the direction of movement of the vehicle. The vehicle classification and direction determination scheme utilizes this feature, and the results are presented in a later section.

Extended Circuit System for Multiple Lanes

The system currently used for data collection from a multiple lane roadway for lane-by-lane detection is as shown in Figure 3*a* (*35*). In this system each loop is attached with an oscillator and a processor for data extraction. This results in complex connections and a costly system as each loop requires its own processor to detect traffic. Figure 3*b* presents the proposed system for a serially connected lane-by-lane detection.

To reduce the complexity of the system in 3*a*, the preliminary focus was to connect all the loops to a single DAQ, which reduces the required connections. Following Ali et al. (*1*), in which a similar system was developed to address the problem of less lane-disciplined traffic conditions, a circuit system as shown in Figure 3*b* was developed. Although Figure 3*b* shows a modified loop design for demonstration, a normal ILD can also be easily used in this design for lane-by-lane detection. The methodology employs a band-pass filtering technique to separate out the information from each lane. Every loop is connected in parallel to a capacitor, forming a resonant circuit as shown in Figure 3*b*. Each loop is connected separately to C_1, C_2, and C_3 capacitors, forming resonant circuits with frequencies f_1, f_2, and f_3. Each of these loops is connected in series electrically with each other and to a DAQ. An input signal V_{in} is generated that contains all the resonant frequencies of the loops f_1, f_2, and f_3, with the help of an arbitrary function generator and power supply unit, which is controlled by a computer. The complete output signal from all the loops is measured using the DAQ. The output signal with all the resonant frequencies is analyzed using band-pass filters. The change in amplitude of the output signal corresponding to each resonant frequency is found using suitable filters. This separates out the information from each loop, giving lane-wise data. This system was implemented in the field for data collection, as discussed next.

Data Collection

In order to implement a multiple-lane system and also to determine the directional information, data were collected from a test bed with a bi-directional traffic flow in a controlled environment at Iowa State University.

An aerial view schematic of the test bed setup is shown in Figure 4*a*, where all the loops are connected in series producing a single output. The test is designed to determine the direction of vehicles. Therefore, on both the lanes, vehicles were passed in both directions producing direction specific signatures. In order to observe a change in signature due to new loop configuration, a conventional rectangular loop is placed on one lane as shown in Figure 4*a*. Data were collected for 4 hours from 10:00 a.m. to 1:00 p.m., and a total of 253 vehicle signatures were collected for vehicle classification and a total of 182 signatures were collected for direction estimation. Three vehicle classes, namely car (Class 2), motorbike (Class 1) and four-tire single-unit truck (Class 3) were used for the data collection. Figure 4*b* shows a snapshot of the experiment, where a car passes over two loops and corresponding signatures were generated on panels

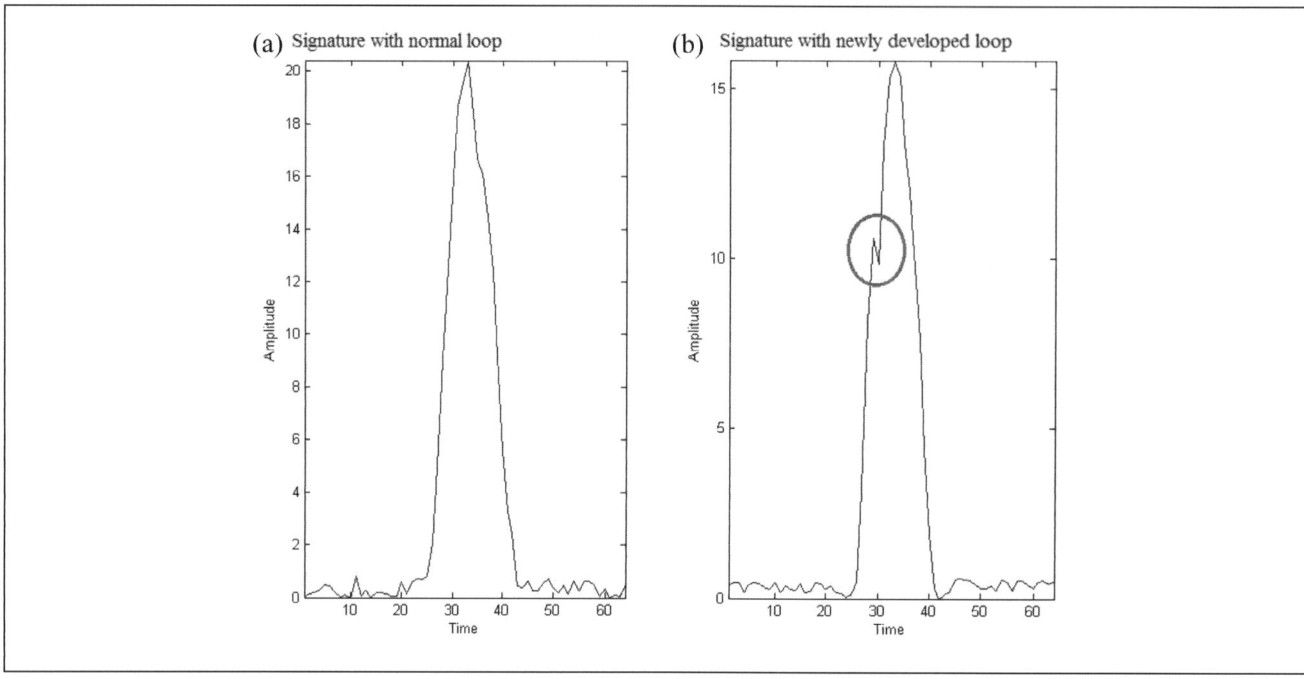

Figure 2. Change in car signature before and after the inner loop shift.

Loop_1 and Loop_2. Since there is no vehicle presence on Loop 3, panel Loop_3 shows random noise.

Vehicle Classification and Direction Detection

To provide an accurate classified vehicle count, and to determine directional information in real time, an efficient and robust four-phased methodology was developed. Two of these phases, signal de-noising and feature extraction, include usage of discrete wavelet transform (DWT). The intermediate segmentation phase used a simple peak-based method to efficiently collect vehicle signatures. Finally, the identified signatures are trained using supervised random forest and support vector machine (SVM) methods. Figure 5 shows the algorithmic steps in the methodology. The techniques used in each phase are detailed in the following sections.

Discrete Wavelet Transformation

DWT is often used to analyze a signal in a time series. DWT can remove instrumental noises, which often occur in higher frequencies (*25*). A wavelet is a wave-like oscillation with amplitude that starts at zero, increases, and then decreases back to zero. Unlike the sine curves used in Fourier transform for decomposition of a signal, wavelets are generally much more concentrated in time. They usually provide an analysis of the signal, which is localized in both time and frequency, whereas Fourier transform is localized only in frequency. The DWT decomposes a signal into different scales with different levels of resolution. The majority of signal information is in low frequency, which is the most important part of the signal, and the information in high frequency may indicate noise. The transform for any signal can be computed efficiently, using Mallat's algorithm (*36*). Figure 6 demonstrates Mallat's algorithm of three-level decomposition, which uses a set of consecutive low- and high-pass filters. At each decomposition level, the filters produce signals with an output down-sampled by a factor of 2. The detail coefficients ($d_i[n]$) tend to have high value in the noisy parts of the signal. Therefore, these coefficients ($d_i[n]$) are nullified, and the signal is reconstructed producing a de-noised signal with smaller error. This algorithm is used in the present study for signal de-noising.

Random Forest

A random forest consists of a number of decision trees, with each tree grown using some form of randomization. Decision tree classifiers divide one large complex question into a series of simple sequential questions, forming a hierarchy of questions. In a single decision tree classifier, a node is created by defining an associated "question" about all the available features; for example, the question could be whether the features are above a certain threshold. Finally, these questions end with classifying the data into leaf nodes or terminal nodes. Since several trees are created in this algorithm with each tree's

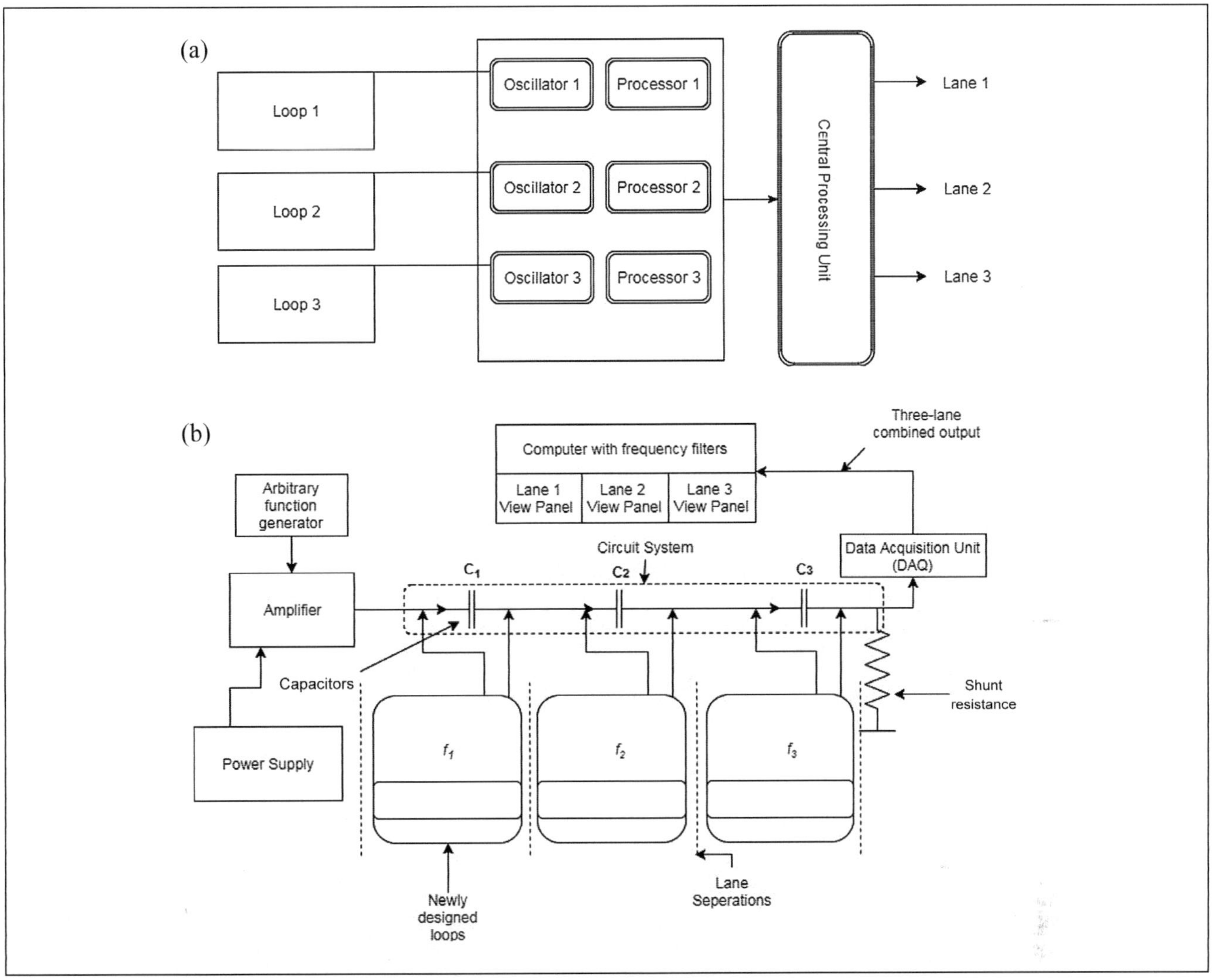

Figure 3. Multi-lane detection systems: (*a*) existing lane-by-lane detection system overview for three lanes (adapted from Minsen et al. [35]); (*b*) proposed lane-by-lane detection using a serially connected modified ILD (the same circuitry can be used with normal ILDs to perform lane-by-lane detection).

nodes formed using only a fraction of randomly chosen features, it is called a "random forest" (*31*). The algorithm starts with creating a decision tree, in which the input data are classified based on some feature *i*. The result from the question asked on feature *i* further generates hierarchical branches. This process goes on till terminal node, where the data are classified as one of the given classes. A number of such trees are generated and the terminal class for a given data vector is determined by majority voting of the trees (*31*).

SVM

Traditional statistical pattern recognition methods are developed for large samples, and they may not work well in practical cases with a limited number of data samples. In order to achieve good performance for limited sample data sets, a new powerful tool called SVM was developed (*37*). Due to its performance and robustness, this algorithm is extensively used in various domains. The working principle of SVM is as follows: it maps the input vector into a high dimensional feature space by suitable transformations. In this space an optimal separating hyperplane is constructed (*38*). Many other supervised learning algorithms, like perceptron, try to find any hyperplane without considering how clearly the hyperplane separates the data, which is not desirable. A hyperplane that is as far away as possible from either class is preferred and is used in SVM. A technical measure of how clearly a hyperplane separates data is its margin. This is the distance of the hyperplane to the closest point in the dataset. A large

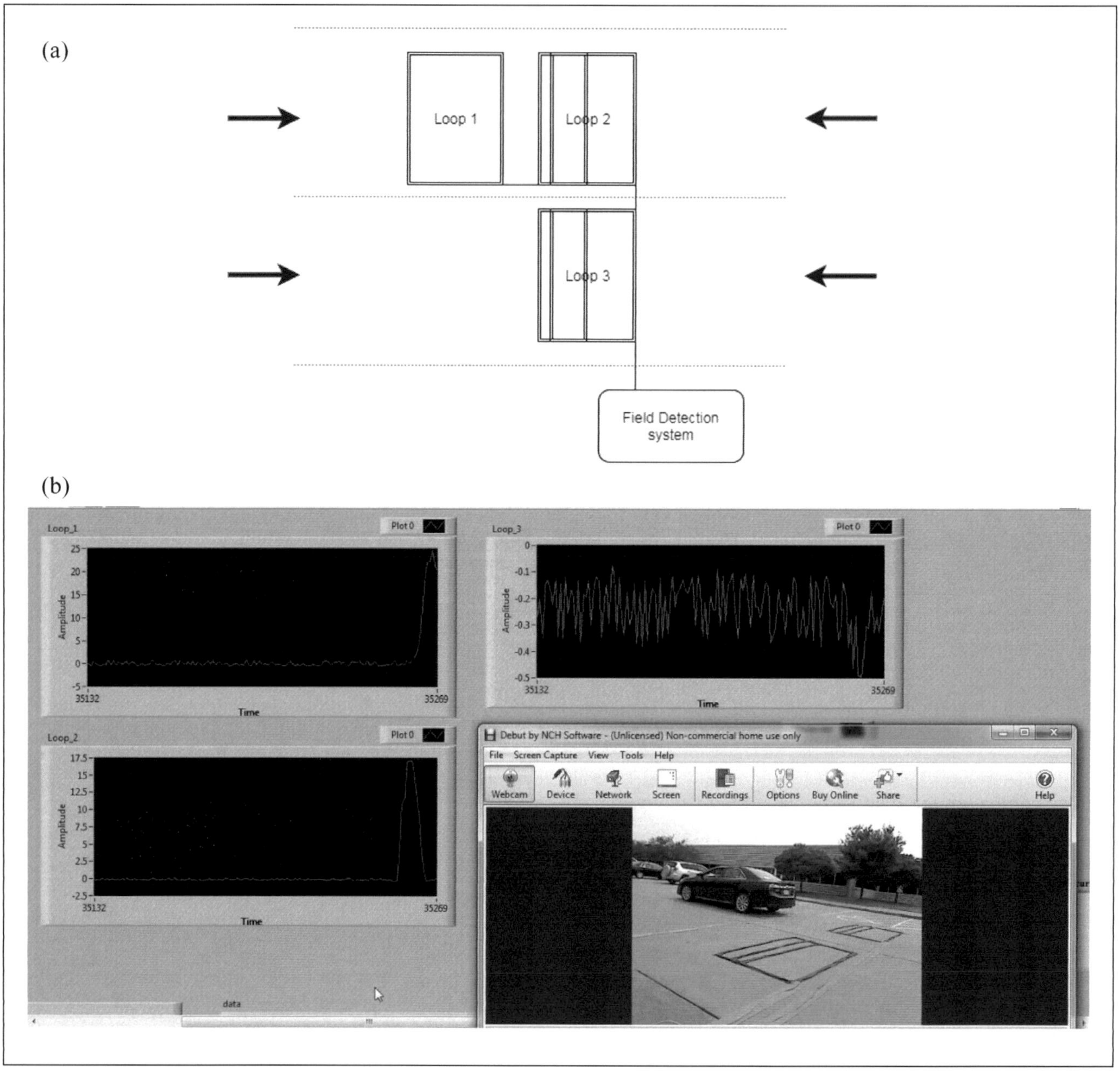

Figure 4. Details of test bed: (*a*) test bed setup for lane-by-lane and direction detection; (*b*) snapshot of the field experiment.

margin means that the hyperplane very clearly separates the data.

Implementation and Results

The proposed methodology of filtering is done in LabView to obtain lane-wise data and is further processed for vehicle classification and direction detection in Matlab. For field implementations, this can be replaced with a simple single board processor.

The accuracy of lane-by-lane detection was tested first, and the proposed modified circuit system with software integration provided 100% lane-by-lane vehicle detection during test runs. These results can be obtained in real time as the vehicle moves without any requirement of off-line data processing. Figure 7 shows the total number of correctly identified vehicle runs and actual vehicle runs in every 10 minute interval.

De-Noising the Raw Data using DWT

A Daubechies wavelet of second order (dB2) with two vanishing points was chosen in this study because the sample signatures and the wavelet chosen have a similar

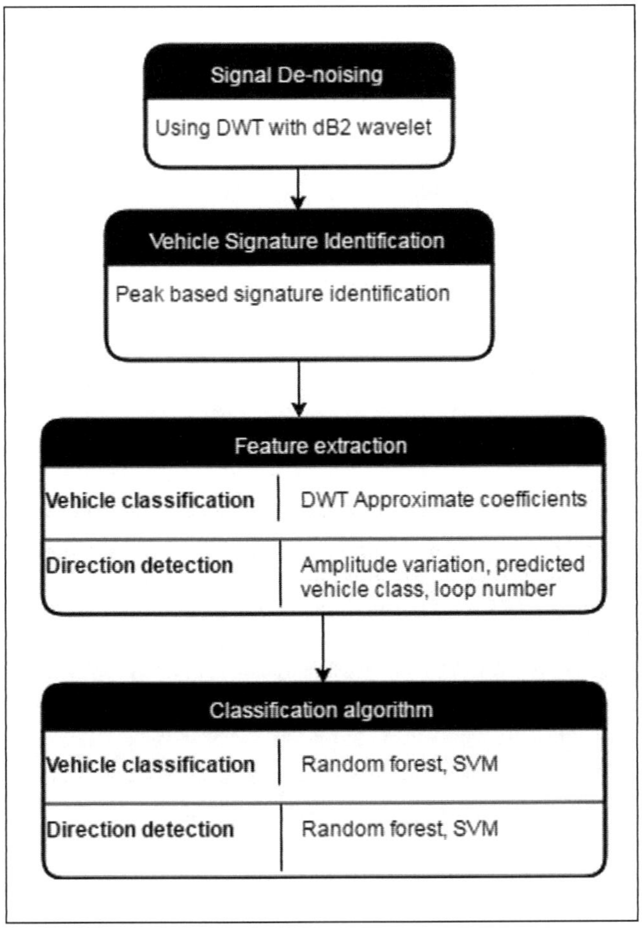

Figure 5. Methodology of direction identification and vehicle classification.

shape. Figure 8 shows a sample of de-noised signal after applying this method.

Peak-Based Signature Identification

One of the unique distinctions between vehicle signature and noise can be identified in the amplitude domain. The amplitude of vehicle signatures is greater than noises generated during no-activity periods. Therefore, a simple peak-based solution was implemented. A sliding window is passed across the raw data which identifies signatures if the peak is greater than a set threshold. With this method all the vehicle signatures generated were identified and were trained for vehicle classification and direction determination.

Vehicle Classification using Random Forest and SVM

The approximation coefficients ($c_i[n]$) obtained from DWT were used as input features for vehicle classification. Here also, Daubechies wavelet with two vanishing points (db2) was used and Level 2 approximate coefficients were used as features. The results are presented in the form of a confusion matrix. A confusion matrix is a summary of prediction results on a classification problem. The number of correct and incorrect predictions are summarized with count values broken down into four categories as "true positive" (correctly predicted event values), "false positive" (incorrectly predicted event values), "true negative" (correctly predicted no-event values), and "false negative" (incorrectly predicted no-event values). Thus, it gives an insight not only into the errors being made by the classifier but more importantly the types of errors that are being made (*39*).

In this study, a random forest classifier with 400 trees and an SVM classifier with radial kernel were built. The random forest classifier achieved an accuracy of 98.42%, while the SVM classifier achieved 99.60%, and corresponding confusion matrices are provided in Table 1*a* and *b*, respectively.

Direction Identification using Random Forest and SVM

In the case of direction identification, in order not to retain the significant information for better identification, the variation in amplitudes of the signatures along with the loop number and the predicted vehicle class are

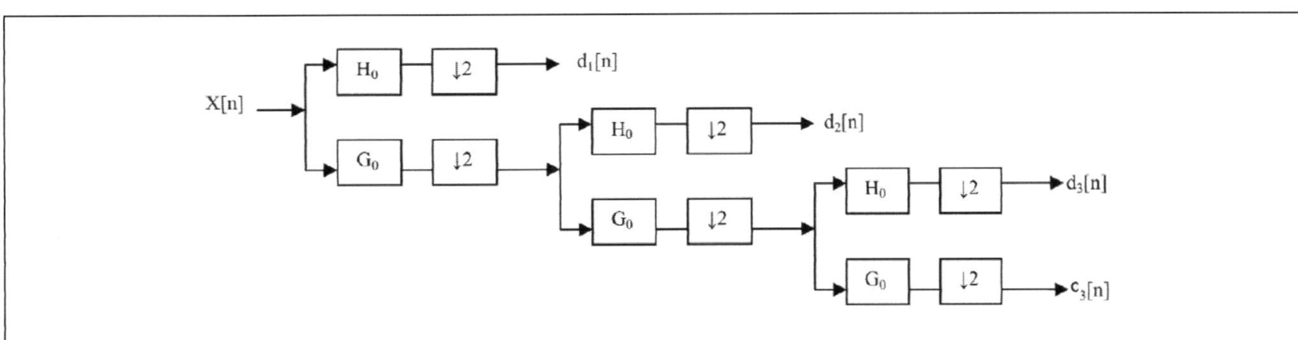

Figure 6. Mallat's algorithm demonstration, where X[n] = original signal; H0 = high-pass filter; G0 = low-pass filter; di[n] = detail coefficients at *i*th level of decomposition; and ci[n] = approximation coefficients at *i*th level of decomposition.

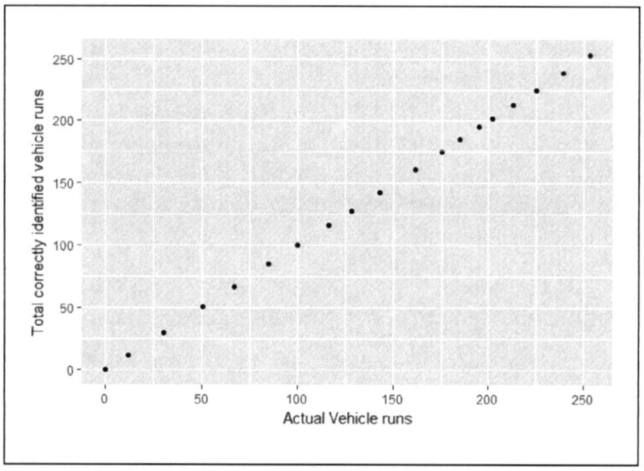

Figure 7. Lane-by-lane classification by the system versus actual runs.

used as input features for both the classifiers. The random forest model achieved an accuracy of 78.57% while the SVM model achieved an accuracy of 74.17%. Their corresponding confusion matrices are given in Table 2*a* and *b*.

Summary and Conclusions

This paper proposed an innovative circuit and inductive loop shape design to enhance its capabilities. The improved circuit design and signal processing algorithm presented in this study can facilitate lane-by-lane detection in serially connected loops. This ensures that the traffic engineer does not have to make a trade-off between operational efficiency and space utilization. The complexity of the system will slightly increase due to the integration of capacitor specialized design and complex waveforms to enable filtering of lane-by-lane detection, and also due to co-located loop design for single application purposes. However, from another angle, the complexity reduces since the number of connections and processors is reduced. In addition, the cost of the system would be low for both direction determination and vehicle classification applications. The improved loop design proposed in this research enhances the vehicle classification accuracies reported in the literature. It is able to achieve 99–100% accuracy under test scenarios compared with 80–94% accuracies reported in the literature. In addition to improving classification accuracies, the improved loop design also facilitates identifying the

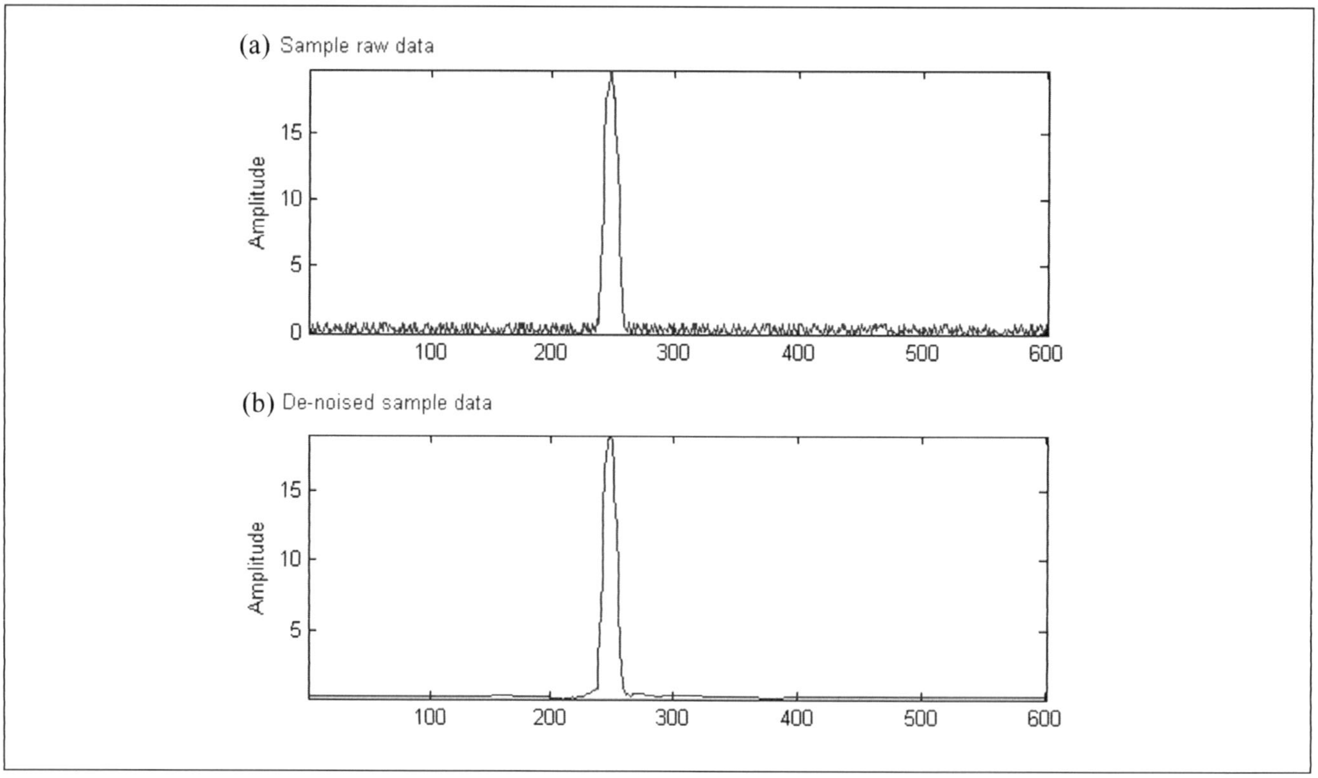

Figure 8. De-noised sample data using Mallat's algorithm.

Table 1. Confusion Matrix for Vehicle Classification

(a) Confusion Matrix for Vehicle Classification with Random Forest Model

		Predicted		
		Motorbike	Car	Truck
Actual	Motorbike	42	0	0
	Car	0	127	5
	Truck	1	1	77
	Accuracy	100%	96.21%	97.47%

(b) Confusion Matrix for Vehicle Classification with SVM Model

		Predicted		
		Motorbike	Car	Truck
Actual	Motorbike	42	0	0
	Car	0	131	1
	Truck	0	0	79
	Accuracy	100%	99.24%	100%

Table 2. Confusion Matrix for Direction Determination

(a) Confusion Matrix for Direction Determination with Random Forest Model

		Predicted	
		Northbound	Southbound
Actual	Northbound	68	24
	Southbound	21	69
	Accuracy	73.91%	76.67%

(b) Confusion Matrix for Direction Determination with SVM Model

		Predicted	
		Northbound	Southbound
Actual	Northbound	58	34
	Southbound	13	77
	Accuracy	63.04%	85.55%

direction of travel using a single loop. The accuracy of direction detection ranged between 74% and 77%. Such accuracies would be consistent in lane-disciplined traffic, where the traffic moves through the center of the loop. The proposed methodology can be implemented for various applications, such as lane-by-lane detection using serially connected loops, enhanced vehicle classification applications, estimated vehicle count on reversible lanes in homogeneous and lane-disciplined traffic, in parking management systems, and for median-less roadways, as well as under heterogeneous and less lane-disciplined traffic conditions to determine classified directional traffic count.

One of the downsides of this proposed loop design is the extra effort required for the installation of the system. However, the cost of the system would be low for both direction determination and vehicle classification applications when compared with a two-loop lane-by-lane detection system. Another possibility for future work is to develop the system at a commercial scale, where it has to be designed for ease of calibration. Similarly, the system has only to be tested for a limited number of vehicle types. The system can easily differentiate vehicle classes which are significantly different from each other, but the system is not to be used if there is a need to differentiate between classes which might be similar in terms of metal content and base height. As a future step, there is a need to develop a prototype field-ready version of the system and test in various real-world test beds. The standardization of the parameters could be potentially done after wide-scale testing of the system. Some of the factors that will affect the design parameter will include number of turns, the wire gauge, number of lanes connected, the loop length, and so forth.

Acknowledgments

Our research results are based on work jointly supported by the Center of Excellence in Urban Transportation at IIT Madras and REACTOR lab at Iowa State University.

Author Contributions

The authors confirm contributing to the paper as follows: study conception and design: Goli Koti Veera Yogesh, Dr. Anuj Sharma, Dr. Lelitha Vanajakshi; data collection: Goli Koti Veera Yogesh, Dr. Anuj Sharma; analysis and interpretation of results: Goli Koti Veera Yogesh, Dr. Anuj Sharma, Dr. Lelitha Vanajakshi; draft manuscript preparation: Goli Koti Veera Yogesh, Dr. Anuj Sharma, Dr. Lelitha Vanajakshi. All authors reviewed the results and approved the final version of the manuscript.

References

1. Ali, S. S. M., B. George, and L. Vanajakshi. An Efficient Multiple-loop Sensor Configuration Applicable for Undisciplined Traffic. *IEEE Transactions on Intelligent Transportation Systems*, Vol. 14, No. 3, 2013, pp. 1151–1161.
2. Shrivastava, A., and P. Arulmohzivarman. Vehicle Direction Detection Using Symmetrical SURF and Centroid Point Calculation. *2016 International Conference on Communication and Signal Processing (ICCSP)*, India, 2017.
3. Zu, X., S. Zhang, F. Guo, Q. Zhao, X. Zhang, X. You, H. Liu, B. Li, and X. Yuan. Vehicle Counting and Moving Direction Identification Based on Small-Aperture Microphone Array. *Sensors*, Vol. 17, No. 5, 2017, p. 1089.
4. Wahlstrom, N., R. Hostettler, F. Gustafsson, and W. Birk. Classification of Driving Direction in Traffic Surveillance Using Magnetometers. *IEEE Transactions on Intelligent*

Transportation Systems, Vol. 15, No. 4, 2014, pp. 1405–1418.

5. Traffic Detector Handbook. *U.S. Department of Transportation*, Vol. 1, 3rd ed. Federal Highway Administration, Washington, D.C., 2006, pp. 1.1–4.61.
6. Anderson, R. Electromagnetic Loop Vehicle Detectors. *IEEE Transactions on Vehicular Technology*, Vol. 19, No. 1, 1970, pp. 23–30.
7. Prucha, M. J. Inductive Loop Vehicle Presence Detector. U.S. Patent 3576 525, Apr. 27, 1971.
8. Koerner, R. J., and C. Park. Inductive Loop Vehicle Detector. U.S. Patent 3989 932, Nov. 2, 1976.
9. Lu, J., M. Rechtorik, and S. Yang. Automatic Vehicle Identification Technology Applications to Toll Collection Services. *Transportation Research Record: Journal of the Transportation Research Board*, 1997. 1588: 18–25.
10. Cherrett, T., H. Bell, and M. Mcdonald. Traffic Management Parameters from Single Inductive Loop Detectors. *Transportation Research Record: Journal of the Transportation Research Board*, 2000. 1719: 112–120.
11. Smaglik, E., D. Bullock, and T. Urbanik. Evaluation of Lane-by-Lane Vehicle Detection for Actuated Controllers Serving Multilane Approaches. *Transportation Research Record: Journal of the Transportation Research Board*, 2005. 1925: 123–133.
12. Smaglik, E., D. Bullock, J. Sturdevant, and T. Urbanik. Implementation of Lane-by-Lane Detection at Actuated Controlled Intersections. *Transportation Research Record: Journal of the Transportation Research Board*, 2007. 2035: 81–87.
13. Tian, Z., and T. Urbanik. Green Extension and Traffic Detection Schemes at Signalized Intersections. *Transportation Research Record: Journal of the Transportation Research Board*, 2006. 1978: 16–24.
14. Sharma, A., D. Bullock, S. Velipasalar, M. Casares, J. Schmitz, and N. Burnett. Improving Safety and Mobility at High-Speed Intersections with Innovations in Sensor Technology. *Transportation Research Record: Journal of the Transportation Research Board*, 2011. 2259: 253–263.
15. Sharma, A., D. Bullock, and S. Peeta. Limitations of Simultaneous Gap-Out Logic. *Transportation Research Record: Journal of the Transportation Research Board*, 2006. 1978: 42–48.
16. Sharma, A., D. Bullock, and S. Peeta. Stochastic Algorithm for Improved Operations at Isolated High Speed Intersections. *IFAC Proceedings Volumes*, Vol. 39, No. 12, 2006, pp. 325–330.
17. Gajda, J., R. Sroka, M. Stencel, A. Wajda, and T. Zeglen. A Vehicle Classification Based on Inductive Loop Detectors. *Proc., 18th IEEE Instrumentation and Measurement Technology Conference. Rediscovering Measurement in the Age of Informatics (Cat. No.01CH 37188) IMTC 2001*, Budapest, Hungary, Vol. 1, 2001, pp. 460–464.
18. Day, C. M., T. M. Brennan, M. L. Harding, H. Premachandra, A. Jacobs, D. M. Bullock, J. V. Krogmeier, and J. R. Sturdevant. Three-Dimensional Mapping of Inductive Loop Detector Sensitivity with Field Measurement. *Transportation Research Record: Journal of the Transportation Research Board*, 2010. 2128: 35–47.
19. Corey, J., Y. Lao, Y.-J. Wu, and Y. Wang. Detection and Correction of Inductive Loop Detector Sensitivity Errors by Using Gaussian Mixture Models. *Transportation Research Record: Journal of the Transportation Research Board*, 2011. 2256: 120–129.
20. Minge, E., and S. Petersen. Sensor Performance in Measuring Vehicle Length. *Transportation Research Record: Journal of the Transportation Research Board*, 2013. 2339: 47–56.
21. Ali, S. S. M., B. George, L. Vanajakshi, V. Jayashankar, and J. K. V. A Multiple Loop Vehicle Detection System for Heterogeneous and Lane-less Traffic. *IEEE Transactions on Instrumentation and Measurement*, Vol. 61, No. 5, 2011, pp. 1413–1417.
22. Ali, S. S. M., B. George, and L. Vanajakshi. A Simple Multiple Loop Sensor Configuration for Vehicle Detection in an Undisciplined Traffic. *Proc., International Conference on Sensing Technology, ICST*, Palmerston North, New Zealand, 2011, pp. 644–649.
23. Sun, C., and S. Ritchie. Heuristic Vehicle Classification Using Inductive Signatures on Freeways. *Transportation Research Record: Journal of the Transportation Research Board*, 2000. 1717: 130–136.
24. Jeng, S.-T., L. Chu, and S. Hernandez. Wavelet-k Nearest Neighbor Vehicle Classification Approach with Inductive Loop Signatures. *Transportation Research Record: Journal of the Transportation Research Board*, 2013. 2380: 72–80.
25. Ki, Y. K., and D. K. Baik. Vehicle-classification Algorithm for Single-loop Detectors Using Neural Networks. *IEEE Transactions on Vehicular Technology*, Vol. 55, No. 6, 2006, pp. 1704–1711.
26. Lamas-Seco, J., P. Castro, A. Dapena, and F. Vazquez-Araujo. Vehicle Classification Using the Discrete Fourier Transform with Traffic Inductive Sensors. *Sensors*, Vol. 15, No. 10, 2015, pp. 27201–27214.
27. Meta, S., and M. G. Cinsdikici. Vehicle-classification Algorithm Based on Component Analysis for Single-loop Inductive Detector. *IEEE Transactions on Vehicular Technology*, Vol. 59, No. 6, 2010, pp. 2795–2805.
28. Ali, S. S. M., N. Joshi, B. George, and L. Vanajakshi. Application of Random Forest Algorithm to Classify Vehicles Detected by a Multiple Inductive Loop System. *Proc., IEEE Conference on Intelligent Transportation Systems ITSC*, Anchorage, AK, 2012, pp. 491–495.
29. Oh, C., and S. Ritchie. Real-time Inductive-Signature-Based Level of Service for Signalized Intersections. *Transportation Research Record: Journal of the Transportation Research Board*, 2002. 1802: 97–104.
30. Oh, S., S. Ritchie, and C. Oh. Real-time Traffic Measurement from Single Loop Inductive Signatures. *Transportation Research Record: Journal of the Transportation Research Board*, 2002. 1804: 98–106.
31. Jeng, S.-T., and S. Ritchie. Real-time Vehicle Classification Using Inductive Loop Signature Data. *Transportation Research Record: Journal of the Transportation Research Board*, 2008. 2086: 8–22.
32. Azami, H., A. Khosravi, M. Malekzadeh, and S. Sanei. A New Adaptive Signal Segmentation Approach Based on Hiaguchi's Fractal Dimension. *Communications in*

Computer and Information Science Emerging Intelligent Computing Technology and Applications, Vol. 304 CC, 2012, pp. 152–159.

33. Azami, H., K. Mohammadi, and B. Bozorgtabar. An Improved Signal Segmentation Using Moving Average and Savitzky-Golay Filter. *Journal of Signal and Information Processing*, Vol. 3, No. 1, 2012, pp. 39–44.
34. Kaur, A., and N. K. Panag. Different Techniques for the Segmentation of the EMG Signal - A Review. *International Journal of Engineering and Management Research*, Vol. 5, No. 6, 2015, pp. 15–19.
35. Minsen, C., J. Ngarmnil, and T. Rongviriyapanich. Embedded Adaptive Algorithm for Multi-lanes-traffic Inductive Loop Detecting System. *Proc., 2005 2nd International Conference on Electrical Engineering/Electronics, Computer, Telecommunications and Information Technology*, Semarang, Indonesia, Vol. 1, 2005 pp. 359–362.
36. Mallat, S. G. A Theory for Multiresolution Signal Decomposition: The Wavelet Representation. *IEEE Transactions on Pattern Analysis and Machine Intelligence*, Vol. 31, 1989, pp. 674–693.
37. Vapnik, V. N. Direct Methods in Statistical Learning Theory. *The Nature of Statistical Learning Theory*, 2000, pp. 225–265.
38. Gunn, S. R. *Support Vector Machines for Classification and Regression*. Technical Report. School of Electronics and Computer Science, University of Southampton, Southampton, England, United Kingdom, 1998.
39. Brownlee, J. *What is a Confusion Matrix in Machine Learning - Machine Learning Mastery*. [online] Machine Learning Mastery, Vermont, Australia, 2017. https://machinelearningmastery.com/confusion-matrix-machine-learning/

The Standing Committee on Traffic Signal Systems (AHB25) peer-reviewed this paper (18-05861).

Article

TRR
JOURNAL OF THE TRANSPORTATION RESEARCH BOARD

Transportation Research Record
2018, Vol. 2672(18) 154–166

DOI: 10.1177/0361198118791380
journals.sagepub.com/home/trr
SAGE

Building Intelligence in Automated Traffic Signal Performance Measures with Advanced Data Analytics

Tingting Huang[1], Subhadipto Poddar[1], Cristopher Aguilar[2], Anuj Sharma[1], Edward Smaglik[2], Sirisha Kothuri[3], and Peter Koonce[4]

Abstract

Automated traffic signal performance measures (ATSPMs) are designed to equip traffic signal controllers with high-resolution data-logging capabilities which may be used to generate performance measures. These measures allow practitioners to improve operations as well as to maintain and operate their systems in a safe and efficient manner. While they have changed the way that operators manage their systems, several shortcomings of ATSPMs, as identified by signal operators, include a lack of data quality control and the extent of resources required to use the tool properly for system-wide management. To address these shortcomings, intelligent traffic signal performance measurements (ITSPMs) are presented in this paper, using the concepts of machine learning, traffic flow theory, and data visualization to reduce the operator resources needed for overseeing data-driven ATSPMs. In applying these concepts, ITSPMs provide graphical tools to identify and remove logging errors and data from bad sensors, to determine trends in demand intelligently, and to address the question of whether or not coordination may be needed at an intersection. The focus of ATSPMs and ITSPMs on performance measures for multimodal users is identified as a pressing need for future research.

In the United States, more than 300,000 traffic signals are currently in operation. According to the Federal Highway Administration, the operation and performance of most of these signals is assessed through citizen complaints (*1*). In these settings, agencies are forced to rely on software and simulation models to develop timings, with the presumption that if there are no complaints, everything is working acceptably, often compromising on performance and efficiency.

Automated traffic signal performance measures (ATSPMs) are designed to equip traffic signal controllers with high-resolution data-logging capabilities which they may use to generate performance measures. These measures allow practitioners to improve operations and to maintain and operate their systems in a safe and efficient manner (*1*). State-of-the-art ATSPM systems primarily present raw data in graphical representations with the goal of providing tools for visual queries to traffic signal experts. The tool has been very useful for data-driven management of traffic signal systems and has been adopted and modified by several agencies. From conversations with several practitioners who use them, however, the authors have identified three main shortcomings of the tool: (i) ATSPM currently uses raw data feeds but has very little data quality control or quality checks in place, (ii) using ATSPM for system-wide management is resource intensive, and (iii) the primary focus of ATSPM is automobile traffic, and it fails to address multimodal aspects of signal operation.

In this study, the current state of the art is extended by the creation of a new tool called the Intelligent Traffic Signal Performance Measurement System (ITSPM). Instead of primarily automating the signal performance calculation from a raw data stream, this tool uses machine learning techniques, traffic flow theory, and data-driven intelligence to provide additional insights for decision makers. In this paper, three primary enhancements are

[1]Department of Civil, Construction and Environmental Engineering, Iowa State University, Ames, IA
[2]Department of Civil Engineering, Construction Management, and Environmental Engineering, Northern Arizona University, AZ
[3]Department of Civil and Environmental Engineering, Portland State University, Portland, OR
[4]Portland Bureau of Transportation, OR

Corresponding Author:
Address correspondence to Anuj Sharma: anujs@iastate.edu

provided to address the identified shortcomings of the existing state-of-the-art tool, namely:

1. Additional measures for data quality control;
2. Machine learning-based intelligence to deliver initial insights into the data, thus reducing the visual querying time, which results in more efficient utilization of personnel resources; and
3. Some of the current graphics in ATSPM are improved to represent operations better at different spatial and temporal resolutions.

Although noted as a shortcoming, multimodal aspects are not addressed in this paper and will be the focus of future research. The remainder of the paper is as follows: a literature review and the state of practice, followed by the methodology, data used, and results. The paper then ends with conclusions and recommendations.

Literature Review

The development of ATSPMs began with the collection of event-based data by researchers at Purdue University in the mid 2000s (*2*) and the identification of tactical methods to control traffic within NCHRP 3-66 (*3*). Since these initial works, researchers have emphasized the development of event-based data acquisition systems that have the capability of generating high-level performance measures as well as sufficient data resolution for them to be capable of being used for fault recreation and signal fine tuning (*4*). Researchers at Purdue University as well as practitioners at the Indiana and Utah Departments of Transportation (DOT) spearheaded the effort to move Highway Capacity Manual operational parameters from the post-processed environment to real-time performance measures in a mainstream operational environment (*5*). Whereas historical performance measure data were limited to hourly volumes, peak hour factors, and volume to capacity ratio (v/c) values over long analysis periods, these measures use event-based data to empower an agency with the ability to make data-driven decisions regarding detector and communication health, traffic signal coordination, and split efficiency (*6–9*). Although this work has been a game changer in the operation and management of traffic signal systems, thus far it has focused mostly on vehicular performance measures with limited investigation into multimodal performance. The natural platform for this would be an extension of ATSPMs from a multimodal perspective, perhaps incorporating the visualization techniques presented by Tufte (*10*). Existing vehicular ATSPMs may also benefit from improved visualization techniques, although the Purdue researchers responsible for much of the ATSPM development work have already spent much effort on visualization (*11, 12*).

The state of practice with respect to ATSPMs involves the use of open source software and continued improvements to visualizations and metrics using advances in data analytics. From conversations with several practitioners who use them, ATSPMs are most often used for troubleshooting, operations, and planning. The AASHTO innovation initiative led by the Utah DOT has led to the adoption of ATSPMs by 26 transportation agencies across the country (*1*). The open source software used by the Utah DOT produces chart usage reports to track which performance measures and visualizations are most used by agency personnel. A usage report for the period from January 1 to July 23, 2017 in Utah showed that the Purdue phase termination (18,000 queries), split monitor (8,000 queries), and Purdue Coordination Diagram (PCD; ˜ 6,000 queries) are the most used metrics. Conversations with engineers at the Utah DOT and Georgia DOT corroborated this report. Turning movement counts and approach volumes are additional metrics that are used frequently by planners for simulation and modeling purposes.

According to agency personnel at the Utah DOT, the Purdue phase termination metric is used from an operations standpoint to address complaints. The split monitor is used for troubleshooting, retiming, and general operations, whereas the PCD is used for assessing if cycle lengths are optimal as well as the need for general retiming. Currently, with the open source software pioneered by Utah DOT, presentation of the PCD is not optimized. However, planned improvements in the near future involve linking the PCD with the link-pivot diagram to study progression quality and improve operations. The link-pivot algorithm was developed by researchers at Purdue University to optimize offsets along signalized arterials (*13*). Additional improvements involve the addition of transit signal priority (TSP) metrics to evaluate transit delays and to study the transition status of the controller when TSP is implemented. According to agency personnel, the least useful measures currently are: approach delay, arrivals on red, and pedestrian delay. With respect to improvements, engineers expressed interest in improving ways to measure delays, which could then be used in decision making. They also wanted the ability to examine the operational performance at the corridor and network levels, as currently they can only do so at the intersection level.

Methodology

The objective of this study is to improve the current state-of-the-art ATSPMs by providing enhancements in diagnosing sensor errors and assessing demands. The paper also emphasizes the need to associate each graphic with a given spatial and temporal resolution, as described below:

1. Stream analytic measures: These performance measures are used to detect quickly any anomalous behavior at any intersection during the course of a day.
2. Batch analytic measures: This historical chart serves to provide trends over time, and an individual day's information can be retrieved for anomaly detection.
3. Spatial resolution: Spatial resolution can be either at the phase/approach level for a given intersection or at a network level, depending on the desired objective. The decision makers might want to use ITSPMs to compare different corridors or to focus on a given intersection.

The graphic used for each resolution should be carefully planned to avoid any visual overload of information and to provide the user with the ability to identify the information that needs to be conveyed. In this paper, an attempt is made to use appropriate graphics to convey information suitable to a given resolution. The designed graphics and alerts in ITSPM are tied to decision support queries, as shown in Table 1. A comparison of how these decision support queries are answered by Utah ATSPM 4.0.1 is also shown in the table (*14*). Decision support for traffic signals can be divided into four broad categories, namely: (i) data quality, (ii) demand assessment, (iii) traffic control, and (iv) level of service. The focus of this paper is on major improvements in assessing the first three categories.

Table 1. Traffic Signal Decision Support Queries that Could Be Better Addressed by ITSPM

Operator queries	Utah Automated Traffic Signal Performance Monitoring (ATSPM 4.0.1)	Proposed Intelligent Traffic Signal Performance Monitoring (ITSPM)
Sensor and communication health		
Are there any failures in logging?	Utah DOT uses email alerts (not a generated report) for the following items: 1. No data 2. Too many pedestrian push button calls 3. Too many max-outs 4. Too many force-offs 5. Low detector count 6. High detector count Georgia DOT is working on an extension that will allow users to query the database to see how long these errors have been occurring.	New performance measure proposed (shown in Figures 1 and 2)
Are there any sensor failures?	Utah DOT uses the Purdue phase termination plot to determine if sensor failures exist. The phase termination plot is used to identify data gaps, too many max-outs (which can occur due to constant calls), too many force-offs and too many ped calls (which can occur due to constant calls due to a malfunctioning detector).	New performance measure proposed (shown in Figure 3)
Demand assessment		
What days have similar demand patterns?	Manually observed from approach volume graphic that has plot for each day with the x axis as time of day and the y axis as volume in vph.	Machine learning (ML)-based algorithms designed (described in Figures 4 and 5)
How many days were abnormal in a specified historical range?	Not available	
What was the potential cause of an anomaly?	Not available	
Is today a typical day?	Manually observed by comparing today's approach volume graphic with the historical volumes.	
What are the temporal variations for timing plan settings	Not available	Not addressed in this paper
Is the demand randomly distributed or is there a need for coordination	Not available	New performance measure proposed based on ML-based algorithm (described in Figure 6)
Control support		
Are the coordination parameters ideal?	Can be manually identified by visually exploring the Purdue Coordination Diagram	Aggregate platoon coordination diagram (APCD) proposed (Figure 7)

Data Used

The data used for this study were obtained from the City of Portland, OR, which recently started the implementation of ATSPMs at five intersections. For the visualizations in this paper, data were obtained from two specific intersections: NE Sandy Blvd. @ 57th Ave. and SE Division St. @ 122nd Ave. Data were available for February, May, and June of 2017. In this paper, the results are demonstrated based on data from different time frames. The Sensor and Communication Health section and the Control Support section used data from June 18 to June 24 (one week of data), whereas the Demand Assessment section used data from May 1 to June 24 for a better demand pattern extraction. The database contains the high resolution logs from each intersection's controller. The high resolution logs record events, such as phase changes, detector calls, power failures, etc. at one 10th of a second resolution.

Results

The enhancements proposed in this paper are divided into three sections: Sensor and Communication Health, Demand Assessment, and Control Support. First, data logging and sensor errors are discussed. Next, demand is analyzed to find the typical patterns that can be used to identify variation in demand and need for coordination. In the last section, enhancement of the PCD for adaptation to a multi-day display is discussed. It should be noted that existing ATSPMs are very strong in the visualizing control and level-of-service parameters, and the measures proposed here are not intended to substitute all the graphics in the ATSPM system but, rather, to augment the existing set of tools.

Sensor and Communication Health

Data quality is important to assure that decisions are being made using accurate information. ITSPMs are intended to monitor two sources of problems: (i) problems occurring due to incorrect logging of the data, and (ii) sensor errors involving false and stuck calls. It should be noted that issues of missed calls are not observable using high-resolution logs unless a redundant sensor is present to validate the missed calls.

Logging Failures. In high resolution logging, it is recommended that a logging flag that triggers a log entry at a known interval (every 10–15 sec) be added. This will ensure that any logging failures can be directly measured by counting the number of missing logging flags. In the absence of such a feature, ITSPMs propose to use surrogate measures to monitor logging failures.

a. **Spurious Inactivity Period** This is defined as a period during which the controller records no entries for any of the event codes. A tick mark appears for each time one of the event codes is logged during a 30-minute period, as illustrated in Figure 1*a*. The periods of inactivity, annotated by "A" and "B," could be due to an absence of any activity at the intersection or some spurious behavior of the logging program. To use a data-driven technique to find a threshold that separates spurious inactivity from normal inactivity, we explored logging gap distributions, which show the duration of time interval between two events observed over a week on the x axis and the number of times they were observed on the y axis, as displayed in Figure 1*b*. It can be seen that for the distributions observed at both Sandy @ 57th and Division @ 122nd, most of the entries are shorter than 300 seconds and then there is a sudden burst of activity after a long gap, at around 450 seconds (annotated by "C") and near 660 seconds (annotated by "D"). For this work, we used 300 sec as the threshold for detecting spurious vs. normal inactivity.

For the study intersections, a spurious inactivity period was defined as any period of time greater than 300 seconds during which no event was recorded in the database. The average spurious inactivity period versus time-of-day plot is presented in Figure 1*c*. The y axis represents the percentage of time during an hour when there was missing data, classified to be spurious activity, and the x axis represents the hour of the day. It can be observed that the performance of Division @ 122nd was poor over the entire day with approximately 70% of the data not being recorded in any given hour. If a dataset with 70% missing values were to be used for performance evaluation, the results would be misguided. The volume distribution reported for a Friday (6-23-2017) by the Portland ATSPM and the volume observed for a Friday (10-23-2015) using another data collection program (*15*) are presented in Figure 1*d*. The stark contrast between 2015 and 2017 volumes highlight the importance of using data quality checks prior to making decisions using automated performance measures.

b. **Missing Event Error** A second form of logging error can occur if only a single event code is spuriously missed for some period of time. In general, the phase status of "green-start" should be followed by "green-end," and these events should repeat in pairs. After removing the spurious logging failures, if there are still instances when such pairings are violated, a missing event status error will then be recorded. Tick marks for each time green and a detector turned on and off are shown

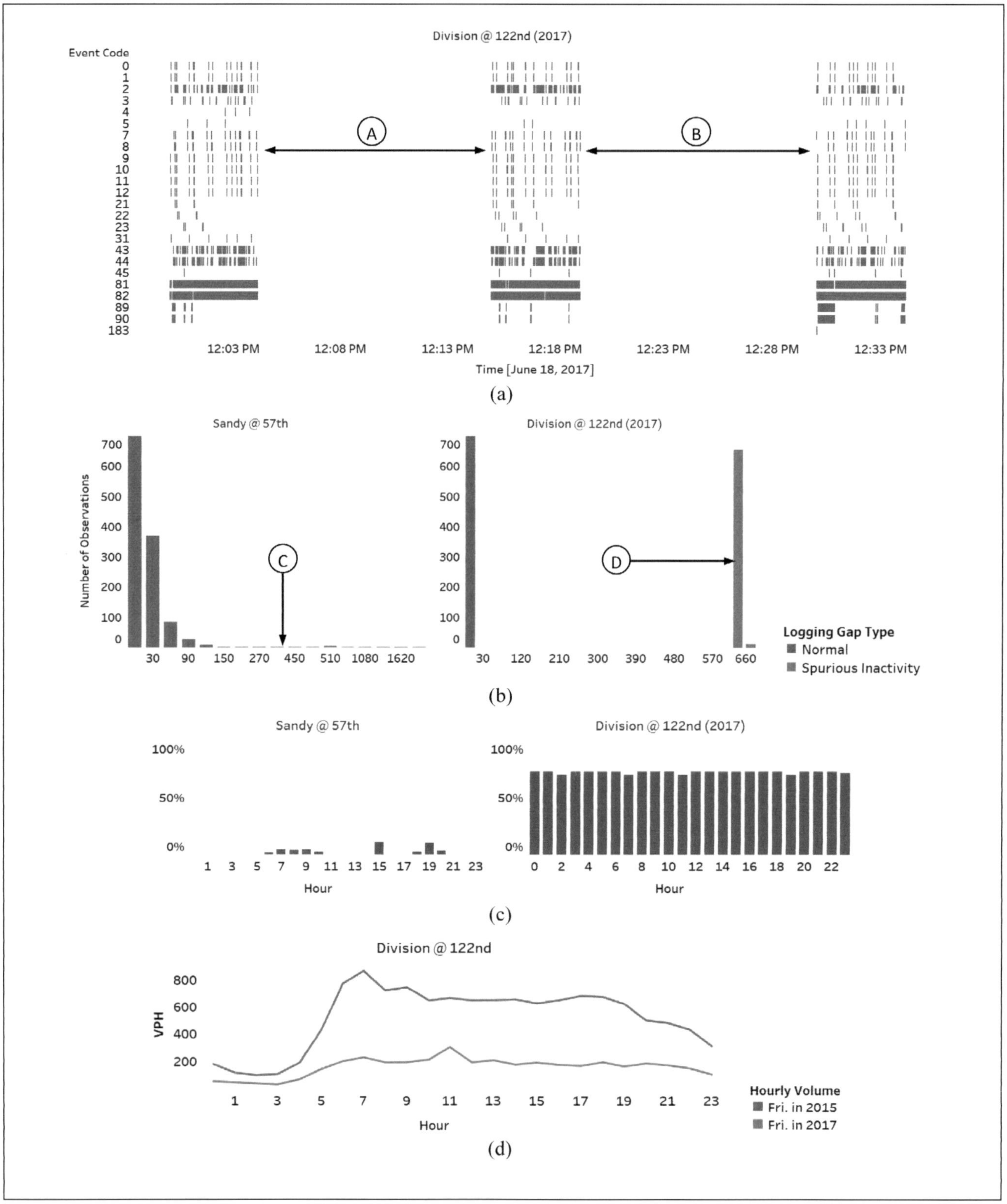

Figure 1. Spurious inactivity assessment procedure: (*a*) Spurious inactivity in logging example, (*b*) Logging gap distribution, (*c*) Summary of spurious inactivity for Sandy @ 57 and Division @ 122nd, (*d*) Impact on volume: Friday with spurious inactivity (6/23/2017) versus normal Friday (10/23/2015).

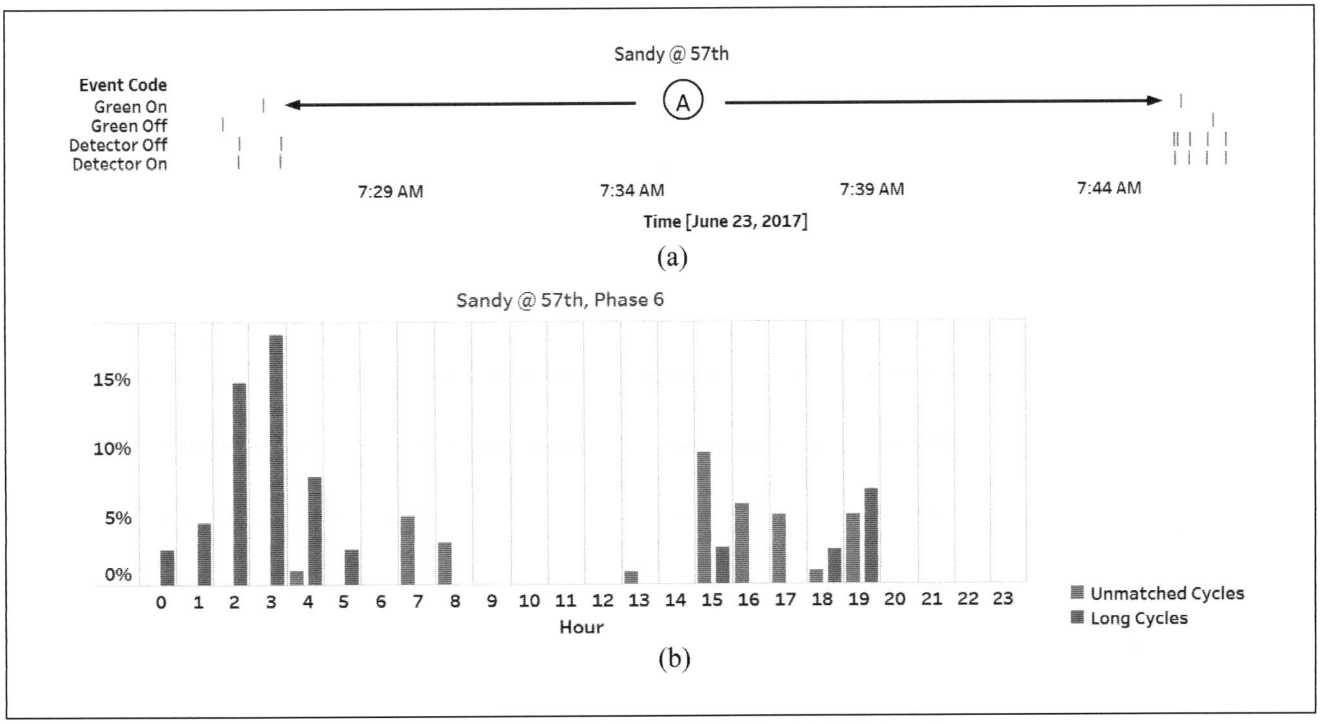

Figure 2. Missing event error assessment: (*a*) Missing phase status example; (*b*) summary of missing phase status for Sandy @ 57th, phase 6.

in Figure 2*a*. An example with phase 2 of Sandy @ 57th, for which the green indication started twice consecutively with no green termination was recorded in between, is annotated with "A." In Figure 2*b*, this data is shown aggregated into a histogram of missing phase 6 status over time of day. Also displayed are the long cycles (greater than 5 minutes) observed for phase 6.

The red bars indicate the percentage of time that phase logging was unmatched, and the blue bars indicate the percentage of time that the cycle length was greater than five minutes in through movements. The long cycles are shown because it is possible that the whole phase pair could be missed because of this logging error. Plotting of very long cycles along with unmatched pairs over time of day can give an indication of how often this might be occurring. Typically, very long cycles are acceptable during night times, but a red flag should be raised if a lot of them are observed during the daytime for movement on a main street. This type of aggregation could also be compiled for detector statuses, as there is an expectation of a certain level of volume by time of day.

Sensor Errors. Stuck call errors can be calculated using Equation 1 with the threshold set as a user-defined parameter. Six minutes implies that a detector was occupied for two or more consecutive cycles (with 180 seconds chosen as a conservatively high cycle length for most jurisdictions), which is highly improbable, especially during non-peak hours.

$$Stuck\ Call = \begin{cases} 1, & \sum_{1min} Occupied\ Time > 6\,min \\ 0, & otherwise \end{cases} \tag{1}$$

A false call error refers to the count that one detector per lane during one minute should not exceed a saturation flow rate. Although the exact saturation flow rate of that approach is not known, a threshold determined from experience can be used. For this research, we used 45 counts per minute (2700 vphpln) as a threshold and any records exceeding that were coded as false call errors.

To identify sensor errors, a scatter plot of the vehicle count per lane per minute is plotted against time occupancy in one-minute bins. A detector example with one month of data is shown in Figure 3*a*. The two shaded regions represent different potential sources of errors: false call errors (annotated by "A") and stuck call errors (annotated by "B"). If the logging error is not removed from the data, the number of stuck call errors will drastically change the data distribution, as in the case of Figure 3*a*. A case of a sensor performing reasonably well is shown in Figure 3*b*. Only one stuck call, annotated by "C," is observed in a period spanning one week. It should be noted that sensor health statistics should be

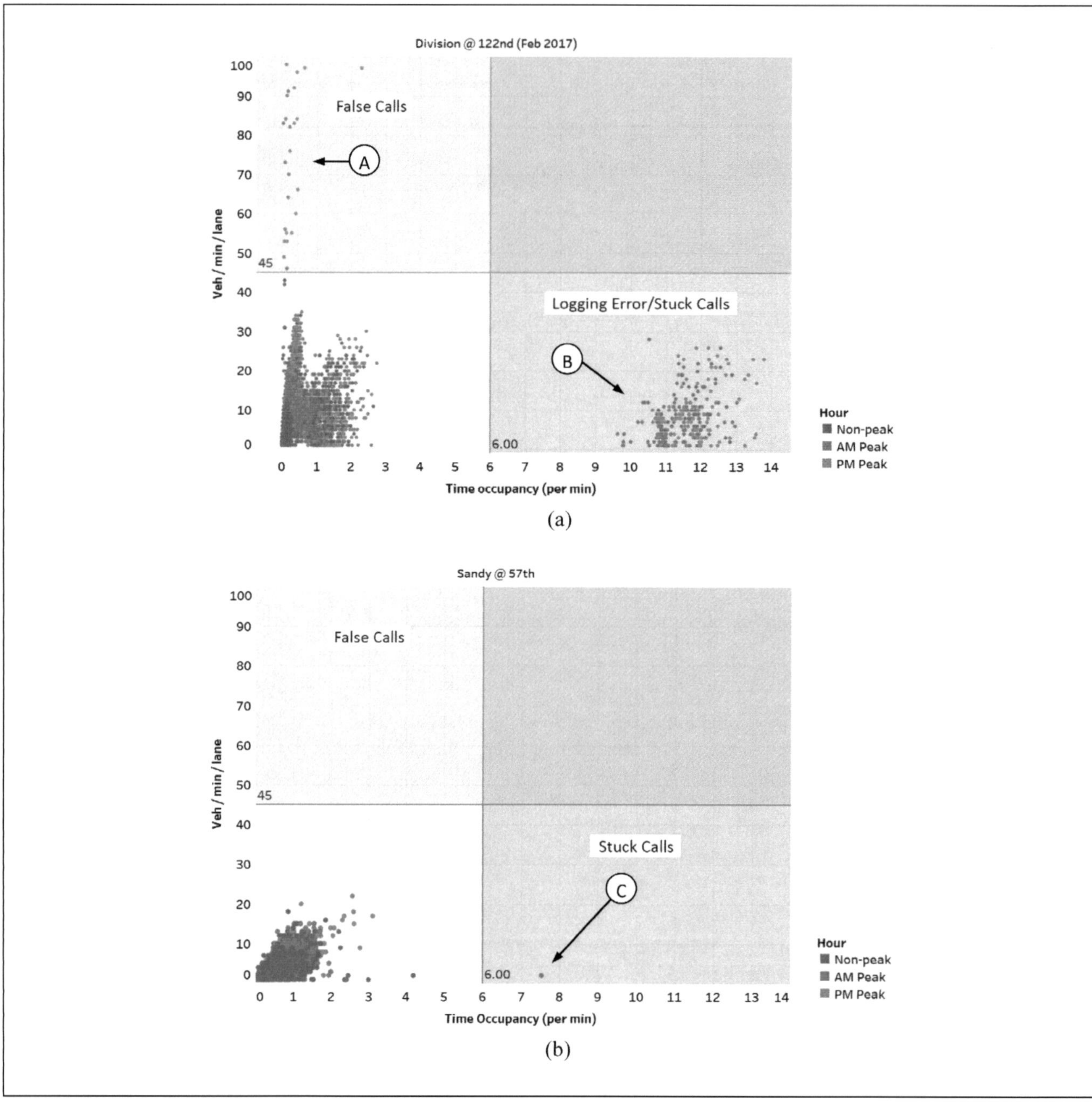

Figure 3. Detector quality performance evaluation: (*a*) Detector count versus total occupied time in 1-min aggregations (diagnostic figure), (*b*) Detector errors after removing spurious inactivity errors.

calculated after eliminating the time periods with significantly high logging errors.

Demand Assessment

Traffic Pattern Estimation. Identifying traffic patterns is critical in designing optimal traffic signal control strategies. Some of the questions that need to be answered by a traffic signal manager include:

1. What does a typical day look like? This is needed to ascertain the base set of timing plans that are designed to meet these conditions.
2. On which days of the week do similar traffic patterns exist? This helps in assigning a given time-of-day plan to certain days of week. The traffic manager can use ATSPMs to look at the volume distribution of different days of the week and make an engineering judgment.

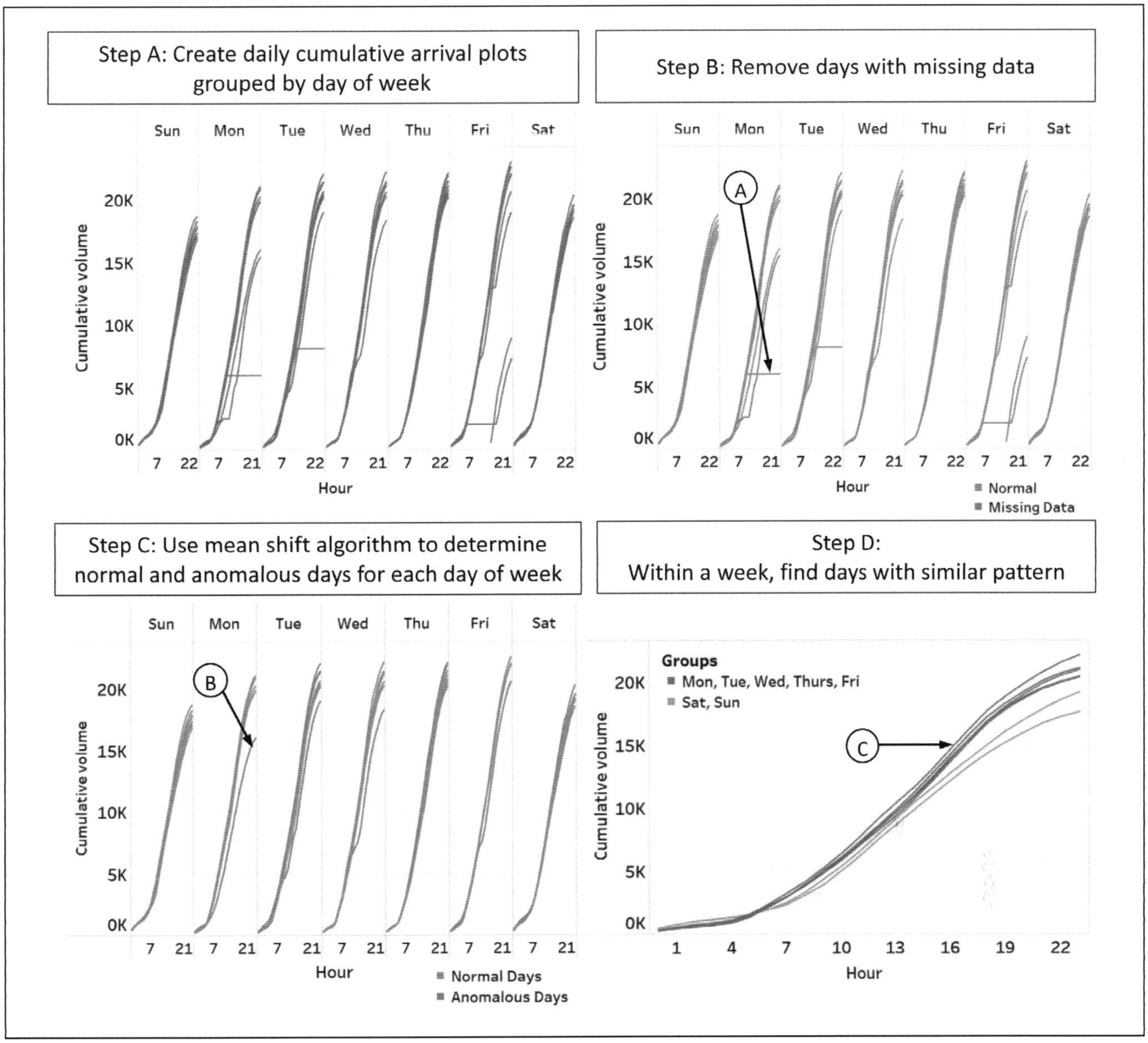

Figure 4. Demand trend analysis on Sandy @ 57th.

3. How many days deviate from the traffic pattern of a typical day? An intersection with a significant number of non-typical days and greater deviation during a typical day may show a greater need for an adaptive control or other unique timing solutions.
4. Are there any intersections today exhibiting an atypical demand? This might help the traffic managers to detect an issue and respond to it in a timely fashion.

An intelligent method of detecting demand patterns to answer the four above questions is presented. As shown in Figure 4, the process uses cumulative demand (all approaches) to determine the days of the week that have similar demand patterns, separates out anomalous demand patterns days, and then aggregates the days of the week having similar demand patterns into groups. This process involves four steps.

- **Step A** involves creating cumulative arrival plots for the total approach volume at the intersection.
- **Step B** identifies days with missing data, determined by instances for which the cumulative volume shows no change over any one-hour period. One such example is shown by annotation "A" in

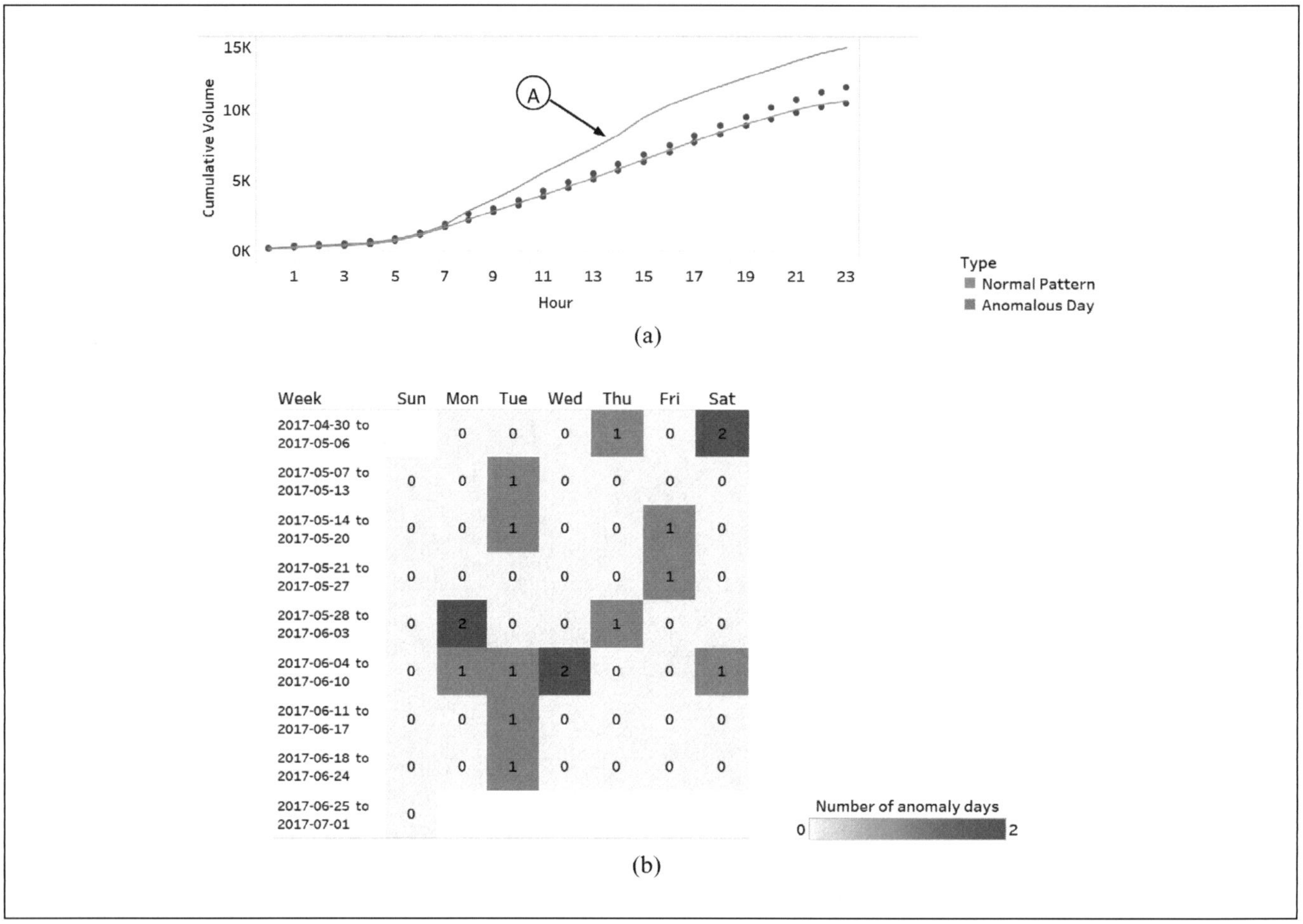

Figure 5. Anomalous days—detection and summary: (*a*) Real-time or daily representation of an anomaly (Division @ 122nd), (*b*) Weekly number of anomalies seen on the network.

Figure 4, as the volume remaining fixed after 7 a.m. After such days are identified, they are removed from further analysis. These horizontal lines can also be used to identify detector issues.

- **Step C** finds typical and atypical days for each day of week. This is done by computing a representative base curve for that day and then computing the deviation of the other curves from the base curve. The base curve is obtained by averaging the demand data points for every hour of the day. These deviation values (means and standard deviations) are clustered by a mean-shift clustering algorithm, and the largest group is identified to be the typical pattern for that day of week. The mean-shift algorithm works on the following principle:
    - First, assume each feature point is a cluster center.
    - Then, take all the points within the bandwidth or radius of the feature center and recalculate the mean of these feature points as the new center.
    - Repeat this for all the points until convergence is achieved; that is, the center points remain unchanged. Further details about mean-shift clustering algorithms can be found elsewhere (*16–19*).

The curve annotated as "B" in Figure 4 was selected as an atypical day by the algorithm. It can be seen that the daily arrivals on that day were much lower than the rest of the cluster.

- **Step D** is to group the days of week with similar demand patterns into one cluster. First, the base day for each day of week is determined. For example, curves for all the Sundays are combined to form one curve for a Sunday and so on. This grouping method is similar to that of Step C, and the representative curves are clustered. As shown by annotation "C" in Figure 4, weekdays show as one cluster and weekends make up another.

It is important to identify anomalous data in real time as well as to present an overview at the network level. After estimating the group of typical days, a 90% confidence bound is generated, and a curve departing from the confidence bound can trigger alarms in real time. An example of such a day is shown in Figure 5*a*, annotated as "A." In addition, anomalous days can also be identified at a network level over a period of time, as shown in Figure 5*b*, with the shading of each cell indicating the level of anomalous days.

Need for Coordination. Whether or not to provide signal coordination for a given time period is a challenging question for traffic operators to answer. This section provides a data-driven methodology to address the following problems:

1. Are there bunches/platoons arriving at an intersection? This provides evidence that the arrivals are not completely random and that the upstream intersection is impacting the arrival pattern at the subject intersection. This in turn points toward exploring the impact of providing coordination.
2. What is the time period for which coordination should be explored? The proximity of an upstream intersection shows only the possibility of creating tightly packed platoons. For coordination to be beneficial, there need to be enough platoons for enough cycles. The percentage of cycles showing platoons during a given time period and the average platoon length can be good measures for answering the question.

Steps used to identify the presence of platoons and calculate surrogate measures to identify a good time period to explore coordination options are presented in Figure 6.

- In **Step A**, the arrivals that can be classified as a platoon of vehicles are identified. To cluster those arrivals, a machine learning technique, density-based spatial clustering of applications with noise (DBSCAN) (*20*), was used to identify the group of vehicles in a single cycle. The following rule was used: A platoon should contain a minimum of five vehicles and not be separated by average headway greater than 1.6 seconds. It results in a parameter setting of five as minimum samples and four seconds as epsilon in the DBSCAN algorithm. These parameters were user-defined and can be chosen as deemed appropriate by the operator. In Figure 6, an example using data from Sandy @ 57th St. and Division @ 122nd St. for a given day is shown in Step A. The black dots are random arrivals that are not clustered into any potential platoons, and the red dots show vehicles classified as platoons.
- In **Step B**, the percentage of cycles in a given hour that have platoons is identified. An example distribution of percentage of cycles containing platoons is shown in Step B of Figure 6. The blue line indicates the median cycle percentages by time of day; the gray shading implies the range from the 25th and 75th percentiles of the data by time of day.
- In **Step C**, the distribution of average platoon lengths by number of vehicles is calculated. The average platoon length distributed by time of day is shown in Figure 6, Step C. The blue line and gray shading represent the median and the range from the 25th to the 75th percentile of each distribution, respectively.

The percentage of cycles with a platoon and average platoon length by time of day are needed to identify the time period for which the impact of providing coordination should be explored. A predefined threshold can be used to identify the time periods to explore. As an example, if a threshold of 60% was chosen for percentage of cycles exhibiting platoons with platoon length greater than eight vehicles then, as annotated by "A" and "B" in Figure 6, the evening peak for Sandy @ 57th might be explored for the effects of coordination.

Control Support

Aggregate Platoon Coordination Diagram (APCD). The PCD is currently used by a number of agencies to identify if most arrivals occur during the green band, among other items. It is an effective tool to identify if there are any occurrences of platoon incursions happening during the start or end of green times. Despite being a very useful tool at a single-day resolution level, the PCD in its current form is cannot be extended to longer time durations. When given a month as a time range, UTAH ATSPM 4.01 generates one PCD per intersection per day. Here, a new tool named Aggregate Platoon Coordination Diagram (APCD) is proposed. Note that the term "Platoon" instead of "Purdue" is used to avoid any claims that this measure has been supported or recommended by Purdue University personnel.

Vehicle distributions using one day and one week of data, respectively, are shown in Figure 7*a* and *b*. In Figure 7*b*, the green and red lines show average red time and average cycle time, respectively. The color map indicates the density of arrivals, which is the average number of vehicle arrivals for each five-second period per cycle.

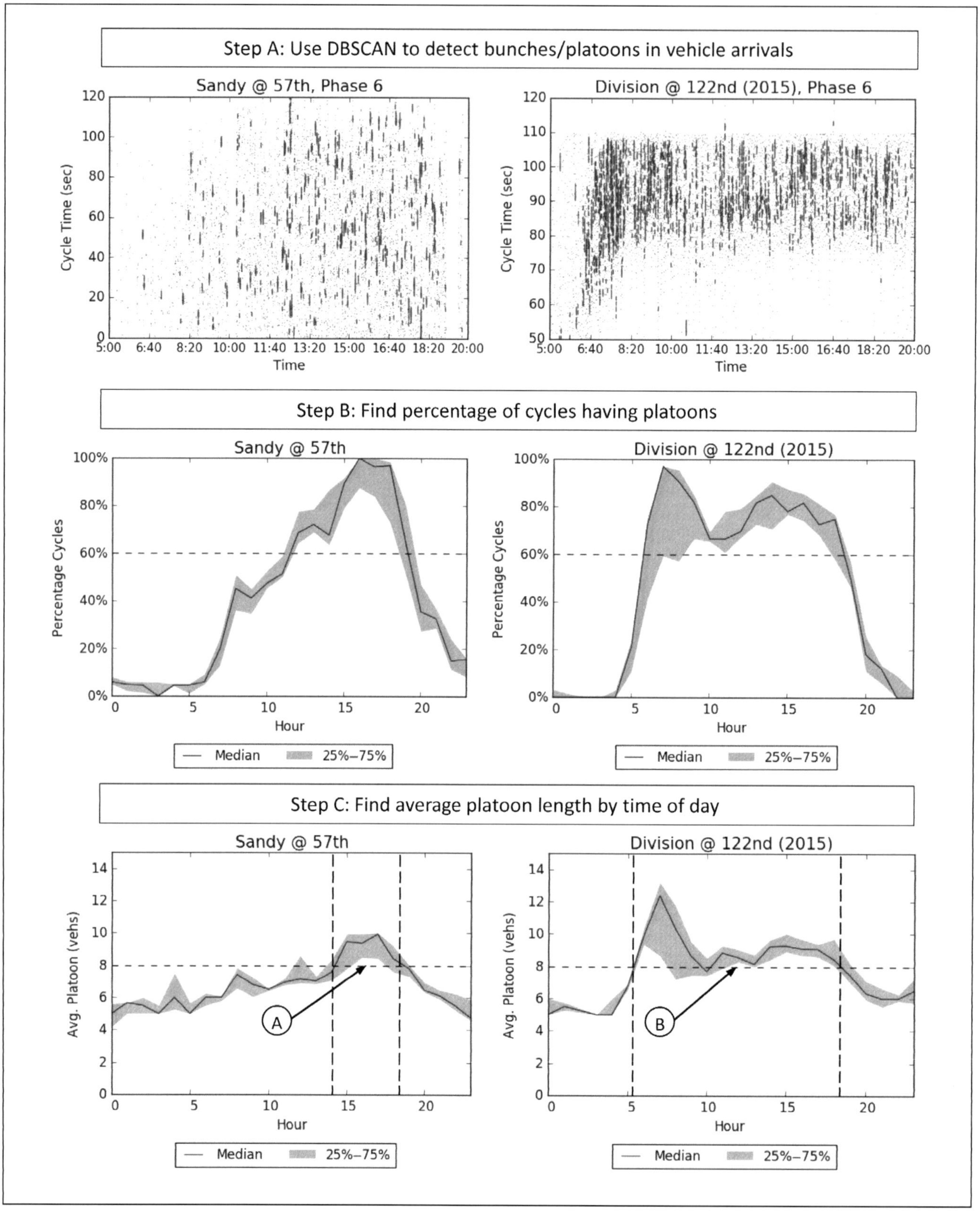

Figure 6. Procedures for determining the need of coordination.

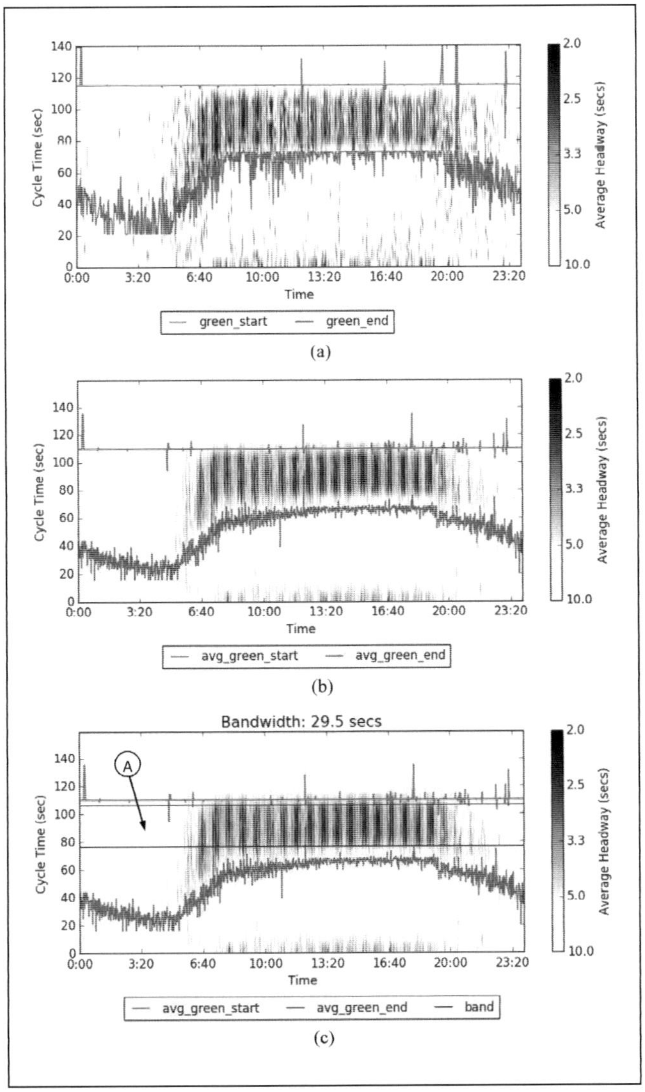

Figure 7. Proposed APCD features on Division @ 122nd (2015), Phase 6: (*a*) single day PCD example, (*b*) APCD over one week, (*c*) APCD with bandwidth detected.

The darker color indicates higher density and thus shorter headways. The color threshold was chosen to populate only densities that can be considered as platoons.

Use of the APCD to find the potential platoon bandwidth is demonstrated in Figure 7*c*, annotated by "A." For each time slot (along the *y* axis), if over 10% of vehicles arrive within a two-second headway, this slot would be considered as part of the platoon band. Taking the lowest and highest values from all the time slots, which are the band boundaries, the platoon bandwidth can be determined. Here, the intersection of Division @ 122nd St. (2015) shows a 29.5-second platoon bandwidth.

This example of an APCD is presented with an average of one week of data. It should be noted that the APCD can aggregate similar days, as identified in the Demand Trend Analysis section.

Conclusion

The ITSPM presented in this paper builds on the concepts of machine learning, traffic flow theory, and data visualization to minimize the human time needed for data-driven traffic signal management systems. Practitioners intimately familiar with ATSPMs reported the existing state-of-the-art systems to have three primary limitations: (i) limited data quality control, (ii) intensive resource requirements, and (iii) falling short in addressing the multimodal aspect of the operations. This paper addressed the first two of these shortcomings.

In this paper, a methodology to identify and remove data-logging errors as well as to identify bad sensors was presented. Eliminating these errors improves the quality of data, which leads to more precise decision making and also results in more efficient human asset management. After providing the methodology for data quality control, machine learning principles, which include intelligence in demand trend identification, were used. The use of machine intelligence will reduce the time taken by human operators, who would otherwise have to detect the same patterns manually, and will allow them to allocate their time more efficiently in making decisions rather than identifying patterns. In addition, ITSPM also includes an intelligent method for identifying whether or not coordination may be needed at an intersection. This can be achieved using two surrogate measures defined in this paper, namely, percentage of cycles exhibiting platoons per hour and average platoon sizes. Finally, in this paper the Aggregate Platoon Coordination Diagram (APCD), which is an advancement over the current PCD, was proposed. The APCD can be used to minimize visual clutter by removing any vehicle that is not in a platoon and can be scaled to include multiple days of data, thus eliminating the need for the traffic signal managers to browse through multiple individual PCDs when looking to improve coordination.

Future work could concentrate on two aspects. With regard to sensor and communication health, besides the methods proposed in this paper, a future research direction would be integrating more data sources like crash and weather data to conduct a causation analysis. In addition, shortcomings in the ATSPM outputs in addressing the multimodal aspect of traffic signal operations were identified, and the authors recommend that efforts be focused on designing these performance measures in future research. The availability of multimodal information through high resolution logs is also limited. At a minimum, there is a need to integrate more multimodal sensors on the roadways to provide data on approach volumes and delays for all modes. New surrogate measures can be calculated to investigate if phase allocations are equitable.

Author Contributions

The authors confirm contribution to the paper as follows: study conception and design: Peter Koonce, Anuj Sharma, Edward Smaglik, Sirisha Kothuri; data collection: Anuj Sharma, Tingting Huang; analysis and interpretation of results: Tingting Huang, Subhadipto Poddar, Anuj Sharma; draft manuscript preparation: Tingting Huang, Anuj Sharma, Edward Smaglik, Sirisha Kothuri, Subhadipto Poddar, Cristopher Aguilar, Peter Koonce. All authors reviewed the results and approved the final version of the manuscript.

References

1. Federal Highway Administration. *Automated Traffic Signal Performance Measures.* Center for Accelerating Innovation, Federal Highway Administration, Washington, D.C.www.fhwa.dot.gov/innovation/everydaycounts/edc_4/atspm.cfm. Accessed March 7, 2018.
2. Smaglik, E. J., A. Sharma, D. M. Bullock, J. R. Sturdevant, and G. Duncan. Event-Based Data Collection for Generating Actuated Controller Performance Measures. *Transportation Research Record: Journal of the Transportation Research Board*, 2007. 2035: 97–106.
3. Urbanik, T., D. Bullock, L. Head, D. Gettman, R. Campbell, M. Ablett, E. Smaglik, S. Beaird, J. Yohe, and S. Quayle. *Traffic Signal State Transition Logic Using Enhanced Sensor Information.* NCHRP 3-66 Final Report. National Academic of Science, Transportation Research Board, Washington, D.C., 2006.
4. Day, C. M., E. J. Smaglik, D. M. Bullock, and J. R. Sturdevant. *Real-Time Arterial Traffic Signal Performance Measures.* Publication FHWA/IN/JTRP-2008/9. Joint Transportation Research Program, Purdue University, West Lafayette, Indiana, 2008.
5. Taylor, M. *Overview of UDOT SPM System.* Utah Department of Transportation. http://docs.lib.purdue.edu/cgi/viewcontent.cgi?article = 1025&context = atspmw. Accessed March 7, 2018.
6. Day, C. M., D. M. Bullock, H. Li, S. M. Remias, A. M. Hainen, R. S. Freije, A. L. Stevens, J. R. Sturdevant, and T. M. Brennan. *Performance Measures for Traffic Signal Systems: An Outcome-Oriented Approach.* Purdue University, West Lafayette, Indiana, 2014.
7. Day, C. M., D. M. Bullock, H. Li, S. Lavrenz, W. B. Smith, and J. R. Sturdevant. *Integrating Traffic Signal Performance Measures into Agency Business Processes.* Purdue University, West Lafayette, Indiana, 2015.
8. Bullock, D., R. Clayton, J. Mackey, S. Misgen, and A. Stevens. Helping Traffic Engineers Manage Data To Make Better Decisions. *ITE Journal*, Institute of Transportation Engineers, Vol. 84, No. 3, 2014. pp. 33–39.
9. Day, C. M., J. M. Ernst, T. M. Brennan, C. Chou, A. M. Hainen, S. M. Remias, A. Nichols, B. D. Griggs, and D. M. Bullock. Performance Measures for Adaptive Signal Control Case Study of System-in-the-Loop Simulation. *Transportation Research Record: Journal of the Transportation Research Board*, 2012. 2311: 1–15.
10. Tufte, E. *The Visual Display of Quantitative Information*, 2nd ed. Graphics Press, Cheshire, Conn., 2001.
11. Day, C. M., R. J. Haseman, H. Premachandra, T. M. Brennan, J. S. Wasson, J. R. Sturdevant, and D. M. Bullock. Evaluation of Arterial Signal Coordination: Methodologies for Visualizing High-Resolution Event Data and Measuring Travel Time. *Transportation Research Record: Journal of the Transportation Research Board*, 2010. 2192: 37–49.
12. Brennan, T. M., C. M. Day, J. R. Sturdevant, and D. M. Bullock. Visual Education Tools to Illustrate Coordinated System Operation. *Transportation Research Record: Journal of the Transportation Research Board*, 2011. 2259: 59–72.
13. Day, C. M., and D. M. Bullock. Computational Efficiency of Alternative Algorithms for Arterial Offset Optimization. *Transportation Research Record: Journal of the Transportation Research Board*, 2011. 2259: 37–47.
14. Utah Department of Transportation. *Automated Traffic Signal Performance Measures.* http://udottraffic.utah.gov/atspm/. Accessed March 7, 2018.
15. Liu, C., A. Sharma, E. Smaglik, and S. Kothuri. TraSER: A Traffic Signal Event-based Recorder. *SoftwareX*, Vol. 5, 2016, pp. 156–162.
16. Fukunaga, K., and L. Hostetler. The Estimation of the Gradient of a Density Function, with Applications in Pattern Recognition. *IEEE Transactions on Information Theory*, Vol. 21, No. 1, 1975, pp. 32–40.
17. Cheng, Y. Mean Shift, Mode Seeking, and Clustering. *IEEE Transactions on Pattern Analysis and Machine Intelligence*, Vol. 17, No. 8, 1995, pp. 790–799.
18. Golubev, A., I. Chechetkin, K. S. Solnushkin, N. Sadovnikova, D. Parygin, and M. Shcherbakov. Strategway: Web Solutions for Building Public Transportation Routes Using Big Geodata Analysis. *Proc., 17th International Conference on Information Integration and Web-based Applications & Services*, ACM, Brussels, Belgium, 2015.
19. Choi, J. Y., and Y. K. Yang. Vehicle Detection from Aerial Images Using Local Shape Information. *Proc., Advances in Image and Video Technology, Third Pacific Rim Symposium*, PSIVT, Tokyo, Japan, 2009, pp. 227–236.
20. Ester, M., H. Kriegel, J. Sander, and X. Xu. A Density-based Algorithm for Discovering Clusters in Large Spatial Databases with Noise. *Proc., Second International Conference on Knowledge Discovery and Data Mining (KDD-96)*, AAAI Press, Portland, OR, 1996, pp. 226–231.

The Standing Committee on Traffic Signal Systems (AHB25) peer-reviewed this paper (18-05800).

Article

TRR
JOURNAL OF THE TRANSPORTATION RESEARCH BOARD

Transportation Research Record
2018, Vol. 2672(18) 167–178

DOI: 10.1177/0361198118794042
journals.sagepub.com/home/trr
SAGE

Data-Driven Ranking of Coordinated Traffic Signal Systems for Maintenance and Retiming

Christopher M. Day[1], Howell Li[2], James R. Sturdevant[3], and Darcy M. Bullock[2]

Abstract

Automated traffic signal performance measures (ATSPMs) have been deployed with increasing frequency. At present, the existing ATSPMs are focused on the performance of individual movements or intersections. As the number of ATSPM users has increased, a need for system-level metrics has emerged. This paper proposes a method of evaluating corridor performance at the system level using high-resolution data. The method is demonstrated for eight signalized corridors in Indiana, including 87 intersections. This method develops five subscores for the areas of communication, detection, safety, capacity allocation, and progression; these five interrelated aspects of performance are each given a category subscore based on quantitative performance measures, with scales appropriate to the context of the operation. An overall score for each corridor is determined from the lowest subscore of each of the five areas. This approach simplifies the analysis process, as opposed to examining several hundred individual movements as currently would be required using ATSPM tools that are commonly available at present. The methodology is presented as a prototype for further development and adaptation to individual agency objectives and data sources.

Automated traffic signal performance measures (ATSPMs) (*1–5*) have seen an increasing number of deployments. In 2017, with the selection of ATSPMs for the Federal Highway Administration's (FHWA) Every Day Counts program (*6*), this trend seems poised to accelerate, with many more agencies seeking to deploy the technology. As more agencies deploy ATSPMs and gain experience with the data, new research needs are emerging.

One need is to provide some context for performance measure data. Currently, a wealth of quantitative data can be obtained for specific aspects of operation, but users must rely on their own knowledge of the system and experience with the data to ascertain whether a given value represents good or poor operation. There is also a need to aggregate the data from individual movements to intersections to a system level, comprising all the corridors and networks managed by an agency. Some inspiration for this aggregation effort can be found in existing methods for pavement management systems. It is possible, for example, to express the percentage of a roadway system that is performing at or above a certain threshold. Similar types of metrics are needed for traffic signal systems.

In the case of signalized arterials, the system has a maintenance aspect as well as an operational aspect, and the latter relies on the former. It would not be reasonable to develop a system grade that focuses entirely on operations and ignores maintenance. Similarly, a "maintenance score" may be helpful for targeting maintenance activities, but does not tell whether the signal operation is providing a good quality of service.

Hierarchical Quality of Service

To begin developing an overall score for the quality of service of a traffic signal system it is important to understand the hierarchical nature of different aspects of its operation and maintenance (*7, 8*). Figure 1 shows a diagram illustrating this concept, with five layers highlighting different aspects of performance:

[1]Iowa State University, Ames, IA
[2]Purdue University, West Lafayette, IN
[3]Indiana Department of Transportation, Indianapolis, IN

Corresponding Author:
Address correspondence to Christopher M. Day: cmday@iastate.edu

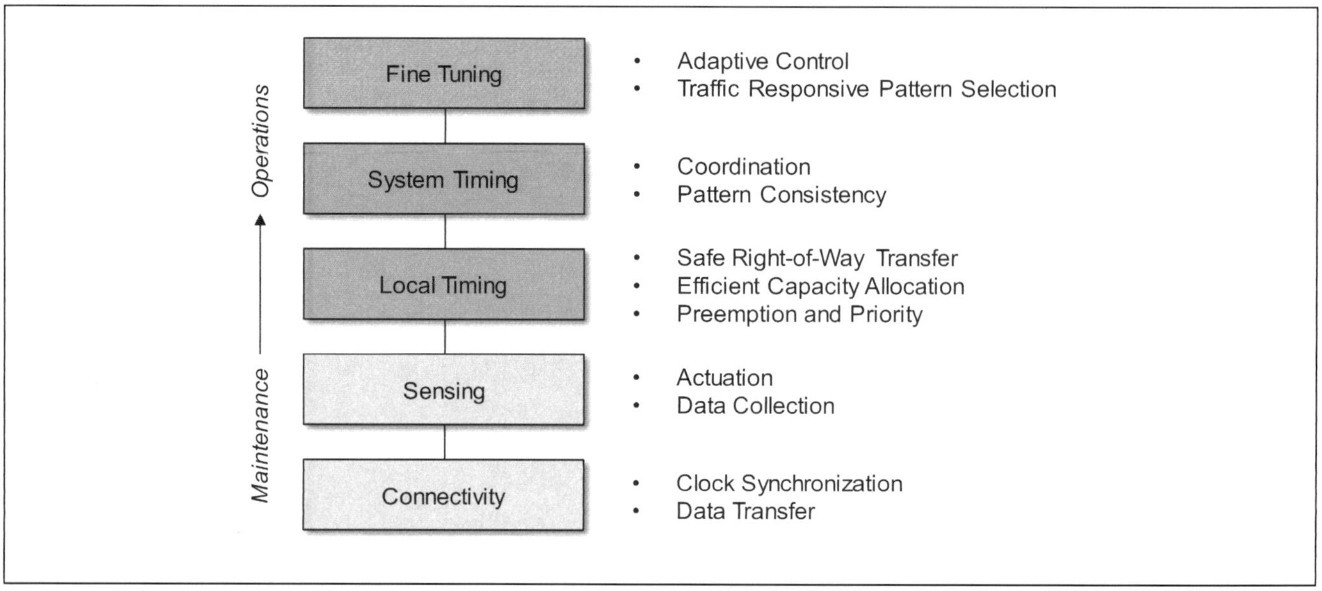

Figure 1. Hierarchical quality of service concept (*8*).

- The top layer, "fine tuning," represents advanced control methods that seek to improve operation by managing system and local timing in reaction to changes in traffic conditions. For this operation to be successful, system and local timing must be appropriately configured, detection systems must be able to measure conditions, and communication is needed to obtain that information and to send back the control decision.
- System timing represents the coordination of groups of signalized intersections by imposing consistent control patterns. A common objective is to provide a smooth flow of traffic. For system timing to be successful, local timing must be configured, detection is helpful, and communication must work to keep controllers synchronized.
- Local timing comes next; this represents basic phase-switching operations of local controllers. Signal timing must facilitate transfer of right-of-way and efficient allocation of capacity. Basic actuation requires working detection, and communication is needed to send out data.
- Sensing, enabled by detection systems, enables basic actuated operation and to measure intersection conditions, greatly improving the potential quality of control when appropriately configured.
- The bottom layer represents communication systems. Communication enables basic coordination functionality through clock synchronization, and transfer of data from the intersection to the central office. Without communication, it is difficult to develop a clear picture of site conditions without a field visit. Probe vehicle data can identify locations where congestion (slow speeds) exists in a network, but cannot determine if it is attributable to signal timing.

In laying out the descriptions of these layers, their interdependence quickly becomes apparent; each layer generally relies on the functionality of layers beneath it. Signal systems are relatively robust and will continue to offer basic function even as components fail. When detectors fail, actuated movements operate as fixed-time, but they still function; when communication is lost, controller clocks drift and coordination degrades, but the intersections continue to go through their cycles and rudimentary service remains. However, the inefficiencies accumulate over time and across systems.

Using the hierarchical view in Figure 1, it is possible to develop the notion of a quality of service "score" that includes several subscores reflecting different aspects of operation. For each, a basic question can be asked expressing the substance of the potential subscore:

- Communication: How many intersections are online, and reporting data?
- Detection: How many detectors are malfunctioning?
- Safety: Are there any issues with right-of-way transfer?
- Capacity: Are there any movements that need more green time (and can it be given to them)?
- Progression: Is coordination achieving the agency objectives, such as smooth traffic flow (and can it be improved)?

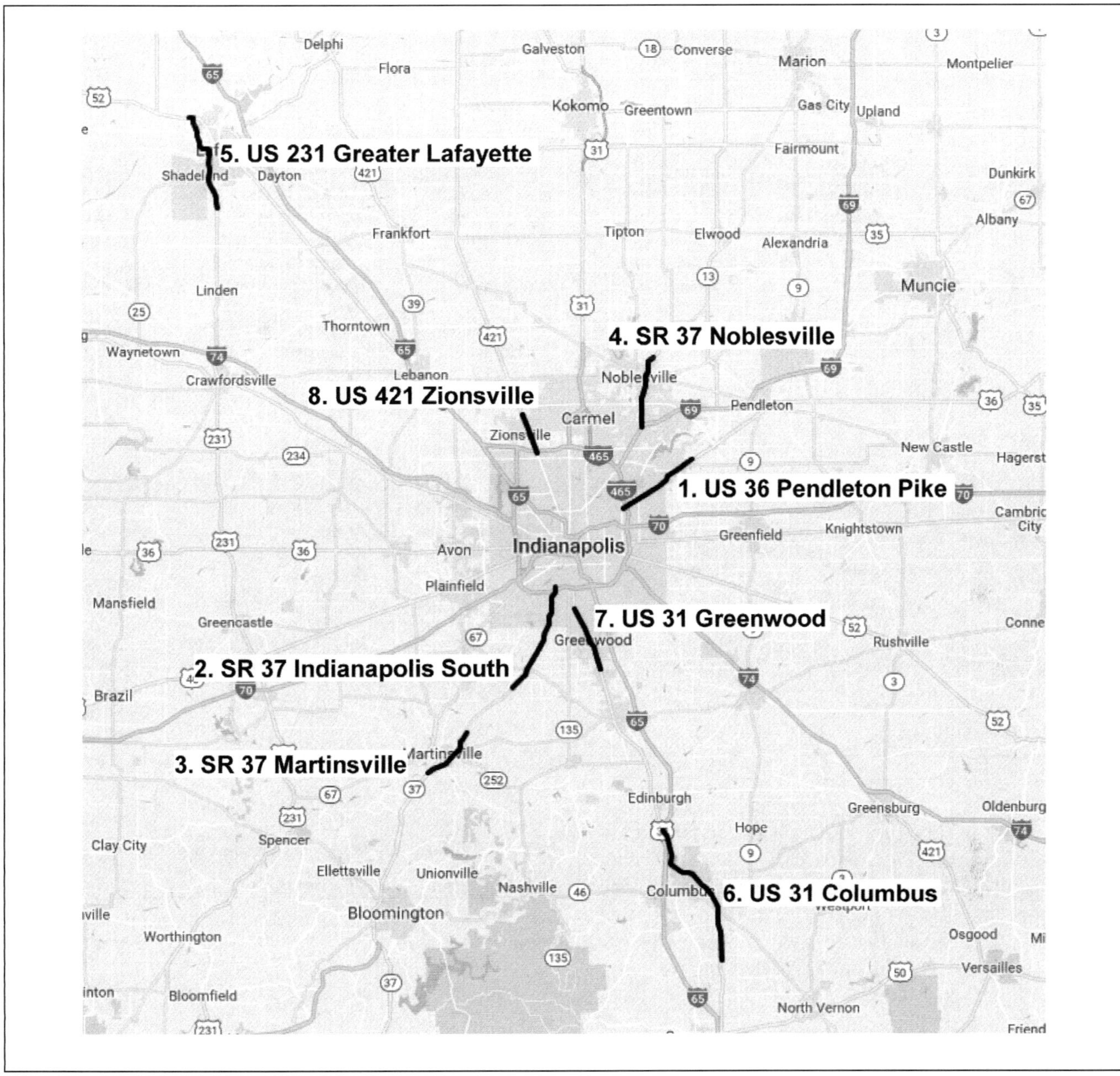

Figure 2. Map of corridors included in the study.

In this framework, the success of "fine tuning" could be determined from the value added to existing control and the prevention of degradation to subscores over time.

This paper focuses on the quality of service for coordinated operations in existing systems, using existing high-resolution data, and offers a demonstration of the scoring concept with real-world data. The following section explains the source of the data used, while the subsequent portions of the paper expand on the subscores listed above. Finally, these are combined to develop an overall score for eight signalized arterials in central Indiana.

Development of an Evaluation Methodology

Field data for this study was collected on eight signalized, coordinated arterial corridors in Indiana. Figure 2 shows a map of their locations. These eight corridors contain a total of 87 intersections. Each intersection had a controller capable of collecting high-resolution data. This study uses data collected in 2015, and is intended to reflect the perspective of an analyst viewing data on Monday, November 23, for the previous five weekdays.

The requirements for signal performance measures are discussed in greater detail elsewhere (*1*, *2*). In brief,

Table 1. Performance Subscore Thresholds

	Communication	Detection	Safety	Capacity	Progression
Evaluation element	Entire corridor (percentage of intersections)	Entire corridor (percentage of detectors/ phases)	Intersection	Intersection	Coordinated approach
Performance measure(s)	Percentage of intersections online	H1 detectors; H2 detectors; H3 phases; H4 ped phases.	Rate of red light violations per 1,000 vehicles	Highest percentage of split failures, F; Degree of intersection saturation, X_C	Platoon ratio, R_p; Volume-to-capacity ratio, x
Method for aggregation to corridor	N/A	Lowest heuristic subscore	Lowest score of any intersection in corridor	Weighted average of all intersections (using intersection volume as weight)	Weighted average of all approaches (using approach volume as weight)
Criteria for "A"	100%	<5%	0–5	Above line $X_C = 20F - 1$	Above line $x = -6R_p + 7.8$
Criteria for "B"	90–99%	5–15%	5–10	Above line $X_C = 5F - 0.5$	Above line $x = -6R_p + 6.9$
Criteria for "C"	80–89%	15–35%	10–20	Above line $X_C = 2.85F - 0.429$	Above line $x = -6R_p + 6.0$
Criteria for "D"	70–79%	35–50%	20–40	Above line $X_C = 1.25F - 0.25$	Above line $x = -6R_p + 5.1$
Criteria for "F"	<70%	>50%	>40	Below line $X_C = 1.25F - 0.25$	Below line $x = -6R_p + 5.1$
Time period evaluated	Previous 24 hours	Previous 5 weekdays	Previous 5 weekdays	Previous 5 weekdays	Previous 5 weekdays
Additional notes					No score given when $x > 0.9$

the intersections in this study mostly had standard detector layouts in Indiana. The standard layout uses advance detection on the mainline through movements, with stop bar detectors for all the minor movements. The exception was US 31 Columbus, which had almost no advance detection, instead having stop bar detection on every lane at every intersection.

Performance Measures

Subscores were assessed for each of the five areas of communication, detection, safety, capacity, and progression. This section presents the performance measures used to evaluate each area. These are metrics available in the ATSPM system developed using high-resolution data for corridors in Indiana. The system was described in greater detail previously (*4*). Each of these metrics are discussed in more detail in subsequent parts of this paper.

Table 1 summarizes the threshold definitions for the various letter grade thresholds for quality of service in each area and a conceptual outline for the subsequent sections of this paper that explain the rational of each column. For the communication, detection, and safety scores, the thresholds are based on one variable, while the capacity and progression scores are based on two variables; the variables in these two-variable formulas will be explained later in the paper. Table 2 shows the conversion of the weighted averages to letter-grade scores.

Table 2. Conversion from Weighted Average to Letter-Grade Score

Letter grade	Range of weighted average values
A	3.5–4.0
B	2.5–3.5
C	1.5–2.5
D	0.5–1.5
F	Below 0.5

Similar to past efforts by, for example, the *Highway Capacity Manual* (HCM), these thresholds are subject to considerable debate. Table 1 is presented to provide an initial starting point for discussion. It is expected that these thresholds may evolve as additional experience applying these techniques are obtained.

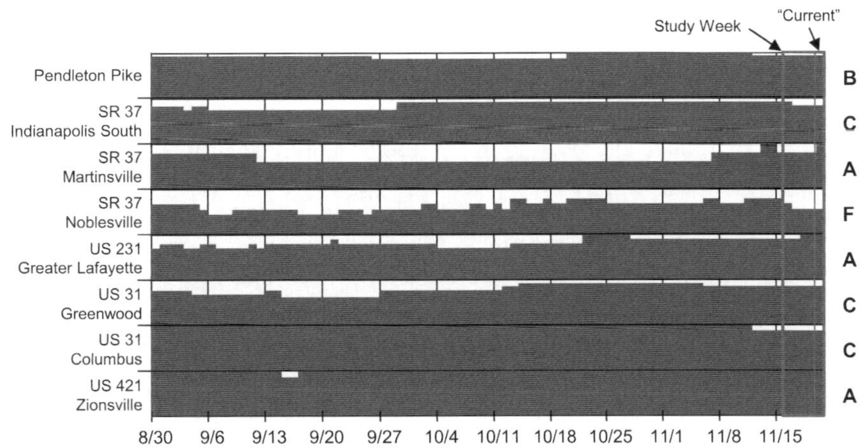

Figure 3. Percentage of intersections that are online in each corridor, by day (8/30/2015–11/20/2015), along with the overall corridor subscore. "Current" conditions are represented by the final day shown in the chart.

Communication

The quality of communication is considered the most basic aspect of service in this perspective, because without it, it is impossible to obtain the ATSPM data in an automated fashion. When long gaps in communication occur, data loss results, varying with the storage space on the local controller. There are several types of controllers in use on the Indiana state highway system, but one common type stores a maximum of about 24 hours of data. For this study, the presence of data in the database was used to evaluate whether a controller was "online" or not during a given time period. A simple count of events per signal ID was used to develop the status for each intersection. There are other nuances of communication quality, such as whether outages are intermittent or not, that could be examined by continuous online-status monitoring (e.g., the results of "pinging" the controller at regular intervals). Such data was not available for the present study.

Figure 3 shows the results of this analysis for a 12-week period. Each row in the chart represents a separate intersection, and the filled space shows the percentage of intersections within the corridor reporting data for each day over time. Ideally, the entire chart would be blue, but occasional losses at individual intersections occur. In Figure 3, the final week is highlighted; this week was selected to serve as the analysis period for subsequent subscores.

For the communications subscore, the online status of all the corridors for the "current" period (here shown as on the day of Friday, November 20) was used to develop the subscore for communication quality. For example, US 421 Zionsville has 100% of its intersections online, so it received an "A" score, while SR 37 Noblesville has only 56%, so it received an "F", applying the thresholds in Table 2.

Detection

Failed detectors degrade the quality of service at an intersection by forcing movements to operate as fixed-time, meaning that they receive an arbitrary amount of green time during every cycle. During fully-actuated ("free") operation, this can be particularly detrimental when a minor phase is always served the maximum green time.

One prerequisite for this analysis is the existence of a database with detector IDs, which is used to establish the channels on which data is to be expected. In Indiana, where most intersections use inductive loops, a "loop tagging" system is used to track detector assignments. Alternately, there are some high-resolution data events related to detector errors (*9*). Those were not used in this study because they have not yet been tested for accuracy.

Four heuristics were developed for identifying detector errors. These are inspired in part by similar error detection methods used in Utah DOT's ATSPM system (*3*), and by prior work by Lavrenz et al. (*10*):

- H1, missing data: This represents any detector defined in the database for which no data was observed. As it is unlikely for any movement to have no traffic for an entire week, this probably represents a faulted detector.
- H2, miscounting: The number of counts per hour was calculated for each detector. To identify miscounting, the count in the 3 a.m. hour was examined; if this exceeded 500 events during that hour, the detector is likely to have a problem. Detector

undercounting was not specifically evaluated in this study, but could be done using a statistical comparison with previous time periods (*11*).
- H3, phases in recall due to detector error: A separate score for this impact of detector error was used, since the number of faulted detector channels could be very different from the number of phases affected. Here, any phase that was called and served its maximum time continually throughout the 3 a.m. hour was considered to need maintenance. Phases in recall by design were excluded.
- H4, stuck pedestrian buttons: Similarly, any pedestrian phase that was served during every cycle in the 3 a.m. hour (except by design) was considered to need maintenance.

Percentage values were computed for each heuristic for each corridor, and thresholds from Table 2 were applied to determine a heuristic score. The lowest heuristic score was then used for the corridor subscore. For example, the scores from Pendleton Pike were as follows:

- H1: 19 of 185 detectors affected, or 10% (heuristic score B)
- H2: 1 of 185 detectors affected, or 1% (heuristic score A)
- H3: 42 of 382 phases affected, or 11% (heuristic score B)
- H4: No phases affected (heuristic score A)

The lowest of any heuristic score was B, so a detection subscore of B emerges for the corridor. Full results for the eight example corridors are presented later in the paper.

The above heuristics are intended to be implementable for any type of detection technology, or detector lane configuration (i.e., lane-by-lane or approach-based). A variety of field configurations were encountered in the data examined in this study. Some adjustments of the thresholds or time periods to be analyzed may be needed to match specific conditions—for example, if a location is expected to have heavy traffic at 3 a.m., the above heuristic for overcounting would probably not work, but different thresholds could be used.

Safety

The first two subscores were maintenance-oriented; the first operational subscore is that of safety. Safety is typically measured by using crash statistics, which require years of data to be compiled since crashes are relatively rare events. Using ATSPMs, however, it is possible to develop a sense of the potential for certain types of crashes by detecting conflicts. One readily measurable conflict is red light violation. Other potential safety metrics include vehicle speeds in excess of the speed limit or the number of dilemma zone vehicles; such metrics were not readily available for aggregation in the present study.

In this study, the red light violation detection methodology proposed by Lavrenz et al. is adopted (*12*). This methodology identifies violations using archived high-resolution data. This method does not consider the presence of conflicting traffic at the time of the violation, but the total number of violations indicates a high potential for such conflicts. The number of red light violations was tabulated for all the intersections during the week of November 14, 2015. The traffic volumes at those intersections were also measured.

To develop a corridor subscore, the lowest score (using thresholds presented in Table 2) of *any individual intersection* was used. For example, on Pendleton Pike, the highest intersection violation rate was 15.2 per 1000 entering vehicles, corresponding to a score of "C", which became the value for the corridor. Full results for all eight corridors are shown later in the paper.

Figure 4 shows a plot of the number of red light violations versus intersection volume for all 87 intersections across the eight corridors. The red lines drawn across the chart represent violation rates, which are used to define thresholds for scores, which are detailed further in the next section. Most intersections had rates between 5 and 20 violations per 1,000 vehicles, a figure similar to other observations (*13*).

Capacity Allocation

The allocation of capacity (i.e., green time) is a critical function of local signal timing. Movements that have insufficient green time are more likely to experience split failures. These are occurrences where a higher amount of traffic is waiting for service than the available green time, so queues form from the residual demand at the end of green.

This study uses previously developed methods of split failure identification. Where stop bar detection was available, the green occupancy ratio (GOR) and red occupancy ratio (ROR) of the first five seconds after the end of yellow were used; split failures were estimated to occur when both values exceeded 80% (*14*). Where advance detection was available, the volume-to-capacity (v/c) ratio was used instead (*15*); split failures were estimated when $v/c > 1.0$. Split failures were calculated for each movement as the percentage of cycles affected; the worst-performing value of any movement was used as the score for the entire intersection.

The rate of split failure by itself does not provide sufficient information to judge the quality of service. As traffic volumes increase, it becomes increasingly difficult to reallocate green time to prevent split failures. Therefore, a series of graduated thresholds are adopted that provide

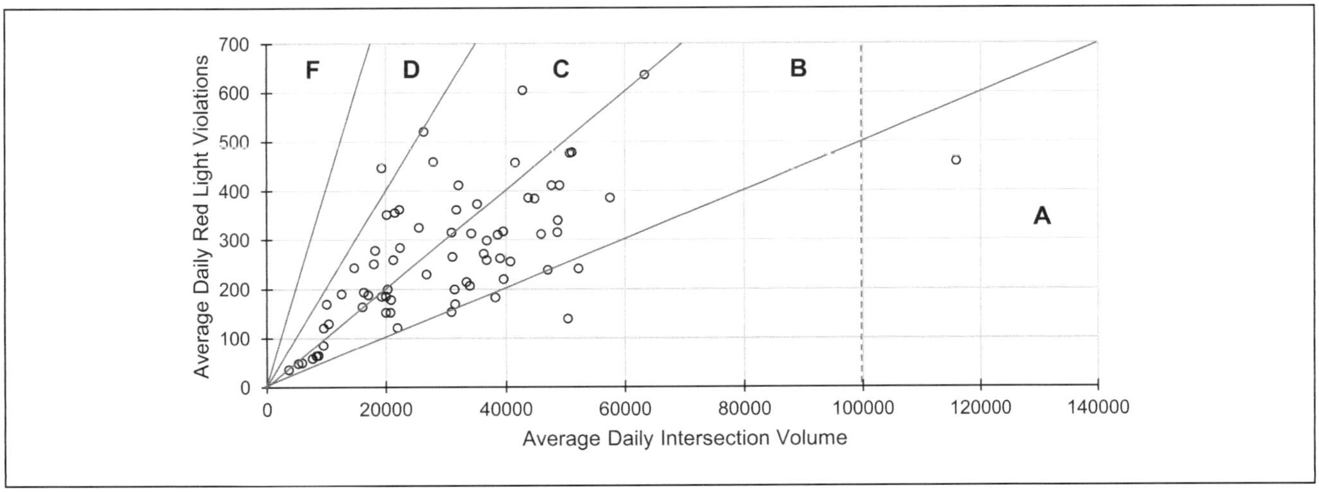

Figure 4. Number of red light violations by intersection and performance ranges. Data is compiled from five weekdays.

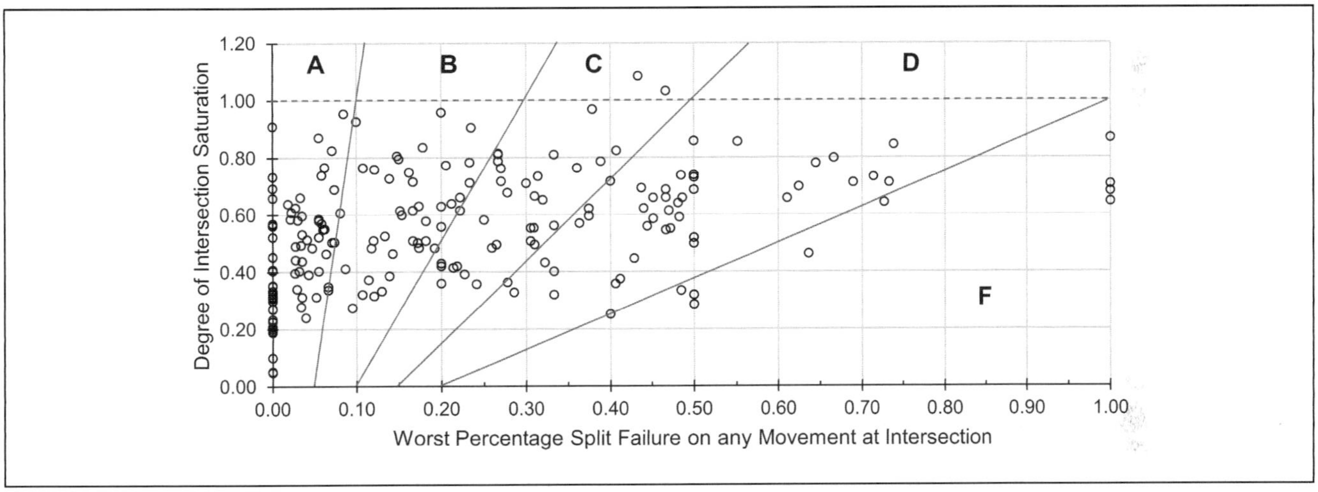

Figure 5. Degree of intersection saturation versus worst percentage of split failure on any movement at each intersection, and performance ranges. Data is compiled from five weekdays.

more lenient scores as the total degree of intersection saturation, X_C, increases. The X_C metric measures the overall intersection saturation by summing volume-to-saturation ratios along the "critical path" through the intersection (*16*). For an eight-phase intersection, the formula is:

$$X_c = [\max(y_1 + y_2, y_5 + y_6) + \max(y_3 + y_4, y_7 + y_8)] \times \left(\frac{C}{C - L}\right) \quad (1)$$

Where y_i, for example, is the volume-to-saturation ratio for phase i, C is the cycle length, and L is the total lost time. Volume-to-saturation flow rates are computed by:

$$y_i = \frac{v_i}{s_i}, \quad (2)$$

where v_i is the volume and s_i is the saturation flow rate of phase i. The units are vehicles per hour for both.

Figure 5 illustrates this concept, showing a plot of X_C versus the highest percentage split failure (F) at each intersection. The red lines show the boundaries for each intersection's score. As the degree of intersection saturation increases, the threshold for each score increases. For example, if $X_C = 0.4$, a "B" grade is acceptable if $F < 0.18$, but for $X_C = 1.0$, a "B" grade is achieved if $F < 0.30$. The rationale is that it becomes more difficult to achieve equitable distribution of green times as X_C increases. For situations where volumes cannot be obtained, the GOR has

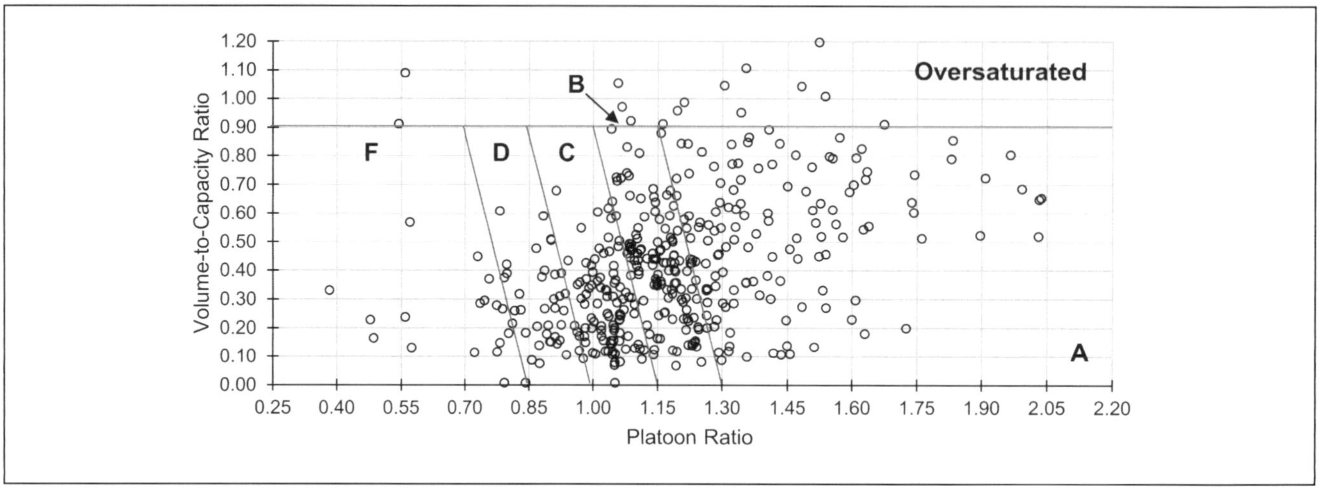

Figure 6. Volume-to-capacity ratio versus platoon ratio, and performance ranges. Data is compiled from five weekdays.

been found to be a conservative surrogate because it reaches 100% at values slightly below v/c or 1.0 and is somewhat sensitive to detector length (*17*).

To compute a corridor subscore, a weighted average was taken of the individual intersection scores. The weighting method used was similar to computing a grade point average; the letter scores for the intersection were converted into numerical values, using 4 for "A", 3 for "B", 2 for "C", 1 for "D", and 0 for "F". The weight was the intersection volume. Table 2 shows the thresholds for converting the numerical weighted average back into a letter grade score for the final value.

Scores were computed for three time-of-day periods (a.m. peak, midday, and p.m. peak). A different weighted-average score was computed for each time period; the lowest of these was used to report the performance of the corridor. For example, for Pendleton Pike, weighted-average scores of "B", "B", and "C" were observed for the three time periods, so the "C" score was passed to the corridor score. Full results for all eight corridors are presented later in the paper.

Progression

The final subscore evaluates the quality of service with regard to signal coordination and the progressive flow of traffic along arterials. There are numerous ways to measure this aspect of signal operation; with high-resolution data, where advance detection is present, it is possible to measure whether individual vehicle arrivals occur during green. The percent on green (POG), in conjunction with the green-to-cycle length ratio, can be used to determine the platoon ratio (R_p), a metric used in the HCM to relate POG to a quantitative score called the Arrival Type (*18*, *19*). The formula is:

$$R_p = \frac{POG}{g/C} \tag{3}$$

where g is the green time and C is the cycle length. The g/C ratio in the denominator can be approximated by the total green time in an analysis period, divided by the total length of the analysis period. The platoon ratio accounts for the fact that high values of POG are naturally obtained when a phase obtains a relatively high share of the available green time. For the same value of POG, a higher value of R_p is obtained if that POG is attained with a smaller g, and vice versa.

Similar to the capacity subscore, good progression is more difficult to achieve as the intersection approaches saturation. To take this into account, the v/c ratio (x) of an approach is measured. The formula is:

$$x = \frac{v}{sg} \tag{4}$$

where v is the volume (number of vehicles), s is the saturation flow rate (veh/s) and g is the total green time during the analysis period (s).

Figure 6 shows a plot of x versus R_p for individual approaches in the eight corridors. As before, the chart is demarcated with lines showing regions that correspond to score values. A horizontal line is drawn at $x = 0.9$; above this value, the approach is considered to be oversaturated, and a score is not given for that approach. Below this value, better scores are granted for higher values of R_p, with the scoring being more lenient as v/c increases. The base score values (at low values of x) are loosely based on the HCM Arrival Type thresholds. A value of $R_p = 1$ indicates random arrivals; the "C" score is centered on that value. Higher values of R_p indicate better progression, corresponding to better scores.

Table 3. Performance Scores

Performance information	Corridor number							
	1	2	3	4	5	6	7	8
Number of intersections total	15	12	5	9	10	13	8	7
Number of intersections online	14	10	5	5	10	11	7	7
Percent online	93%	83%	100%	56%	100%	85%	88%	100%
Communication subscore	B	C	A	F	A	C	C	A
Number of detectors	185	138	75	85	142	133	100	97
H1 detectors	19	11	42	9	4	3	6	8
H1 rate (% of detectors affected)	10	8	56	11	3	2	6	8
H1 heuristic score	*B*	*B*	*F*	*B*	*A*	*A*	*B*	*B*
H2 detectors	1	0	0	0	4	0	0	6
H2 rate (% of detectors affected)	1	0	0	0	3	0	0	6
H2 heuristic score	*A*	*A*	*A*	*A*	*A*	*A*	*A*	*B*
Number of phases	382	242	129	183	199	253	209	148
H3 phases	42	31	123	2	12	3	31	42
H3 rate (% of phases affected)	11	13	95	1	6	1	15	28
H3 heuristic score	*B*	*B*	*F*	*A*	*B*	*A*	*B*	*C*
H4 ped phases	0	0	0	0	0	0	0	0
H4 rate (% of pedestrian phases affected)	0	0	0	0	0	0	0	0
H4 heuristic score	*A*	*A*	*A*	*A*	*A*	*A*	*A*	*A*
Detection subscore	B	B	F	B	B	A	B	C
Highest red light violation rate per 1,000 entering vehicles	15.2	8.6	(a)	12.8	17.3	23.1	8.8	16.4
Safety subscore	C	B	(a)	C	C	D	B	C
a.m. peak capacity subscore	B	B	(a)	C	A	B	C	C
Midday capacity subscore	B	B	(a)	C	A	C	C	C
p.m. peak capacity subscore	C	B	(a)	C	B	C	C	D
Capacity allocation category subscore	C	B	(a)	C	B	C	C	D
a.m. peak progression subscore	C	B	(a)	C	C	(b)	B	C
Midday progression subscore	B	B	(a)	B	C	(b)	A	C
p.m. peak progression subscore	B	B	(a)	B	C	(b)	A	C
Progression category subscore	C	B	(a)	C	C	(b)	B	C
Overall corridor score (worst subscore)	C	C	F(c)	F	C	D(c)	C	D
Average of subscores (for comparison)	C	B	C	C	B	C	B	C

[a]Not enough functioning detection to evaluate operational scores.
[b]No advance detection in this corridor.
[c]An "incomplete" score, with one or more missing subscores.

Similar to the capacity subscore, a progression subscore was also computed as a weighted average. Rather than intersections, the weighted average for progression was computed across individual approaches, with the approach volume serving as the weight. Letter scores were determined for the approaches using the thresholds in Table 2; the same method of converting the weighted average to a category subscore was followed as for progression, using the thresholds in Table 2.

Again, as in the capacity subscore, progression subscores were calculated for three time-of-day periods and the lowest of these was used to represent the overall corridor subscore. In addition, for progression, approaches with $x > 0.9$ were excluded in these computations; any corridor for which more than 50% of its approaches were oversaturated would not be given a progression subscore. However, this did not occur for any of the corridors in this study. As an example, the Pendleton Pike corridor had time-of-day scores of "C", "B", and "B", with the lowest score of "C" being used as the subscore. Full results for all eight corridors are shown in the next section.

System Level Performance

The final result is to compile subscores for each corridor, and return an overall score across categories. The lowest subscore for any individual performance category was reported as the overall corridor score. This approach was used rather than taking an average across categories, to highlight any corridor with severe deficiencies in any aspect of operation. Taking an average resulted in all of the corridors having middling scores, with these weaknesses masked by other areas of acceptable performance.

In this final compilation, in some cases certain categories may be impossible to score for various reasons. In that case, the category would be ignored in determining the overall score. For example, if there is no working detection, it is impossible to determine safety, capacity, or progression subscores. If there is no communication, none of the other subscores can be determined. If there is working detection, but only stop bar detection and no advance detection, the progression subscore would not be available. If overall scores are presented without the accompanying subscores, it would be advisable to mark such scores with an asterisk indicating that they are not based on a complete picture.

Table 3 presents the full results for the eight study corridors. Some observations can be drawn from this table:

- Communication: Three of the eight corridors have all their intersections online, while the others typically range between 80% and 90%. Corridor 4 (SR 37 Noblesville) was missing nearly half of its intersections during the study period, likely due to an ongoing signal modernization project.
- Detection: Most of the corridors scored a "B"; there was one "A" and one "C" score. Corridor 3 (SR 37 Martinsville), received an "F"; there was a very high number of detectors with missing data, and nearly every phase was running in recall due to detector errors. In this case there is an explanation. This section of SR 37 is currently being upgraded to become a section of I-69; since the traffic signals are being removed, their maintenance has been deferred.
- Safety: The violation rates are found to vary from 8 to 23 violations per 1,000 vehicles. Most of the corridors received "B" and "C" scores, with Corridor 6 (US 31 Columbus) receiving a "D."
- Capacity allocation: Here, the overall scores vary from "B" to "D." Unsurprisingly, the p.m. peak tends to have the lowest score of the three time periods considered. However, the a.m. peak tended to operate better than the p.m. peak, and better than the midday for one corridor. A score is not provided for Corridor 3 (SR 37 Martinsville) because too many detectors have failed for a meaningful score to be tabulated.
- Progression: The overall scores are "B" or "C;" in general, progression performance is relatively strong. No time-of-day scores are lower than "C," and a few "A" scores are present for certain times of day.
- Missing subscores: As mentioned earlier, Corridor 3 was missing a high percentage of detectors; as a result, it was not possible to adequately evaluate its safety, capacity, or progression performance. In addition, a progression score is not provided for Corridor 6 (US 31 Columbus) because that corridor did not have advance detection.

The overall results show that all of the corridors received a "C" or lower rating; the two "F" scores present are due to maintenance problems. One of the "D" scores reveals a potentially high number of red light violations at Corridor 6 (US 31 Columbus), while the other "D" score shows a high number of split failures at Corridor 8 (US 421 Zionsville). The others are generally middling scores; in some cases, the worst category was a maintenance issue, and operation is generally good (Corridor 2, SR 37 Indianapolis South), while for others maintenance is relatively good, but operation slightly worse (Corridor 1, US 36 Pendleton Pike).

For comparison, an *average* score is shown in the final row of Table 3, to contrast with the "worst-subscore" method. In this case, it can be seen that every corridor has an overall score of either "B" or "C", and it is not immediately clear which of the corridors is experiencing a severe operational or maintenance deficiency.

Conclusion

This paper proposes a method of evaluating corridor performance using high-resolution data in ATSPM systems, demonstrating the concept for eight signalized arterials in Indiana. The concept is developed to respond to comments the authors have heard from early adopters of ATSPMs, namely that current techniques require "searching" through lots of performance measures to identify problems they can act on. Early adopters have expressed a need for aggregation methods to provide a simplified metric to judge high level performance of their system in the areas shown in Table 1 (communication, detection, safety, capacity allocation, and progression).

The method presented develops five subscores for communication, detection, safety, capacity allocation, and progression. The first three scores are based on overall observations over an analysis period (in this example, one week was used); the others are based on the lowest subscore of time-of-day periods. The overall score is returned as the lowest of the five subscores. The overall results (Table 3) summarize the outcomes for 87 intersections, most with eight phases, for a week's worth of data. An analyst would need to have examined up to 696 individual phase results (for multiple days) to have covered the same scope.

This paper is intended to present a demonstration of the concept, rather than to be prescriptive about the score thresholds and other specific details. These finer points would be better developed through broader application of the concepts, to see whether a reasonable filtering of problem areas occurs on a large scale. It is likely that these considerations would need to be tuned for the objectives and expectations of individual agencies.

There remains considerable room for further refinement, beyond the individual scores. Additional aspects of operation could be further considered, and brought into the individual subscores. For example, in addition to counting red light violations, a measure of dilemma zone performance could be included in the safety subscore. In this study, each failed detector was generally considered to be of equal importance, but this could be changed to give greater weight to failed detectors considered to be serving critical purposes. Analysis of the changes in scores over time will reveal further insights and likely provide a means of benchmarking performance similar to methods currently used in monitoring pavement condition. Other data sources beyond high-resolution data could also be incorporated. As additional agencies implement ATSPMs and consider tools such as these, such refinements will likely emerge.

Acknowledgments

This work was supported in part by the Joint Transportation Research Program administered by the Indiana Department of Transportation and Purdue University.

Author Contributions

The authors confirm contribution to the paper as follows: study conception and design: Day, Bullock; data collection: Li, Sturdevant; analysis and interpretation of results: Day, Li; draft manuscript preparation: Day, Bullock, Li, Sturdevant. All authors reviewed the results and approved the final version of the manuscript.

References

1. Day, C. M., D. M. Bullock, H. Li, S. M. Remias, A. M. Hainen, R. S. Freije, A. L. Stevens, J. R. Sturdevant, and T. M. Brennan. *Performance Measures for Traffic Signal Systems: An Outcome-Oriented Approach.* Report No. TPF-5 (258). Purdue University, West Lafayette, IN, 2014.
2. Day, C. M., D. M. Bullock, H. Li, S. M. Lavrenz, W. B. Smith, and J. R. Sturdevant. *Integrating Traffic Signal Performance Measures into Agency Business Processes.* Purdue University, West Lafayette, IN, 2016.
3. *UDOT Signal Performance Metrics.* http://udottraffic.utah.gov/signalperformancemetrics/. Accessed July 28, 2017.
4. Li, H., L. M. Richardson, C. M. Day, J. Howard, and D. M. Bullock. Scalable Dashboard for Identifying Split Failures and Heuristic for Reallocating Split Times. *Transportation Research Record: Journal of the Transportation Research Board,* 2017. 2620: 83–95.
5. Wu, X., and H. X. Liu. Using High-Resolution Event-Based Data for Traffic Modeling and Control: An Overview. *Transportation Research Part C,* Vol. 42, 2014, pp. 28–43.
6. FHWA. *EDC4 Innovations.* https://www.fhwa.dot.gov/innovation/everydaycounts/edc_4/. Accessed July 28, 2017.
7. Urbanik, T., A. Tanaka, B. Lozner, E. Lindstrom, K. Lee, S. Quayle, S. Beaird, S. Tsoi, P. Ryus, D. Gettman, and S. Sunkari. *NCHRP Report 812: Signal Timing Manual: Second Edition.* Transportation Research Board of the National Academies, Washington, D.C., 2015.
8. Day, C. M., and D. M. Bullock. Integrating Outcome Oriented Performance Measures Into Traffic Signal Operations Business Processes. *Proc., 18th International Conference on IEEE Intelligent Transportation Systems Conference (ITSC),* IEEE, Las Palmas, Spain, 2015, pp. 131–136.
9. Sturdevant, J. R., T. Overman, E. Raamot, R. Deer, D. Miller, D. M. Bullock, C. M. Day, T. M. Brennan, Jr., H. Li, A. Hainen, and S. M. Remias. *Indiana Traffic Signal Hi-Resolution Data Logger Enumerations.* Indiana Department of Transportation and Purdue University, West Lafayette, IN, 2012.
10. Lavrenz, S., J. Sturdevant, and D. Bullock. Strategic Methods for Modernizing Traffic Signal Maintenance Management and Quantifying the Impact of Maintenance Activities. *Journal of Infrastructure Systems,* Vol. 23, No. 4, 2017, p. 05017004.
11. Grossman, J., A. Layton, J. Krogmeier, and D. Bullock. Traffic Signal Detector Error Identification Using Kolmogorov-Smirnov Test. Presented at 95th Annual Meeting of the Transportation Research Board, Washington, D.C., 2016.
12. Lavrenz, S. M., C. Day, J. Grossman, R. Freije, and D. M. Bullock. Use of High-Resolution Signal Controller Data to Identify Red Light Running. *Transportation Research Record: Journal of the Transportation Research Board,* 2016. 2558: 41–53.
13. Retting, R. A., S. A. Ferguson, and C. M. Farmer. *Reducing Red Light Running through Longer Yellow Signal Timing and Red Light Camera Enforcement: Results of a Field Investigation.* Insurance Institute for Highway Safety, Arlington, VA, 2007.
14. Freije, R., A. M. Hainen, A. Stevens, H. Li, W. B. Smith, C. M. Day, J. R. Sturdevant, and D. M. Bullock. Graphical Performance Measures for Practitioners to Triage Split Failure Trouble Calls. *Transportation Research Record: Journal of the Transportation Research Board,* 2014. 2439: 27–40.
15. Smaglik, E. J., D. Gettman, D. M. Bullock, C. M. Day, and H. Premachandra. Comparison of Alternative Real-Time Performance Measures for Measuring Signal Phase Utilization and Identifying Oversaturation. *Transportation Research Record: Journal of the Transportation Research Board,* 2011. 2259: 123–131.
16. Day, C. M., J. R. Sturdevant, and D. M. Bullock. Outcome Oriented Performance Measures for Management of Signalized Arterial Capacity. *Transportation Research Record: Journal of the Transportation Research Board,* 2010. 2192: 24–36.
17. Smaglik, E. J., D. M. Bullock, D. Gettman, C. M. Day, and H. Premachandra. Comparison of Alternative Real-Time Performance Measures for Measuring Signal Phase Utilization and Identifying Oversaturation. *Transportation*

Research Record: Journal of the Transportation Research Board, 2011. 2259: 123–131.

18. *Highway Capacity Manual 2010*. Transportation Research Board of the National Academies, Washington, D.C., 2010.
19. Smaglik, E. J., D. M. Bullock, and A. Sharma. Pilot Study on Real-Time Calculation of Arrival Type for Assessment of Arterial Performance. *Journal of Transportation Engineering*, Vol. 133, 2007, pp. 415–422.

The Standing Committee on Traffic Signal Systems (AHB25) peer-reviewed this paper (18-00115)

The contents of this paper reflect the views of the authors, who are responsible for the facts and the accuracy of the data presented here, and do not necessarily reflect the official views or policies of the sponsoring organizations. These contents do not constitute a standard, specification, or regulation.

Sage Publications USA
sagepub.com

CURRAN ASSOCIATES INC.
proceedings.com

ISBN 978-1-5108-7823-5